W9-CUK-511

LIQUID CRYSTALS,

LAPTOPS

AND

LIFE

SERIES IN CONTEMPORARY CHEMICAL PHYSICS

Vol. 3 Quantum Mechanical Irreversibility and Measurement
by P. Grigolini

Vol. 4 The Photomagneton and Quantum Field Theory: Quantum Chemistry, Vol. 1
by A. A. Hasanein and M. W. Evans

Vol. 5 Computational Methods in Quantum Chemistry: Quantum Chemistry, Vol. 2
by A. A. Hasanein and M. W. Evans

Vol. 6 Transport Theory of Inhomogeneous Fluids
by L. A. Pozhar

Vol. 7 Dynamic Kerr Effect: The Use and Limits of the Smoluchowski Equation and Nonlinear Inertial Responses
by J.-L. Dejardin

Vol. 8 Dielectric Relaxation and Dynamics of Polar Molecules
by V. I. Gaiduk

Vol. 9 Water in Biology, Chemistry and Physics: Experimental Overviews and Computational Methodologies
by G. W. Robinson, S. B. Zhu, S. Singh and M. W. Evans

Vol. 10 The Langevin Equation: With Applications in Physics, Chemistry and Electrical Engineering
by W. T. Coffey, Yu P. Kalmykov and J. T. Waldron

Vol. 11 Structure and Properties in Organised Polymeric Materials
eds. E. Chiellini, M. Giordano and D. Leporini

Vol. 12 Proceedings of the Euroconference on Non-Equilibrium Phenomena in Supercooled Fluids, Glasses and Amorphous Materials
eds. M. Giordano, D. Leporini and M. P. Tosi

Vol. 13 Electronic Structure and Chemical Bonding
by J.-R. Lalanne

Vol. 14 The Langevin Equation: With Applications to Stochastic Problems in Physics, Chemistry and Electrical Engineering, 2nd Edition
by W. T. Coffey, Yu. P. Kalmykov and J. T. Waldron

Vol. 15 Phase in Optics
by V. Peřinova, A. Lukš and J. Peřina

Vol. 16 Extended Electromagnetic Theory: Space Charge in Vacuo and the Rest Mass of the Photon
by S. Roy and B. Lehnert

Vol. 17 Optical Spectroscopies of Electronic Absorption
by J.-R. Lalanne, F. Carmona and L. Servant

Vol. 18 Classical and Quantum Electrodynamics and the B(3) Field
by M. W. Evans and L. B. Crowell

Vol. 19 Modified Maxwell Equations in Quantum Electrodynamics
by H. F. Harmuth, T. W. Barrett and B. Meffert

Vol. 20 Towards a Nonlinear Quantum Physics
by J. R. Croca

Vol. 21 Advanced Electromagnetism and Vacuum Physics
by P. Cornille

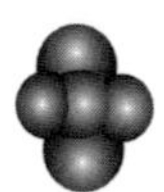

World Scientific Series in Contemporary Chemical Physics – Vol. 23

Michael R Fisch

Kent State University, USA

LIQUID CRYSTALS, LAPTOPS AND LIFE

NEW JERSEY • LONDON • SINGAPORE • BEIJING • SHANGHAI • HONG KONG • TAIPEI • CHENNAI

Published by

World Scientific Publishing Co. Pte. Ltd.
5 Toh Tuck Link, Singapore 596224
USA office: Suite 202, 1060 Main Street, River Edge, NJ 07661
UK office: 57 Shelton Street, Covent Garden, London WC2H 9HE

British Library Cataloguing-in-Publication Data
A catalogue record for this book is available from the British Library.

Cover image: Courtesy of Oleg D. Lavrentovich

LIQUID CRYSTALS, LAPTOPS AND LIFE
World Scientific Series in Contemporary Chemical Physics — Vol. 23

ISBN 981-238-901-6

Printed in Singapore by World Scientific Printers (S) Pte Ltd

For my family: Mary Alyce, Bob, and Andy,
and Fr. Lawrence J. Monville, S.J

Preface

A number of years ago, I was on a university committee that was evaluating student life and how to change or improve it. After a couple of meetings, I stopped by a wise colleague's office and excitedly told him about all the great things that we were planning. He listened politely, and then, taking his pipe from his mouth, asked, "What exactly is the problem you are trying to solve?" After thinking for a moment, I said I didn't know. The committee had talked at length about what we wanted to do, but never asked the key question: Why do we want to do it? Since that time, I have always tried to have the answer to that question and the arguments for why something should be done close at hand.

So, I hope the reader is asking, "Why did he write this book? What was the problem he was trying to address? Why and how is this book part of the solution to this problem?"

In recent years, I have become increasingly concerned with the separation of the more quantitative and technological knowledge of the sciences and engineering and the more qualitative ways of knowing of the arts. Of course, this view is not new, as Snow's classic book,[1] and more recent works by Edward Wilson[2] and the reply to Wilson by Wendell Berry[3] demonstrate. To my mind, it is clear that both ways of knowing are essential to our modern way of life, yet far too many people are turned off to science and mathematics at an early age, and they never really understand all but the simplest of ideas and concepts. Thus, an overriding goal of this book is to take at least a small step towards bringing these cultures together by discussing topics of common interest.

[1] *The Two Cultures: and a Second Look*, C. P. Snow, 1964.
[2] *Consilience: The Unity of Knowledge*, Edward O. Wilson, 1998.
[3] *Life is a Miracle: An Essay Against Modern Superstition*, Wendell Berry, 2000.

Deep in the recesses of my mind, I have a memory of an idea attributed to St. Thomas Aquinas. The gist of the idea is that if you want someone to come over to your point of view, walk over to them, take their hand and lead them to your position. Thus, this work is first addressed to the non-scientist. The material was chosen to explain a most common everyday item, the laptop computer. The very same ideas are then used to help understand life and art. Certainly other examples abound, but these areas most closely associated with my personal interests.

My goal for these non-scientists is to write a book *of* science and technology as opposed to a book *about* science and technology. Far too often, when trying to reach non-scientists, we, the scientists, stop short of giving them hard science and focus on history or society or other less technical areas. I hope the readers who identify with non-scientists will find a clear, united presentation where ideas build on ideas and thereby lead them to better understand and appreciate the work of scientists and engineers, the intricacies of life, the interconnectedness of the sciences to themselves and to technology, and the basic workings of a computer.

I have tried to use simple models and explanations with a minimum of mathematics to reduce the non-scientist's discomfort. At the same time, mathematics is undeniably part of the language of science. I have long felt that part of being a well informed and active citizen is being scientifically and technologically literate, and mathematics must come along for the ride. On occasion, just as when a scientist studies Faulkner or Shakespeare, the reader will be required to "eat their spinach." I have also, possibly to the consternation of these individuals, used Greek letters to represent certain constants and quantities. I have used the same letters in the same manner as more advanced books. My hope is that the reader will consult such books and, because they already know the notation, they will not have the burden of connecting two different notations.

After writing the first draft, I discovered a second audience for this book – the scientific specialist who has not seen some of the material and the connections between the topics. In the time of William Bacon, a single person could know and understand literally *all* of science. This has certainly changed. Centuries ago chemistry, biology, and physics split and went their separate ways. In more recent decades, electronics and biochemistry split off from branches and combinations of these three "major" disciplines. Now interestingly, these "major" fields are starting to come back together. Certainly, this reunification is not total, nor does it encompass all of the areas of these disciplines. Nevertheless, an increasing number of scientists and engineers are working in cross-disciplinary areas. We have, for example, a physicist that needs to know biology, or a biologist that needs to understand polymers. These individuals need to "get up to speed" rather quickly and they need to know and understand the language before they can proceed on to more advanced works.

Liquid crystal science and technology, life and the arts, and computers incorporate many different areas of science and technology and have been a breeding ground for this type of cross-discipline general science education for many decades. Thus, my goal for the scientist who knows some areas but not others is to provide a reasonable introduction to some of the essential ideas and how they connect to other ideas. This should provide an interesting course at the second year level for students in a variety of scientific disciplines. It is hoped that this book will serve as a readily accessible first step into areas which these readers are unfamiliar. The middle chapters, where this is most likely to be the case, have a large number of references for just this reason.

I have tried to convey my excitement, wonderment, and awe of nature in this book. The text is by no means self-contained. Nevertheless, I have tried to include sufficient background material so the reader will not have to consult other references except to obtain greater information and insight. I have started with the most basic ideas and then shown how they allow us to understand more complicated material. The level is deliberately uneven. Life is not "Science 101," and all problems are not equally difficult. Most chapters end with two sections entitled Exercises and Research Questions. The purpose of the Exercises is to give the reader an opportunity to test their knowledge of the material in the chapter. This section also, on occasion, presents applications of the material and tries to illustrate how scientists apply these ideas. The Research Questions tend to be more difficult and, in many cases, are rather open ended. Their purpose is to provide the student the opportunity to extend themselves and the material.

No work is developed in a vacuum. This book has been the beneficiary of help from many people. Dr. John West, former director of the Liquid Crystal Institute at Kent State University first suggested I develop this material into a course and found the financial means to allow me to work on this project. Professor Oleg Lavrentovich, also of the Liquid Crystal Institute, suggested the title.

Several people read all or parts of the drafts and offered concise comments. I particularly wish to thank Dr. Mary Ann Simpson of the Lahey Clinic, my sister Judy Laston (a microbiologist), Mr. John Mishic of Analex

Corporation, and Mr. Glen Novotny of Beachwood High School for reading and commenting on various drafts. Julie Kim of the Liquid Crystal Institute reviewed the chapters that had appreciable chemistry and made several valuable suggestions. David Cohen, Ph.D., M.D. of the Einstein School of Medicine reviewed the discussion of gallstones, his area of expertise. A friend and former student, Dr. Daniel Harrison, taught me enough LaTeX to prepare this document, and was a ready source of good cheer and help when needed. He also reviewed several parts of this book.

A special thanks is reserved for Dr. Jonathan Ruth, also of the Liquid Crystal Institute. Jon served as my in-house editor and read and commented on *every* figure, word, and idea (sometimes in mind boggling detail!). He found unclear statements that others had missed, confusing aspects of figures, and made suggestions too numerous to mention. I have no doubt that this book is better for his voluntary efforts.

Joseph Collura, a former student and current friend, neighbor, and colleague read the "final version" and offered significant suggestions that made it the final version minus one.

Finally, I wish to thank my wife, Mary Alyce Mooney. She read every chapter multiple times and suggested far too many changes for me to count. She also kept me honest by asking those tough questions that forced me to clarify discussions, add footnotes, change figures, and read (and reread) just about every book listed in the footnotes and references.

Of course, all errors, unclear statements, and omissions are the author's.

M. R. Fisch

Contents

Chapter 1

Introduction

1.1 Introduction

What do liquid crystals, laptops and life have in common, besides that they all begin with the letter "l"? At first glance, the answer might well be "not much." Rather surprisingly, they have a great deal in common. Liquid crystals are essential to life and are used in many practical devices that simplify and reduce the physical labor of our lives. Liquid crystals are also a medium that artists use to enrich our lives. The same technology that helped bolster the semiconductor industry has also helped advance the liquid crystal industry.

Liquid crystals are also an essential component of our bodies. Every cell membrane in every living organism is in a liquid crystalline state, the breakdown of this liquid crystalline structure can lead to many diseases. An understanding of the relationship between liquid crystals and diseases may help in finding new medicines and treatments for these diseases.

There is also a less critical, though more easily noticed connection between liquid crystals and our daily lives. In the course of a typical day, we encounter many different kinds of information displays. One of the more ubiquitous is the liquid crystal display that is common in such "low tech" applications as watches, clocks, radios, and household appliances, as well as "high tech" cell phones, automobiles, laptop computers,and electronic instruments. Liquid crystals also suffuse our lives in a less "techie" manner. They are common in simple "toys" such as mood rings and they are being incorporated into many works of art.

The relationship between liquid crystals and laptops is quite clear. In fact, laptop computers could not exist without a light weight, low power, low cost display. The microprocessor and the operating system get more ink, but the liquid crystal display is the lynchpin technology that makes the laptop possible. Does the hardware run a little slow? (Ok, we can wait.) Does the operating system hang from time to time? (<Ctrl><Alt><Del> . . . Ok, nothing's perfect.) Is the display hard to read unless you tilt it at just the right angle? Does it wash out in sunlight? We can tolerate quite a few shortcomings in a laptop, but if the display, the primary and most heavily used of all computer output devices, is difficult and irritating to use, we'll trash the whole unit.

This juxtaposition of display and computer masks a less apparent but more fundamental connection between the semiconductor and the liquid crystal industries. The growth of the electronics industry created much of the infrastructure that allowed larger and more complex displays to be made. Cleanroom manufacturing techniques developed in the electronics industry have been used and modified by the liquid crystal display industry. This synergistic relationship continues to grow and both industries are prospering.

Computers are becoming increasingly common in our lives. The speed and cost of computing is so low that computers are "buried" or embedded inside many appliances, automobiles and other tools of everyday life. The personal computer has become a common accessory in our lives. We use it as a typewriter, a calculator, a terminal for information exchange and a source of entertainment. Much of today's fast paced, high-density communication is based on computers.Unfortunately, in this author's opinion, there are many questions about computer technology that too few people ask: For example, what are the basic science and technologies behind a computer? How does a computer work? How does a computer display work?

Arthur C. Clarke, the famous science fiction writer, once observed that "Any sufficiently advanced technology is indistinguishable from magic."[1] (at least to the ignorant person seeing it for the first time.) Are computers magic? Are the liquid crystals found in displays studied in the biophysics of life magic? Should the average person treat them as magic? Put another way, can the average person understand the basic and applied science and technology behind these three converging topics? Should they try to understand this convergence, and if so, why?

[1]From *Profiles of the Future: An Inquiry into the Limits of the Possible*

1.2 A few preliminary answers

At this point you may be thinking "That's interesting; now let me move on with my life." Why are the connections between liquid crystals, laptops, and life important and worthy of further study? The answer to this question lies in the whole of this book, but we'll begin with a few general, preliminary answers.

1.2.1 *The relationship between technology and life*

Let's start by addressing the importance of a basic understanding the computer and how it functions. You might begin by asking, "Isn't a computer a tool just like a car, a bicycle or a pencil? Can't I just use it and let others worry about the details?" You could if you want to rely on "the others" to always have your best interests at heart. An educated person should want to understand science and technology at a basic level. If educated people do not understand technology, how can they control its use? How can they ensure that appropriate technologies are developed and inappropriate ones left to wither or even be banned? Should the direction of science and technology be determined by the specious arguments of charismatic speakers or by an educated populace? In the time of Galileo,[2] an authoritative body decided what was good science and what was bad science. Four centuries later the authorities have changed, but an uninformed populace, by default, will still let the authorities decide science policy.

History provides examples of people who understand technology controlling people who don't. Author Anthony Walton, in the January 1999 Atlantic Monthly, argues persuasively that technological innovations have traditionally made life worse for African-Americans. He asserts that since the discovery of the New World, African-American encounters with technology have devastated "their hopes, dreams and possibilities." He presents several examples of this devastation. The first was the development, by the Portugese, of the Caravel ships used in the slave trade. Later, the cotton gin hurt these people, because, in his view, it made slavery economically viable. Still later, the mechanical cotton picker threw workers off the land and led to the northward urban migrations after World War II. The increase in au-

[2] Galileo Galilei 1564-1642, invented the microscope and other devices, and built a telescope that he used to discover the moons of Jupiter. He was an early proponent of the theory that the earth revolved around the sun. He was compelled to abjure the theory that the earth was not the center of the universe.

tomation in factories that has occurred in the past 30 years put those who moved north out of work. Walton asserts that these developments all led to "Black folkways," a "consciousness of the race" that he believes is, at best, ambivalent towards interactions with technology. This ambivalence, coupled with the traditional poor education of too many African-Americans, has essentially shut this group out of the benefits of technology. Walton further states, "As a group, they have suffered from something that can loosely be called technological illiteracy."

Increasing technological developments will affect the destiny of all peoples. Therefore, all people must develop an understanding of technology so that they will benefit from it and not be left behind.

The "information age" is predicted to have profound effects upon our lives. James Burke of "Connections" fame has written that we are in the middle of a surge of information. This surge is a powerful agent for change that will change the whole world, including the way we think. He compares this new change to the fundamental changes brought about by the Greek invention of the alphabet[3] and the later development of the printing press by Gutenberg. He argues that the alphabet gave us the step-by-step reductionist manner of thought that has characterized western thinking for the past three millennia. The printing press greatly expanded the availability of information that allowed the nation-state to form.

The newly evolving information systems, which include the Internet and wireless phones, are already starting to break down the barriers of nation-states and may well lead to one world-state. If information is the ultimate commodity, then the ease with which information can be spread, used, and critically analyzed will lead to great economic growth and changes in governments, education, and society itself. Unfortunately, there is a dark side to this transformation. If we do not understand the technology we use, who then is in control? The ignorant end user or the knowledgeable supplier? Once more, only an educated populace that understands the basis

---

[3]This was questioned by one reader. It appears that the Egyptians were the first to represent each consonant phoneme (a member of the set of the smallest distinct units of speech in a language or dialect. For example, the \p\ in *pat* and \f\ in *fat* in English) with a single sign. In this way, typically fewer than thirty 'letters' were needed to convey the consonantal phonemes of a language. A difficulty was some languages required equal representation of both vowels and consonants. The Greeks provided the simple and adaptable innovation of adding vowel phonemes. The Greeks thus invented the first 'complete' alphabet. This is discussed in **A history of writing**, Steven R. Fischer, 2001; **The alphabet**, David Diringer, 1968, and **Alpha to omega**, Alexander and Nicolas Humez, 1981.

and basics of science and technology will be able to control it.

1.2.2 *The relationship between life and the arts*

Paintings, sculptures, woodcuts, silk screens, photographs, movies, videos, and computer images are all visual works of art. We are able to see and appreciate these works because in some way, the artwork interacts with light and we then see the light. How does a typical paint produce the colors we associate with it? How can paints be changed and why might we want to change them? How are liquid crystals and computers changing the traditional manner in which we see art? These are a few of the questions that will now be discussed.

Many types of visual artwork are static. By this it is meant that the composition (the paint, lighting (including shadows and shades of colors), position and relationships of objects) and other details of the artwork is expressed using the materials selected by the artist when the work is formulated and are, to a large extent, never changed except by aging and other similar phenomena. To follow an example from Lovejoy[4], consider the Mona Lisa, painted by Leonardo da Vinci. It is housed in a climate controlled glass case at the Louvre in Paris. It is viewed in artificial light, and is far removed from the Medici palace where it was first displayed. Physically, it is not the same painting it was originally and, it appears different than it would if it were still in the Medici palace under the original lighting conditions. This particular painting has also acquired a great deal of social meaning (or baggage depending on your point of view). In this interpretive sense, it is not and never has been a static work. Nevertheless, except for possible restorations, this work has not been creatively altered since da Vinci completed it.

This "static" work of art should be contrasted with other works, such as mobiles, that, by their very nature, are temporally changing. Similarly, a statue in an outdoor setting may appear very different in winter covered by snow than in summer surrounded with a skirt of flowers. Some contemporary artists have produced works that are ever changing such as Moholy-Nagy's Light Space Modulator.

In recent years, artists have begun to address how to make a traditionally static work of art dynamic by the use of new media. Liquid crystals, molecules that form oriented liquids, and some types of polymers have the

[4] *Postmodern currents: art and artists in the age of electronic media, Prentice Hall 1977.*

potential to achieve this goal. These materials introduce dynamic aspects into the work that cause it to interact with both the environment and the observer. These materials produce colors by selective reflection and differ in both saturation and quality from colors produced by more traditional pigments that depend on absorption of certain wavelengths of light.

David Makow, an artist and physicist, has explored liquid crystals as a medium for painting. He observes that by using liquid crystals, the colors and mood of a composition can change as a viewer approaches, pauses in front of, and backs away from a work. These colors may also be made temperature sensitive.

Traditionally, the artist has had an intimate knowledge of his materials. The sculptor knew the mechanical properties of marble. The painter mixed his own paints, knew good pigments from bad, and knew the effects of grinding the pigments more coarsely or finely. In general, they knew the thousands of little details that were part of the artist's craft. It is no accident that art and science were so closely connected during the Middle Ages and the Renaissance. In more modern times, it more difficult for an artist to develop an intimate knowledge of his materials. Nevertheless, it is possible to develop this understanding. Later in this book, after studying some of the science of light and liquid crystals, we will return to this topic and discuss the differences between liquid crystals and traditional pigments, as well as other aspects of this new art.

The computer has significantly changed the nature of the image. Before computers, even a simple picture was the product of human hands or arts, such as photography, that produced an image through chemical means. The data in a computer is stored as tiny charges of electricity, pits on plastic or tiny magnetic domains. A picture no longer needs to be a physical object – it can be nothing more than a series of ones and zeros recorded on any one of a large number of different storage media. A computer can create images by dutifully following an algorithm, a recipe, that states "Go here and plot color one" and then "Go there and plot color two" and so forth. Using a single keystroke a person can change all the colors to their complementary colors or make a color picture black and white, or vice versa.

The computer has changed the very nature of art. Formerly simple questions such as "What is an original?" have now come to the forefront. Other issues, including ownership of a piece of art, are now opened anew. The computer and digital art may represent the greatest challenge to art since the photograph. We may anticipate that the art forms that this medium spans and those that grow in response to it will be as invigorating and exciting as those spawned by the Daguerreotype. While one does not need to know ones media to be an artist, such an understanding is important, and allows one to fully utilize the potential of the medium. A computer is little more than plastic, metal, and sand, yet it has profoundly affected all of us. Yet another goal of this book is to understand how this technology works so that we may use it to its fullest.

1.2.3 *Liquid crystals and life*

The fundamental importance of liquid crystals to the cells of all organisms might be reason enough for a physical scientist to want to study them. On the other hand, a layperson might feel that the biophysics of cell membranes is excessively complex and difficult to understand, and question the importance of such studies to the non-technophile. Satisfactory answers to this question can be hard to come by. Nevertheless, a few answers will be provided.

Certainly, one of the big questions that we all ask is, Where and how did life originate? Can life be created in a test tube? Should such an experiment be performed? Another fundamental question is, "Did bilayer membranes, a liquid crystalline structure, form in the 'primordial soup'?" Similarly, why are liquid crystalline structures so common in biological structures? Why does a breakdown of this structure lead to disease? These questions require an understanding of how cellular life functions. This is the realm of biochemistry and biophysics, highly technical areas filled with specialized terminology. Nevertheless, ethical issues raised by recently publicized work in the areas of cloning and the study of DNA strongly suggest the importance of learning the terminology and relationships of these areas. (The educated populace argument once more rears its beautiful and bountiful head.)

While it is true that yesterday's lecture may well be out of date by next year, the ability to make the connections between disciplines and a primary knowledge of the relevant terminology, combined with the ability to teach oneself, will always allow a citizen to make educated choices.

A further motivation for studying the biophysics of liquid crystals is that the relationship between liquid crystals and life is at the very frontier of science where ideas and approaches from many disciplines are important and help to move the field forward. All of the "prefixed" sciences (bio-x, physical-x, structural-x, etc.) grew from the frontiers of the past. This makes studying such areas an exciting intellectual experience.

Too often, science is misrepresented as a well-known set of facts (most likely needing to be memorized) that have been known for ages. Usually, these "laws" were discovered by a bunch of dead white people who are completely unlike you. This author believes non-technical people should at least catch a glimpse of a frontier area of science so that they can understand why science is so fascinating and a discipline worthy of study and support. They should also learn about scientists who work at the frontier and see that scientists and non-scientists are much more alike than different.

The biophysics of membranes and the relationship between liquid crystals and life and disease is certainly not the only frontier in liquid crystal science. However, it is an ever more important one that naturally ties into our study of molecules and their interactions.

1.3 Overview - the road ahead

This book is a tool to help you start to understand the relationship between basic science, technology, everyday high tech devices, and life. There are no easy shortcuts that allow one to go from start to finish without climbing each hill and stopping at each summit to observe and understand the surroundings. Modern high tech devices rely heavily on many basic and applied technologies, each of which depends on basic science. There is no single basic science that dominates these areas. A healthy dose of chemistry, physics and biology will be needed to understand this material.

One of the amazing features of contemporary science is its reliance on scientific models. Often, the results of an experiment are explained in terms of a scientific model. The use of models to guide science and suggest experiments has been critical to the growth of science and technology. The explanations in this book are almost exclusively based on models. Thus, the first topic will be a discussion of scientific models.

One of the tools that scientists and engineers use to simplify problems, and to unify different aspects of problems is symmetry. Symmetry deals with similarities in patterns, and an understanding and knowledge of the use of symmetry is one of the techniques used by many scientists. We will have occasion to use symmetry to simplify discussions; for this reason a chapter on symmetry comes next.

With this as background, we will study the basic science that is the foundation of the technologies that we will investigate. One of the fundamental forces of nature is the force between two charged bodies. This force is called the electric or electrostatic force. We will study it and learn how more complicated charged objects interact with an electric field. This is important in biophysics, liquid crystal displays and all of electronics. The nature of light, how to control it, and its behavior at interfaces and in matter is crucial to understanding displays, sight, and colors. This will form the next chapter of this book. We will then turn our attention to the chemical nature of materials and discuss atoms and molecules. The systems that we observe and study consist of large numbers of molecules acting in concert. Thus we must start to understand how large numbers of molecules behave in order to discuss solids, liquids and gases. This study will form a backdrop for us to explore and understand semiconductors, polymers, liquid crystals and related phenomena.

The next four chapters will explore several types of molecules and materials. The first will discuss polymers, long chain or branched molecules. These molecules are important to life in the form of proteins and DNA and important to technology as a class of materials that can be made with a wide variety of properties. The second chapter in this group will discuss liquid crystals. Liquid crystal molecules act in concert to form structures that are not only useful in technological applications but fundamentally important to life itself. Semiconductors, the electronic building blocks of computers and microelectronics, will then be studied. You will learn about the unique properties of these materials that make them so technologically useful. only useful in technological applications but fundamentally important to life itself. The last chapter in this group will discuss the molecules that are important to life. These molecules are often polymers and amphiphiles and particular and important examples of the types of molecules in the previous chapters.

We will then discuss how these materials are put to use in technologically useful configurations. We will next investigate digital devices and microcomputers. This chapter will present and demonstrate new ideas that are needed to understand and explain these devices. It will end by discussing a fully functional, bare bones computer. We will then discuss liquid crystal displays, and bring together almost every topic that we've considered and introduce a few new twists that are exciting all by themselves. Finally, the combination of these devices into a laptop computer will be summarized.

The last part of this book contains two topics of contemporary interest. These chapters can be taken in any order and will explain how the basic science and technologies discussed earlier in the book relate to these topics. The next chapter explores some of the relationships between art and science mentioned earlier. In particular, we will look at some of the scientific aspects of pigments and the use of liquid crystals as an artistic medium. The last chapter is a rather detailed discussion of liquid crystals and life.

The language of quantitative physical science is mathematics. This does not mean that one must love math or be highly proficient at mathematics to understand how these technologies work. What it does mean is that we must go beyond a simple word description of the sort, "I do A, and B results." One must in some way associate numbers with A and B so we can discover how much a change in A causes a certain amount of change in B. For this reason there is an appendix with a very brief review of math you already know.

The more technical reader will already know some of the material presented in this book. It is hoped that there is sufficient new material and that the material is presented in such a way as to satisfy their curiosity.

Chapter 2

Scientific Models

2.1 Overview

Much of the analysis of science is performed on models and not on an actual physical object. Such analysis is what allows science to make predictions. One important feature of scientific models is that they are provisional; a scientist may propose a model and, after testing it, disclose that it does not explain all the data.

The author prefers to relate the goals of a chapter to questions. In this way the student will see "the problem we are trying to solve" and not become lost in the details. The following questions will be answered in this chapter.

(1) What is a scientific model?
(2) Are models correct or incorrect or simply adequate or inadequate?
(3) What are some examples of scientific models?

2.2 Introduction

The goal of science is to explain and predict natural phenomena. One of the primary tools used in this process is the scientific model. Almost of all of our discussions and analysis will be performed using models and not real objects. For this reason you should understand what a scientific model is and what its limitations are.

2.3 What is a scientific model?

A scientific model is a substitute for a real object or system. It emphasizes *some* features of a real object while de-emphasizing or even ignoring other features. The kind of features that are included or not included is determined by the purpose and audience of the model. Thus, a model which is considered appropriate for this book may be too simple for the detailed understanding needed to make a liquid crystal display that conforms to military specifications.

Generally, a scientific model can be written down. And while it can be described with words, it can frequently and more usefully be represented by a picture or series of pictures, a collection of formulas or mathematical equations, or a graph of the relationship between two variables. Sometimes one uses combinations of the above. The important feature which makes all of these representations models and not a random collection of pictures or equations is that they are used to describe a real entity and, for the purposes of the scientific discussion, represent the object.

Good models follow Occam's Principle,[1] which states that the simplest possible model that explains all the desired phenomena is the model of choice. A model includes the **minimum** number of features necessary for its desired purpose. A model should be as simple as possible within this constraint and, consequently, **all models are incomplete**.

Scientists perform calculations and make predictions based on models. If the predictions do not agree with experiment, the model is too simple or has left out some essential feature. **The ultimate judge of the validity of a model is experiment.**

Why do we use models? The basic answer is that nature is so complex that to make progress one must simplify until one obtains a model that is tractable, yet still closely approximates nature. Thus a model is best thought of as an aid in understanding a system or situation. Some properties are exaggerated while others are de-emphasized or even neglected. The simplicity of a model facilitates understanding relationships that may be obscured by the complexity of the real system.

[1]This is also called the the Law of Parsimony. It seems that the the statement normally attributed to him is found nowhere in his writings and is a more of an urban legend than fact. Some version of the "razor" can be found as early as Aristotle and certainly in Thomas Aquinas (SUMMA THEOLOGICA, Part I, Q. 2, a. 3, obj. 2). It was also referred to (but probably also violated) by Duns Scotus of dunce cap fame. Simply put, this principle states that the number of assumptions should always be a minimum. Those that are unnecessary are "shaved off."

The following points should not be forgotten.

(1) All models have limited utility.
(2) All relationships that hold for a model do not necessarily hold for the real object or system the model represents.
(3) Models are *adequate or inadequate* for tasks, not right or wrong.

These ideas can be illustrated by the following example.

Example 2.1

The primary component of natural gas used for heating and cooking is methane. Suppose one wants to discuss chemical structure of natural gas. Several scientific models are available. Some of these models are shown in Fig. 2.1.

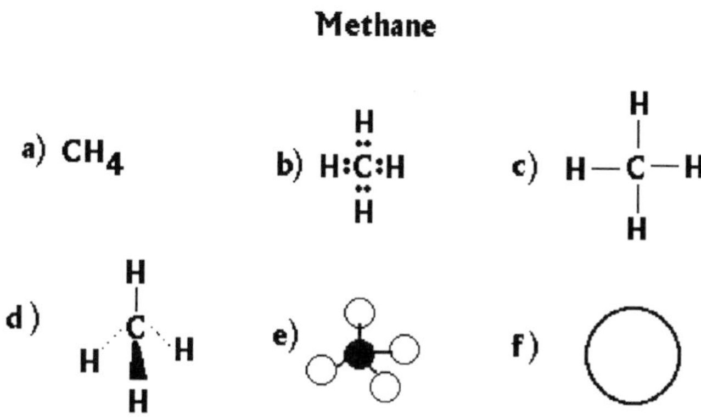

Fig. 2.1 Models of methane

The first model, labeled a), is the chemical formula for methane. It tells us that methane is a molecule made of one carbon atom and four hydrogen atoms. This model tells us what the composition of methane is at the atomic level. The next two models are similar. Model b) is the electron-dot formula. This model uses dots to represent electrons involved in the binding of the carbon and the hydrogen. Such a model is useful in discussing chemical reactions with other molecules and atoms. Model c) is the structural formula. It indicates that there are four single bonds coming from the carbon. Each line represents a single bond.

Models b) and c) may lead one to believe that methane is a planar molecule. This is not the case. Models d) and e) try to emphasize the three dimensional structure of this molecule. Model d) is an example of a perspective formula. In this representation the lines (solid, broken and triangular) represent single bonds. The solid line is in the plane of the page, the dotted line is behind the page and the triangular line is coming out of the page. Thus, one sees that methane has a tetrahedral structure. Model e) is a drawing of a ball and stick model of methane. This model could be constructed from colored gumdrops and toothpicks. Models are not restricted to lines drawn on paper. Here the atoms are represented by gumdrops, say red for carbon and green for hydrogen and the bonds are represented by the toothpicks. This toothpick and gumdrop representation is an example of a scientific model that is also a physical model.

The last model, f), is the simplest. It represents methane by a sphere and ignores all of the internal details of the molecule. It emphasizes that a molecule takes up space, and that to a good approximation, the molecule is rotating rapidly. Thus, it can be well represented by a sphere, which is drawn as a circle in a plane. Model f) is useful in discussing the physical properties of liquid and gaseous methane.

2.4 References

This chapter was developed from notes that were written by Professor Klaus Fritsch at John Carroll University, University Heights, Ohio. A good reference for this material is **Projects and Investigations The Practice of Physics**, by R. E. Vermillion (Macmillan Publishing Co. 1991).

2.5 Exercises

(1) In what sense is a map a model?
(2) Many children play with toy cars and trucks. Discuss how and why these toys are models. For example: What features are de-emphasized? What is its intended purpose? ...
(3) In the early 17th century, Kepler concluded, by analysis of Tycho Brahe's data, that the orbit of Mars was an ellipse with the sun at one foci. This result was theoretically explained by Sir Isaac Newton in 1687.[2] Earlier Copernicus constructed a model of the solar system

[2]This result is in the Philosophiae Naturalis Principia Mathematica, "The Principia."

with the sun at the center and the planets revolving around the sun in circular orbits. However, to explain the observed motion of the planets this model had to invoke "epicycles." The epicycle model consisted of small circular orbits about the larger circular orbit of the planet. An example of an orbit with four epicycles is shown in Fig. 2.2. An orbit according to Newton's theory is also shown.

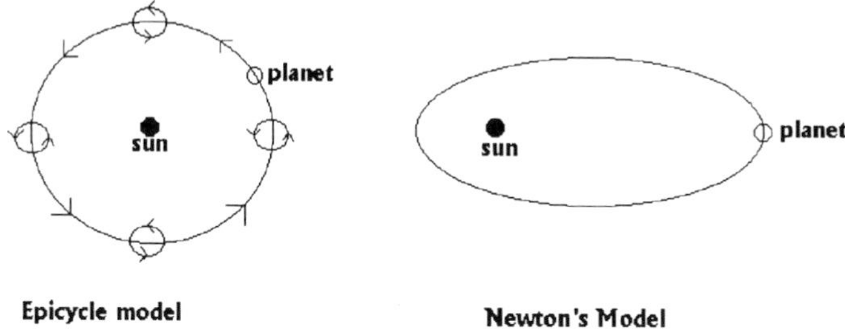

Fig. 2.2 Epicycle and elliptical orbits

The epicycle model explains the motion of the planets as well as the Newton model, and no one has ever traveled to a point well above the ecliptic, the plane containing the orbits of most of the planets, and verified Newton's prediction. Why then is Newton considered correct and the epicycle model a historical footnote?

(4) The motion of a thrown ball is described by a set of three equations that describe its position as a function of time: $x(t)$, $y(t)$, and $z(t)$. Here, the symbols $x(t)$, $y(t)$, and $z(t)$ are equations that express the x, y, and z positions of the ball (relative to a reference point) as a function of time, t. $x(t)$, $y(t)$, and $z(t)$ are called the dependent variables and t is called the independent variable. A skilled physicist could envision many different sets of equations which have different mathematical complexities, various types of forces, include or do not include the spin of the ball, and so forth to describe the position of the ball.
Do these equations represent a model? If a set of equations predicts the trajectory of a thrown ball, is it correct? Suppose a set of equations only explains the trajectory approximately. Is that set of equations

However, Newton showed this before or during 1684 and communicated this idea to Sir Edmund Halley, who later predicted the return of the comet that bares his name.

useless? Explain.

2.6 Research question

(1) What makes a model scientific? For example, a person claims, ("has a model") that quartz crystals can cure headaches. Would such a model be scientific? How do you decide? Explain and defend your reasoning. *You are strongly encouraged not to use this example.*

Chapter 3

Symmetry

3.1 Overview

Most people have a sense of what symmetry is and can easily recognize a symmetric object. The ability to identify symmetry in various situations is a useful skill that is used in many arts and professions. In science, symmetry goes beyond simple geometry and plays a major role in the understanding of many phenomena. We will use symmetry later in this book to understand liquid crystals and some of the molecules of life. This chapter will introduce some concepts of symmetry and answer the following questions:

(1) What is symmetry? Why should we care?
(2) What are the basic symmetry operations—how does one repeat a pattern in a symmetric manner?
(3) Can simple symmetry operations be combined into more complex symmetry patterns? How does one do this?
(4) How does one develop the skills to observe and discern symmetry in objects?

3.2 Introduction

You are familiar with some aspects of symmetry. For example, you know that, to a very good approximation, a mirror placed along a line extending from the middle of your nose through your navel will reveal that the left and right sides of your external body are symmetric. Similarly, you know that there is something symmetric about your left and right hands. Many of us admire the petals of flowers and their seemingly symmetric organization, and see a certain beauty in a well-built fence.

An understanding of symmetry will be of assistance in understanding solids and liquid crystals. In fact, one of the fundamental postulates of crystal physics is *Nuemann's Principle*. This may be stated as follows: "The symmetry elements of any physical property of a crystal must include the symmetry elements of the point group of the crystal." At this stage this has many undefined terms, as this chapter unfolds this statement will become clearer. It is also essential to understanding their various molecular organizations or phases. Many of the molecules of life have a certain *handedness*. This chapter will explore some aspects of symmetry that we will need later in this book.

The Greek root of the word symmetry is *metron*, which means to measure, and the prefix *sym* means together. Combined, they convey the sense of an object having the same measure or "look." The equality of certain dimensions or measured quantities is the essence of symmetry.

3.3 Representation of a pattern

To illustrate symmetry we must first pick a motif or pattern that is unsymmetrical. For example, consider the shoe shown in Fig. 3.1

Fig. 3.1 A shoe, an example of an asymmetrical object.

Of course, shoes come in two forms, one for the right foot and one for the left foot, as illustrated in Fig. 3.2.

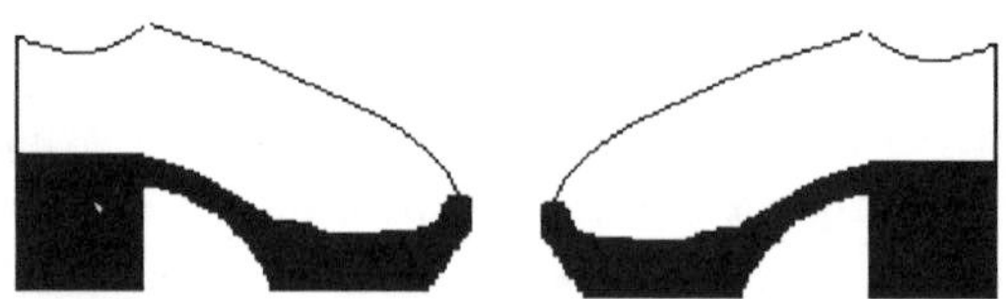

Fig. 3.2 Left and right shoes, examples of enantiomorphic objects.

To distinguish between these two forms, we call one *left-handed* and the other *right-handed.* Any asymmetric object can exist in both left-handed and right-handed forms. Pairs of objects that are related by having right and left-handed forms are said to be *enantiomorphous*. Enantiomorphic objects are mirror images of each other. If you hold your right hand up to a mirror, it looks like your reflection is holding up its left hand. Thus your left hand is said to be the mirror image of your right hand and vice-versa. Hence the terms left-handed and right-handed.

It is important to note that enantiomorphic objects are not identical to each other. One test of the identicalness of two objects is the ability to superimpose one on top of the other. If the two objects can be made to exactly superimpose one on top of the other (for instance by rotation), they are not enantiomorphic.

Suppose one tries to superimpose exactly, the $<$ onto the $>$. This can be done by rotating one of these symbols in the plane of the page. This is in contrast to superimposing the same symbols with the lower leg colored green. In this case superposition of the objects may only be obtained by rotating the object out of the plane of the page. Similarly, a left hand-print and a right hand-print (of the same person) can only be made to superimpose by turning one over. For a planar object this is a rotation in three dimensions and is not allowed. Thus, we can distinguish two kinds of objects: those that can be superimposed onto their mirror images and those that cannot be superimposed onto their mirror images.

Objects that are not superimposable on their mirror images are said to be *chiral.* This definition goes back over 100 years to Lord Kelvin, who wrote in 1893: "I call any geometric figure, or any group of points, *chiral,* and say it has *chirality,* if its image in a plane mirror, ideally realized, cannot be brought to coincide with itself."[1]

Chirality is the necessary and sufficient condition for the existence of enantiomers. An object (or, as we will see later, a molecule) that is chiral can exist in two forms, left-handed and right-handed. Hence there are two enantiomers of the object. An achiral (without chirality) object has only one form and cannot possess enantiomers. Thus a symmetric two-dimensional object which exists in two forms (*e.g.* '(' and ')') is not chiral since by rotating after the mirror reflection one of these, they can be made to coincide. However, an object need not be three-dimensional to be chiral, as the example of hand-prints illustrates. The test of chirality is to see, either in your mind's eye or in an actual drawing, if the object and its

---

[1]Chiral is the Greek word for hand.

mirror image can be superimposed exactly upon one another. If the object and its mirror image are **not superposable**, the object is chiral.

To see how one might do this, consider the object in Fig. 3.3:

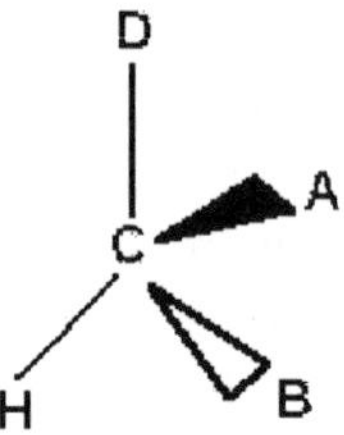

Fig. 3.3 A chiral object.

and its mirror image in Fig 3.4. In these figures the two thin lines are in the

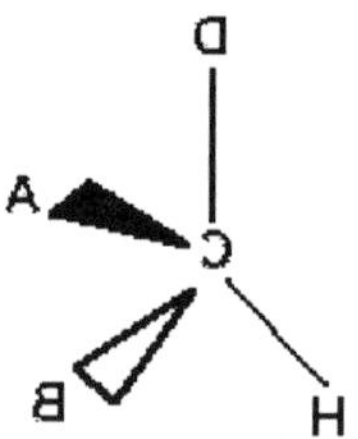

Fig. 3.4 The mirror image of the object in Fig 3.3.

plane of the page, the dark triangle is considered to be going into the page and the open triangle is considered to be coming out of the page. Placing these objects on opposite sides of a mirror gives us Fig. 3.5

You can test that these two objects are mirror images by noting that the parts of the object closest to the mirror have corresponding parts on the image that are also closest to the mirror, and vice-versa. Now imagine moving the image on the left past the mirror and placing it next to the object on the right. This situation would look something like Fig. 3.6. The two objects cannot be superimposed. Therefore, they are enantiomers, and there are two forms this three dimensional object.

As another example, consider a case with a non-chiral (or achiral) ob-

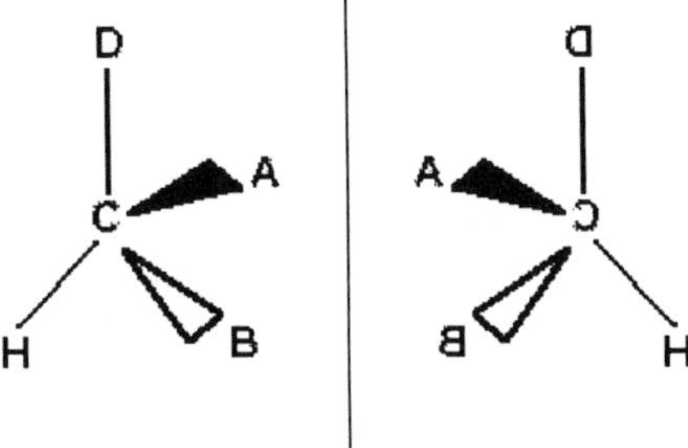

Fig. 3.5 The original object is on the left, a mirror is in the middle and the mirror image of the original object is on the right.

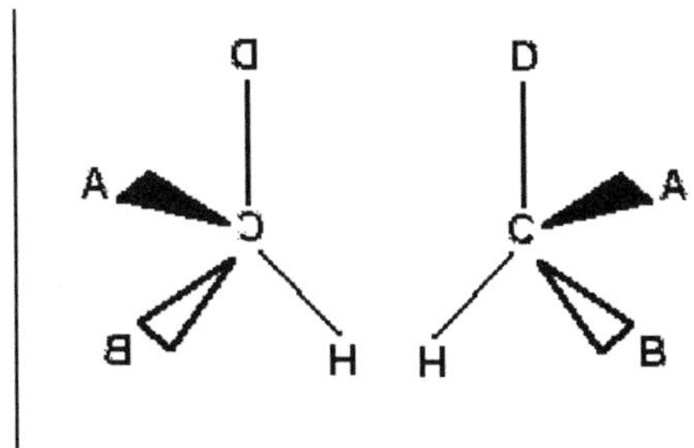

Fig. 3.6 The object translated past the mirror.

ject, the previous object with two of the five letters the same. To perform the chirality test, we follow the same procedure as before. First, draw the two objects and a mirror, as shown in Fig. 3.7. Ask yourself, "Is the image correct?" [Yes.] Then translate the object on the left across the mirror to obtain Fig. 3.8.

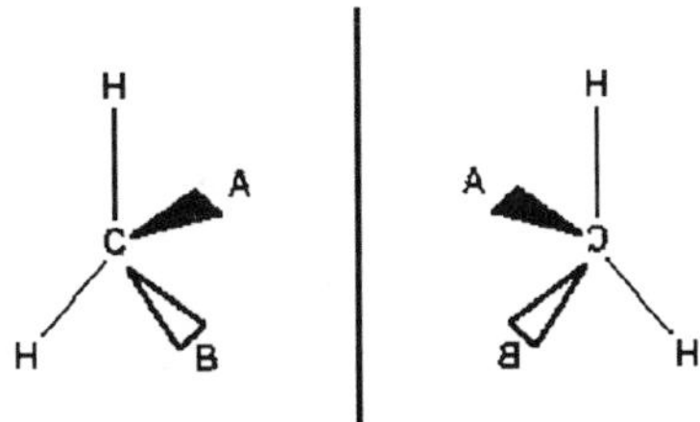

Fig. 3.7 A non-chiral object, mirror and its reflection.

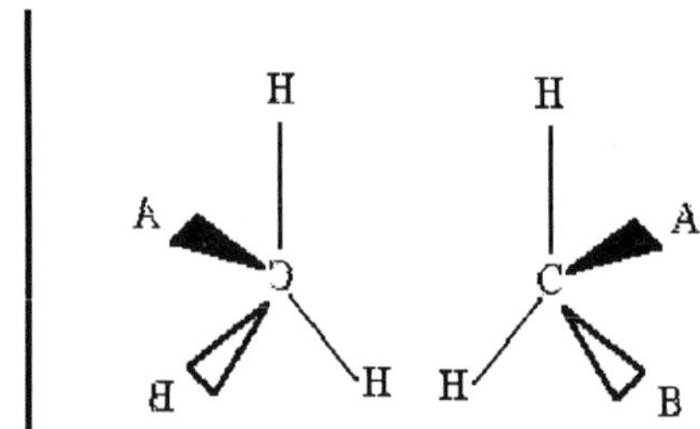

Fig. 3.8 Non-chiral object to the left of the mirror moved to the right of the mirror.

These two do not look identical. However, they can be made to exactly superimpose by rotating one of the objects, in the plane of the page, so that both pairs of H's point in the same direction. The direction of rotation is shown and how the objects appear after rotation are shown in Fig. 3.9.

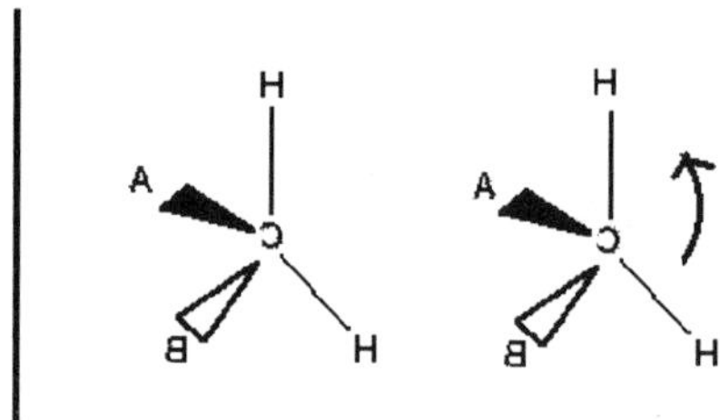

Fig. 3.9 Demonstration by rotating the translated object in Fig. 3.8 that the two non-chiral objects are in fact, the same.

The objects can now (in your mind's eye) be moved and made to exactly superimpose. Since these two objects are superimposable, the original object is not chiral.

3.4 Why is chirality important?

The previous section strongly suggests that chirality is important. However, the reasons were not discussed. Molecules can be chiral, and the single crystals that grow from these molecules can exhibit this symmetry. In fact, this is how chirality was discovered in 1848 by Louis Pasteur. More recently, it was predicted (and found) the a certain class of chiral liquid crystal materials would exhibit interesting and useful electric properties when the long range order of the molecules consisted of a layered structure with the molecules tilted with respect to the layers.

At a very fundamental level, living organisms use an extremely large number of carbon containing molecules that are sometimes chiral. For example, the building blocks of proteins, the amino acids (except for the simplest, glycine) are chiral and left-handed. The building blocks of DNA, the nucleotides are always right-handed. Interestingly, humans only metabolize right-handed glucose (table sugar). Left-handed sugar tastes sweet but passes unchanged through the body. Similarly many pharmaceuticals are chiral: one handedness can be useful and the other useless or even dangerous. An example of this is Thalidomide.[2] Another common chiral substance is aspartame (Equal^{TM}): S- or left-handed aspartame is bitter while R- or right-handed aspartame is sweet. We will explore these areas in greater detail later.

3.5 Repetition of an object

Earlier we stated that symmetry had to do with an equality of measure. Another way to state this is to say that symmetric objects, in some way, possesses repeated patterns. Recalling that the essence of symmetry is a repeated pattern, we now ask "How can one symmetrically repeat an object?" In answering this question, the equivalence of "equality of measure" and "repeated patterns" will also be demonstrated.

How can we symmetrically repeat a pattern? The first thing we must do is pick an unsymmetrical object or pattern to repeat. The reason for this is that a repeated symmetric pattern may have symmetries that go beyond the simple symmetries that we intend to illustrate. Consider the number seven (7). It is sufficiently unsymmetrical to use to illustrate the basic principles.

Figure 3.10 shows an enantiomorphous pair of sevens, one on each side and horizontally equidistant from a vertical line. Observe that while a single seven is asymmetrical, the figure shown is symmetrical. Each side of the figure looks like a mirror image of the other reflected across the vertical line. Figure 3.10 demonstrates one of the more basic symmetry 'operations,' ***reflection* of an object across a *symmetry line***. Symmetry resulting from the reflection of an object across a symmetry line is sometimes called

[2]The S isomer has antinausea effects, while the R isomer is teratogenic, causing serious birth defects. However, in the body, the S isomer is partially changed to the R isomer so that a 50-50 mixture results.

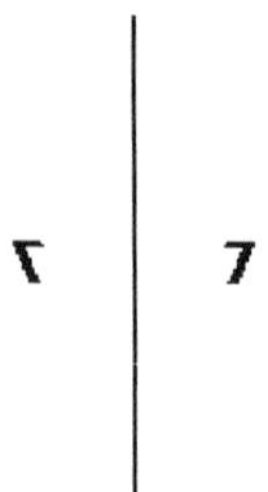

Fig. 3.10 An enantiomorphous pair of sevens.

mirror plane symmetry because the symmetry line can be thought of as the intersection of a *mirror plane, MP* with the page.

Another basic symmetry operation, is **repetition by *translation***. This may be illustrated in the following manner. Place a copy of the object at some point. Then move a distance **t** in some direction and place a copy of the object, in exactly the same orientation, at this new point. Then move again a distance **t** *in the same direction* and place another copy of the object, once more in the same orientation. Continue forever to place copies of the original object after moving, always in the same direction, a distance **t**, while leaving the orientation of the object unchanged. Of course, in real situations the repetition does not go on forever. Translation is, of course, familiar in fences, railroad ties, telephone poles and the like. Here is an example of translation in one dimension:

7 7 7 7 7 7 7 7 7 7 7 7 7

→ (the direction of translation)

t ↔ (the translation distance).

You may have noticed that the translation distance, **t**, is in bold type. Bold type is traditionally used to tell the reader that this quantity has both size (how large the translation is) and direction. In the above example, the size is approximately 1/4 inch and the direction is to the right. Physical quantities that are characterized by both magnitude and direction are called vectors. Thus, **t** is called "the translation vector."

One can also symmetrically repeat an object by rotating the original object. To generate such a rotationally symmetric pattern one must rotate the original object by a certain fixed angle about an axis that is perpendicular to the object and then place a copy of the original object in this position. This process is repeated until a rotated copy superimposes upon the original. The check that the pattern has rotational symmetry is that the final object appears unchanged by rotation by the constant rotation angle.

Once more, the number 7 will be repeated to generate a pattern with rotational symmetry. In the examples that follow, the number 7 will be a distance "a" from a central point which will be represented by a +. Imagine a thin rod pushed through the + so that it is perpendicular to the page. This rod represents the rotation axis. We can then symmetrically repeat a seven by rotating the seven about the rod as shown in the two examples in Fig. 3.11.

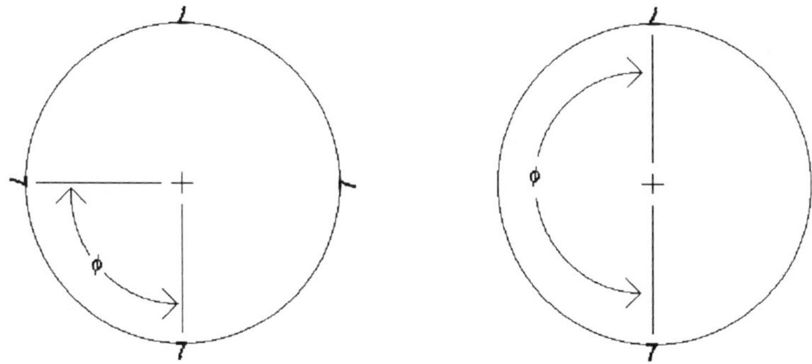

Fig. 3.11 Examples of rotational symmetry.

Example 3.1 Consider the left side of Fig. 3.11 which has four sevens. The way to understand this picture and the operation that formed it is to imagine a rotation axis passing through the + at the center of the circle and perpendicular to the page. Now, pick the seven at the bottom of the circle as the original object. Then rotate the axis and this seven clockwise by 1/4 of a turn clockwise, and draw another 7. This is the 7 that is on the left of the circle. Then rotate the axis by the same amount once more by 1/4 of a turn clockwise and draw another seven. Notice that because of the rotation, this seven is upside down from the original, that is to say, it is an upside down seven. Repeating this process one more time leads to the seven on the right of the circle. Finally, rotating 1/4 turn clockwise for the fourth time leads to a seven that is exactly superimposed on the original 7.

Now consider the whole object consisting of four sevens. If you rotate this object by 1/4 turn, clockwise or counterclockwise, it will appear the same as the unrotated object. This means that the object exhibits symmetry of some sort. This symmetry is called ***rotational symmetry***, or, more precisely, ***proper rotational symmetry***.

Note that if one of the sevens carries a little tag, the pattern will return to its original position and orientation after 4 partial rotations. Generally, any pattern of sevens will return to its original orientation after an integer number n (n = 1, 2, 3, 4, ...) of partial rotations, (*i.e.* n partial rotations = one full rotation). This means that the angle ϕ (the Greek letter phi; this letter is often used by scientist to represent angles) between each repeated object must be an integral submultiple of one revolution: $\phi = \frac{360^o}{n}, n = 1, 2, 3, ...$ To see this, recall that a full rotation corresponds to 360°. The example just discussed corresponds to n=4 and $\phi = 90^o$.

Example 3.2 The object on the right side of Fig. 3.11 has two sevens. Note that if one makes a copy of this object and rotates it about the axis by 1/2 turn, it will superimpose on the original. A second partial rotation by 1/2 turn will return the object to its original orientation. The reader should convince themselves that this situation corresponds to n=2, and $\phi = 180^o$.

The three symmetries that have just been discussed, reflection, translation, and rotation, are called *point group symmetries*. This is because they all have **at least one special point** that differs from all others. This point is unique because it remains unchanged no matter what symmetry operation, **consistent with the symmetry of the object**, is performed. Consider the rotational symmetry that we just discussed. The center that we rotated about, the +, is a unique point. This point is unique because i) it is the center, ii) it is not repeated elsewhere in the pattern, and iii) it does not change or move when we rotate about an axis through it. Similarly, the mirror plane is a member of a point group because each point along the mirror plane is unique, not repeated elsewhere and unchanged by the action of the mirror plane. The translation clearly has many special points. For example the center of all of the first object, the second object, ...

The reader should also be aware that translational symmetry is not limited to one direction. It is also possible to have translational symmetry in two and three dimensions. This will be discussed in greater detail later when we discuss solids and liquid crystals. For now simply observe the following 2-d translational pattern:

g g g g g g g g g g ...

g g g g g g g g g g ...

In this pattern the translation distance, $\mathbf{t}_{horizontal}$, is about 1/4 inches and the vertical translation distance, $\mathbf{t}_{vertical}$, is 1.5 lines.

We will now look at rotational symmetry more carefully. When rotational symmetry is present, **successive rotations** by an amount ϕ (phi) will lead to superposition of the rotated object upon the initial object. After a sufficient number, n, of rotations by the angle ϕ, the rotated pattern will be superimposed upon the original. This is illustrated by the pinwheel in Fig. 3.12. One petal of the pinwheel has been marked so that it is easier

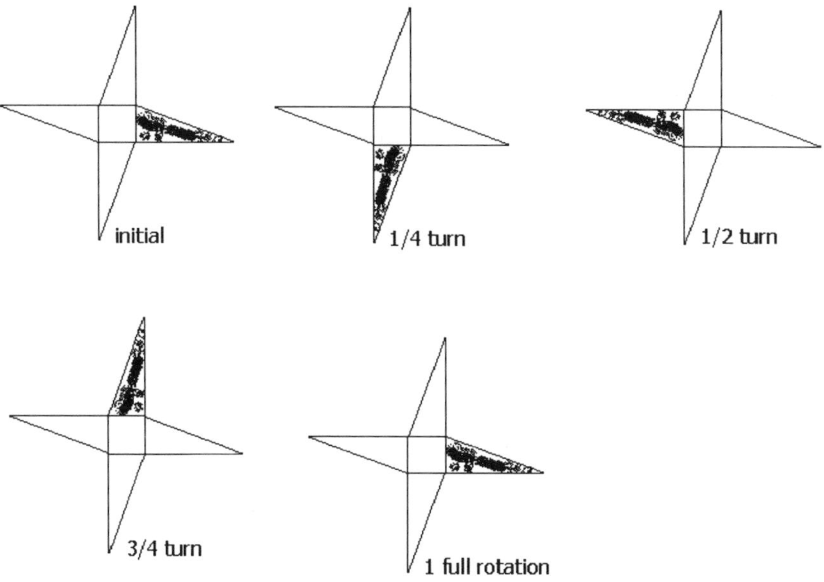

Fig. 3.12 Rotational symmetry of a pinwheel.

to follow. **This marked petal does not exist in the real symmetric object; it has been added as a visual aid.** The important point is that except for the marked petal, the object is the same after each quarter turn as it was initially. This object, without the marked petal, has rotational symmetry about an axis through its center.

The angle of rotation, ϕ (phi), is called the *throw of the axis* and the number, n, of successive rotations needed to achieve superposition is called the *fold of the axis*. The pinwheel has a 4-fold rotation axis and a throw of 90°.

A rotation of the pinwheel by 1/2 turn and a full turn also leave it unchanged. These are in fact, examples of the generally observed rule that if a n-fold axis exists then folds which are obtained by dividing n by an integer that leaves no remainder also exist. Thus a 4-fold axis will generally also be consistent with a 2-fold and (always) a 1-fold axis. A 6-fold axis will also have 3-, 2- and 1-fold axes. When quoting the symmetry of an object the highest fold consistent with the symmetry of the object is quoted. Thus the object with 4 7's has 4-fold symmetry.

A rotational axis is said to exist when *all of space around the axis* has the property of rotational symmetry about this axis. An important characteristic of isolated objects is that there is no limitation on the allowed throws. This is not the case for crystals. **Crystals can only have throws of 360°, 180°, 120°, 90°, and 60°** because translational symmetry puts an added constraint on the allowed rotations.

3.6 Combined symmetries

It is common to observe objects with combinations of symmetries. While the pinwheel has only the 4-fold rotational symmetry, many objects combine more than one of these point symmetry operations.

Figure 3.13 illustrates an object possessing both rotational and mirror symmetry. Such a combination of symmetries is found in many objects and naturally in flowers. Once more, the small partially shaded area near the rotation axis is just meant as a visual aid. Consider the top row of this figure. The original object, rotated by 1/2 turn about an axis perpendicular to the page and passing through the +, is shown on the right. The rotated object appears, except for the small shaded area, to be the same as the original object. Thus, the original object has a 2-fold rotation axis. There are also the two mirror planes shown in the bottom of the figure. Notice for the vertical mirror plane that the part of the object to the left of the mirror plane is the mirror image of the part to the right, and vice-versa. Similarly, the part of the object above the horizontal mirror plane is a mirror image of the part below the mirror plane and vice-versa. There are also two 2-fold axes in the plane of paper along the two mirror planes. If these mirror planes are treated as rotation axes, then rotation about these axes by 1/2 turn will leave the object unchanged.

One should not infer that mirror planes necessarily have rotation axes associated with them. Consider the a rectangular table as shown in Fig. 3.14. This table has a two-fold rotation axis running vertically through

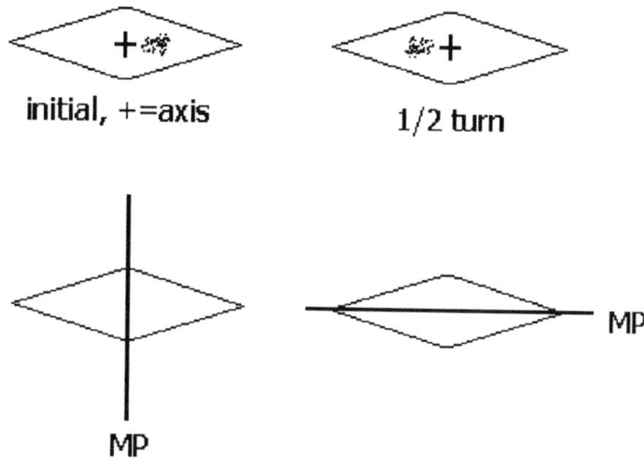

Fig. 3.13 An object with rotational and mirror symmetries.

the plus sign. If one rotates the table by 180° then the table appears unchanged — the hallmark of a symmetry operation. Similarly, the table appears unchanged when reflected across the two mirror planes shown. However, rotating around either one of the mirror planes would make one flip the table over. Clearly, an overturned table does not look like an upright table!

3.7 Observing symmetries

You are now aware of the three basic symmetry operations: reflection across a mirror plane, translation, and proper rotation about an axis. However, the observation and identification of these symmetries in real objects is somewhat more difficult and requires some practice. First, at a fine enough level almost no object has symmetry since there is always a small scratch, change in color or the like. These small imperfections have to be ignored to observe symmetry in real objects. This section will present and discuss several examples of identifying symmetries in objects.

First, observe that every object has a one-fold rotation about any axis. This means that if you rotate the object by a full rotation about any axis it will appear the same as it did before the rotation. (This is not very

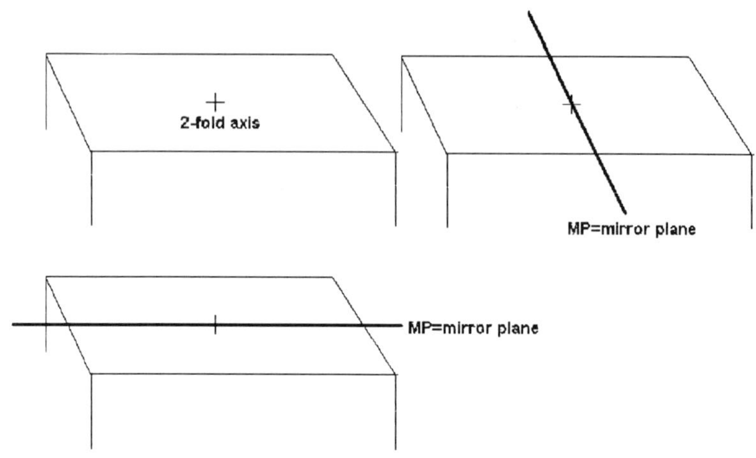

Fig. 3.14 A rectangular table and its symmetries.

interesting.) Thus, in searching for rotation symmetries the goal is to find higher fold (smaller angle) rotations that leave the object unchanged. Furthermore, an object may possess both two-fold (the object is unchanged by a rotation by 180^0) and a four-fold (the object is unchanged by a rotation by 90^0) rotation about the same axis. In this case, the object is said to possess the higher fold symmetry axis — the smaller 2-fold and 1-fold axes are inferred as discussed above.

The author knows of no "best" approach to listing symmetries. However, a systematic approach with checks will always work. Also, it is often easy to observe or eliminate translation as a first step, by asking, "Does the motif (pattern, object, ...) repeat itself exactly after being moved a fixed distance in a given direction?" If the answer is no, then the object lacks translational symmetry. An answer of yes leads to the question, "What is the translation distance and direction (translation vector)?" Another key to remember is that **mirror planes must pass through the center of the object**. If they did not, the two sides could not be mirror images and the plane would not be a mirror plane.

Example 3.3 Consider the planar object shown in Fig 3.15 It is an isolated object. Therefore, it **does not possess** translational symmetry. What about a mirror plane? Note that the left and right sides of this object appear to be the same. This is a clue. If one places a mirror plane down the center of the object as shown, the left and right hand sides of the object turn

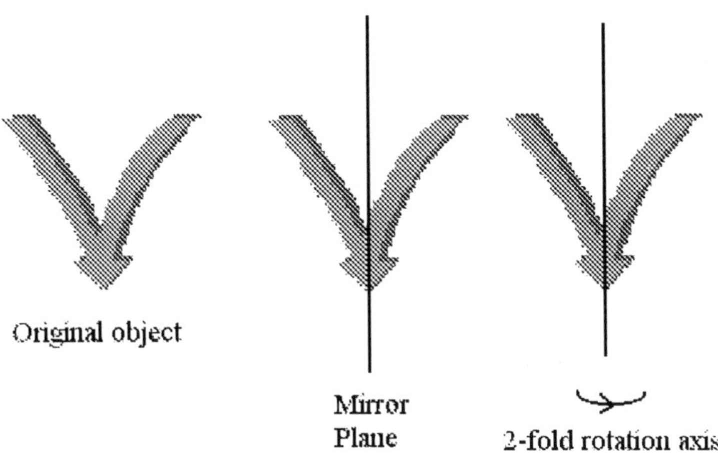

Fig. 3.15 A planar object and its symmetries.

out to be mirror images of each other. Is there any rotational symmetry? First, ask "Is there anyway I can rotate this about an axis perpendicular to the page that will leave the objects appearance unchanged?" It appears not. Then ask, "Are there rotations whose axis is on the page (in the plane of the page)?" The mirror plane is a good place to start looking, especially for planar objects. If I rotate this object by 180° along an axis as shown, the object is unchanged. Thus, this is a two-fold axis. There are no other symmetries to this object.

Example 3.4 Consider now the planar star shown in Fig 3.16. Again, this is an isolated object and there is no translational symmetry. Next consider mirror planes. The second star shows one of five equivalent mirror planes that exist in this object. The other four mirror planes go through the center of the star and the end of each of the other points of the star. If the top and bottom of the star are the same, this mirror plane is also one of five 2-fold axes that lie in the plane of the page. The other four 2-fold axes are along the other four mirror planes. Lastly, ask, "Is there a rotational axis perpendicular to the figure?" Yes, it goes through the center of the star, and since rotation by 1/5 of a turn leaves the object unchanged, it is a 5-fold axis.

Example 3.5 Sometimes the symmetries are more difficult to observe. Consider the Chevrolet^{TM} "Bow Tie" symbol shown in Fig. 3.17.

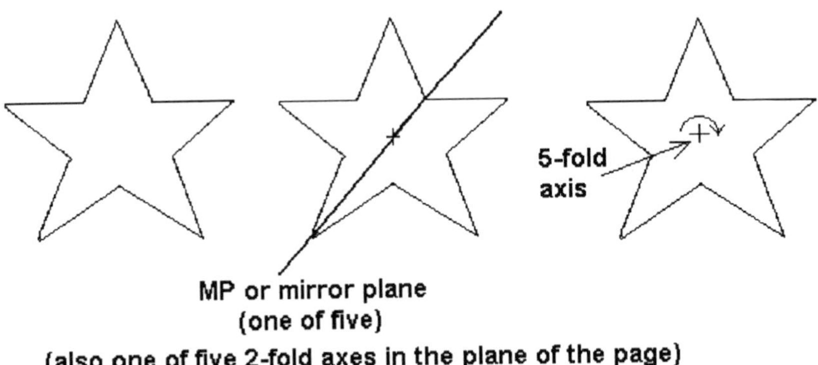

Fig. 3.16 A star and its symmetry operations.

Fig. 3.17 Outline of the Chevrolet^{TM} "Bow Tie" symbol.

Once more, there is only one object so there is no translational symmetry. Are there any mirror planes? Since this object is oblong, the best places to look are along the object's long and short axes. These mirror planes also must pass through the geometric center of the object thus two possibilities exist.

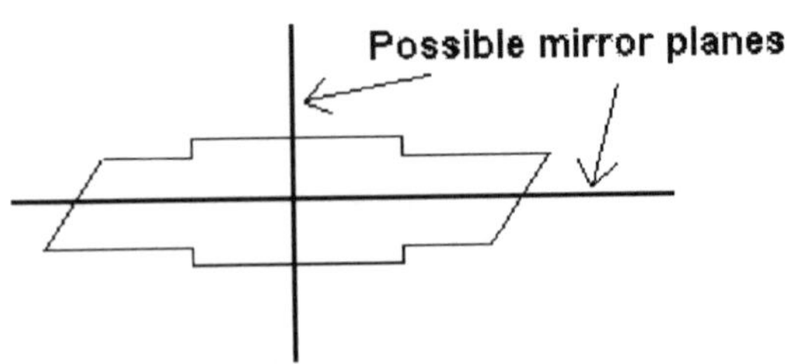

Fig. 3.18 Possible mirror planes for the Chevrolet^{TM} symbol.

Consider the vertical mirror plane shown in Fig 3.19. Does the right part of the object appear to be a mirror image of the left part? In particular, ask,

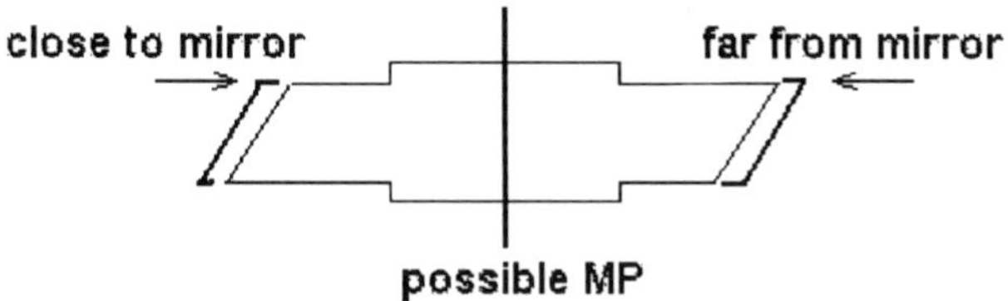

Fig. 3.19 Possible vertical mirror plane in Chevrolet symbol.

"Are there any places where there is not an exact one to one correspondence between two parts of the figure?" If the answer is yes, then the two indicated sections are not mirror images. Therefore, a vertical mirror plane does not exist.

The reader should draw the corresponding picture for the possible horizontal mirror plane. The same two "end sections" will not be mirror images of one another. Therefore, there are no mirror planes in this object.

Does a rotation axis perpendicular to the page exist? Such an axis should have a fold > 1 or it is not considered rotational symmetry. A good place to look for such an axis is at the geometric center of the object. This is indicated in Fig. 3.20 by the plus sign.

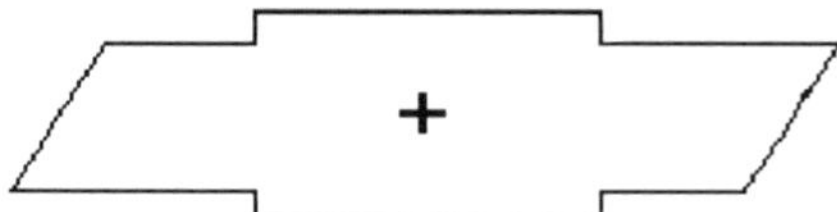

Fig. 3.20 The center of the Bow Tie.

Clearly, rotation by less than 1/2 of a turn will not leave the horizontal lines horizontal and the vertical lines vertical. This means that there cannot be throws (rotation angles) of this object of less than 180°. Thus, try 1/2 turn. Figure 3.21 shows the original object with the "r" on the left and the rotated object with the upside down "r" on the right. The "r" is used so you can see that the object was indeed rotated. Now erase the r's (in your mind) and observe that the object and the rotated object are identical. Therefore, there is a 2-fold rotation axis perpendicular to the object through the +. There are no other symmetries in this object.

Translational symmetry is never as ideal in practice as in theory. No

Fig. 3.21 A Bow Tie and a rotated Bow Tie illustrating the 2-fold rotation axis.

object is repeated without change for an infinite distance. So, in practice, when one looks for translational symmetry one looks for identical objects that are repeated many times. The word 'many' is deliberately vague. It generally means four or more. However, neither the minimum nor the maximum number of repeated objects is specified. Thus, can four identical equally spaced objects in a row (for example: ? ? ? ?) be said to exhibit translational symmetry? This is a tough call. In a pinch, say, on an exam, one may say yes. However, in general this is too few. Even so, one must be careful when one looks at translation not to pick too small of a motif.

Example 3.6 Consider the motif shown in Fig. 3.22 and the object shown in Fig. 3.23.

Fig. 3.22 Motif for a repeated object.

Fig. 3.23 A pattern with symmetry constructed from motif in Fig 3.22.

This object clearly has translational symmetry at least for a few steps; a square is repeated after a fixed distance in the same orientation. Note however, that the repeating pattern is not a single square, but a combination of an empty and a filled square.

Observe in Fig. 3.24 that this pattern also has a horizontal mirror plane and a 2-fold axis in the plane of the page along this mirror plane. However,

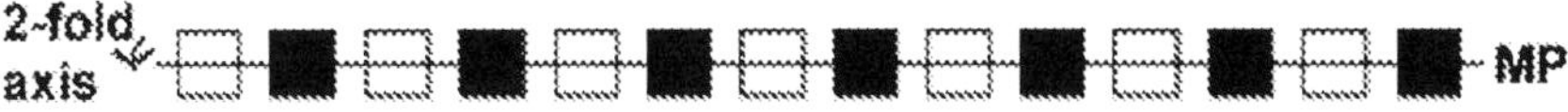

Fig. 3.24 Mirror plane and 2-fold axis of this object.

there is no rotation axis perpendicular to the page because the object is asymmetric in the sense that it starts with an empty square and ends with a filled square. This illustrates that one must sometimes look very carefully to eliminate symmetry.

Example 3.7 Suppose we cut off the rightmost black square from the object in Fig. 3.24. We are then left with the object in Fig. 3.25.

Fig. 3.25 Symmetric object with approximate translational symmetry.

One could still argue that there appears to be translational symmetry, but now there are two mirror planes as shown in Fig. 3.26.

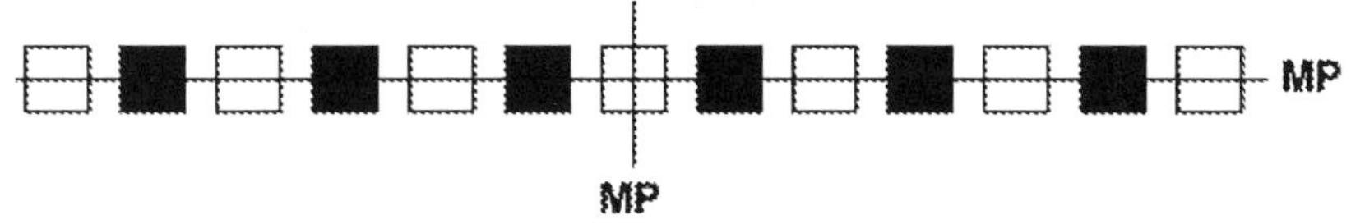

Fig. 3.26 New mirror planes object of Fig 3.25.

These mirror planes also correspond to 2-fold rotation axes in the plane of the paper as shown in Fig. 3.27.

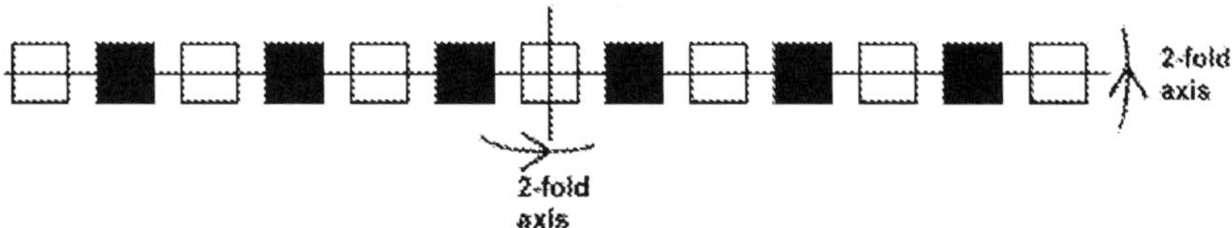

Fig. 3.27 Two 2-fold rotation axes of the object in Fig. 3.25.

These are not all of the symmetries of this new, slightly modified object. There is now a 2-fold rotation axis perpendicular to the page, through the intersection of these two mirror planes. This is illustrated in Fig. 3.28.

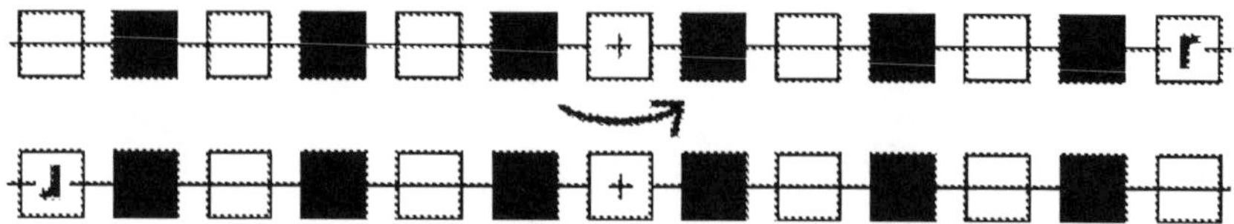

Fig. 3.28 Additional symmetry axis of object in Fig. 3.25.

Once more, an "r" has been appended so that the rotation is clear. The object looks the same after rotation as it did before once the "r" is removed. Therefore, the object has a 2-fold rotation axis through the + in the center. This extra symmetry demonstrates how tricky it can be to discover subtle symmetries in an object.

Example 3.8 As a final example, consider the Ying-Yang symbol shown in Fig 3.29.

Fig. 3.29 Ying-Yang symbol.

This object has a harmonious appearance, yet does it possess any symmetry? First, there is only one symbol, so there is no translational symmetry. There is no clear up or down, left or right, or any version of a mirror plane, so there are no mirror planes. Does this object exhibit rotational symmetry? The best bet would be to rotate it by 1/2 turn through an axis perpendicular to the page. The symbol then looks like Fig. 3.30.

At first glance, Fig 3.30 appears similar to Fig. 3.29. Note however, that the filled part is now empty and the empty part is now filled, *i.e.* the light and dark sections have been interchanged. Thus, this 2-fold axis is by itself not a symmetry axis of this object. This rotated object can be restored to its original appearance by another operation. Thus, **this object exhibits compound symmetry**. There are two possibilities for

Fig. 3.30 Rotated Ying-Yang symbol.

the second operation. One is to interchange empty part and filled part. This is a new type of symmetry operation. The other possibility is to rotate the object by 1/2 turn through an axis in the page and oriented as shown in Fig. 3.31. Note that by performing both rotations the doubly

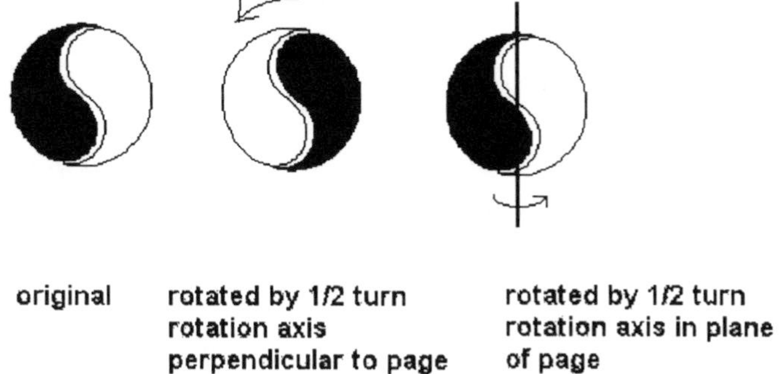

Fig. 3.31 Two rotations along different axes show the compound symmetry of the Ying-Yang symbol.

rotated object and the original are identical.

3.8 References

More examples of finding and identifying symmetries are in the exercises. A particularly excellent book on symmetry is **Symmetry A Unifying Concept** by I. Hargittai and M. Hargittai, Shelter Publications, Bolinas, Ca. This book has an amazing collection of fascinating photographs that illustrate symmetry in the natural and man-made world. Another useful book is **Introduction to Solids**, L. Azároff, McGraw-Hill, 1960. The second chapter on symmetry was useful in in preparing this chapter. Finally, Neumann's Principle and symmetry in crystals is well discussed in **Physical properties of crystals their representation by tensors and**

matrices, J. F. Nye, Oxford University Press, 1955.

3.9 Exercises

(1) Some readers may not be familiar with how a mirror works, and by extension, since many mirrors are planar, mirror planes. The purpose of this exercise is to familiarize the reader with mirrors.

   (a) Stand in front of a mirror and stretch out one of your arms towards the mirror so that your hand is close to the mirror. Does the image of your hand appear to be of the same type (right or left) as you held up or different?
   (b) Study your image. Does the image of your hand appear closer to the plane of the mirror than your face or further away?
   (c) Is your image upside down or right side up?
   (d) Imagine now that your outstretched arm and body are (very schematically) represented by a 7. Also, imagine that you can see a side view of this seven, the mirror and its mirror image. Sketch all three objects.
   (e) Summarize, including examples how distances between objects and the mirror images are related.

(2) Find and/or create 5 enatomorphic pairs of objects that are not mentioned in this chapter.
(3) Identify, list and sketch at least three objects that have symmetry lines (mirror planes). Show the symmetry lines (mirror planes).
(4) Identify, list and sketch three objects that have translational symmetry. Identify the basic translation and the pattern that is repeated after each translation.
(5) Identify, list and sketch several objects that have rotational symmetry. (Try cut open fruits, vegetables and the like). Identify the symmetry (fold and axis); also identify any reflection symmetries.
(6) Sketch the location and orientation of a collection of "3"s in such a way that they are consistent with rotational symmetry about a 2-fold, 4-fold, and 6-fold axis.
(7) Identify the location and type of symmetries in the object in Fig 3.32.
(8) Suppose that I were totally bald. When facing you do I have a symmetry line that goes through the center of my nose and my navel? Explain.
(9) Suppose you walk in a straight line through newly fallen snow. Your

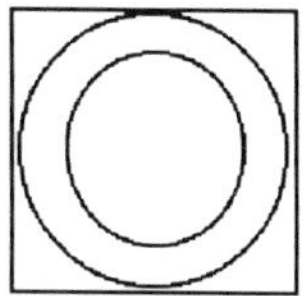

Fig. 3.32 Find the symmetries of this object, problem 7.

footprints would like something like Fig 3.33 What **compound sym-**

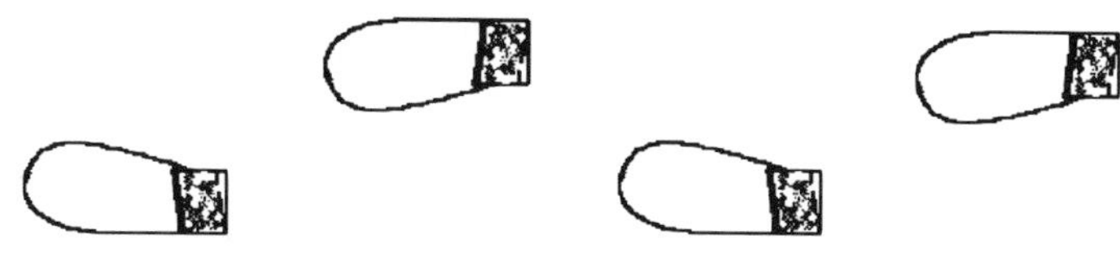

Fig. 3.33 Footsteps in the snow, problem 9.

metries are present in this pattern? Hint: Assume you walk for a long distance. Consider all the right footprints first. What basic symmetry do they show? Consider then the left footprints. Finally, observe where and how the left footprints relate to the right footprints in the picture.

(10) Consider a standard square card table. This table has one four-fold rotation axis and four mirror planes. Where are they located? Draw them.

(11) Consider a square piece of cardboard that is the same on the top and the bottom and has no markings. What symmetries are consistent with this object?

(12) Identify the types and locations of all symmetries in the planar objects in Fig 3.34. [ALL OBJECTS COPYRIGHTED BY THEIR RESPECTIVE OWNERS.]

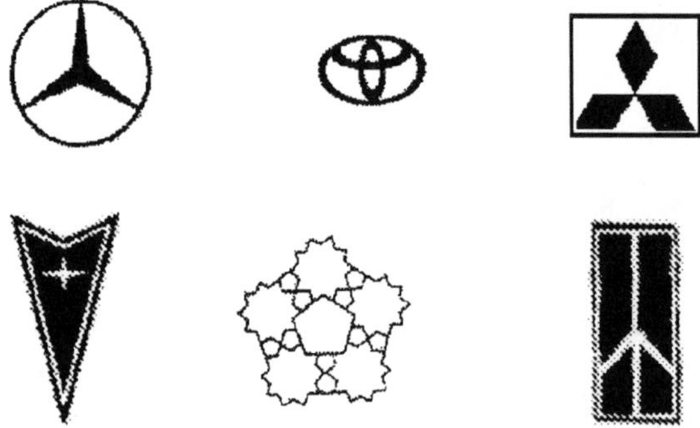

Fig. 3.34 Planar objects with symmetry, problem 12.

Chapter 4

Forces and Fields

Force is a basic concept in physics. An unbalanced sum of forces, a "net force," will cause a body to accelerate. That is, the object's velocity will change magnitude or size, in this case how fast it is moving, and/or the direction it is traveling. Forces are required for objects to change from a state of rest to a state of motion and back to rest, or to change their state of motion.

About 150 years ago Michael Faraday proposed that, in addition to the forces that exist because objects come into contact with each other (contact forces for short), there are also forces caused by fields. The reasoning goes like this: an object creates a field, another object interacts with this field, and this interaction leads to a force. While this seems a bit contrived, it has turned out to be a very fruitful concept. This chapter will briefly review forces in general and then focus on electrostatic forces since they are of primary importance in later chapters. The following questions will be addressed:

(1) What is a force and how is it related to acceleration?
(2) How are entities such as forces that have both magnitude and direction represented?
(3) What is the nature of the electric force between two objects and what is the electric field?
(4) What is the electric field in some technologically important geometries?
(5) What happens when material objects are placed in an electric field?

4.1 Force

We experience forces every day. Examples include:

(1) Gravity as seen in the ocean tides and the acceleration of an object as it travels down a hill.
(2) Electrical forces found in TV's and computer screens and in lightning.
(3) The mechanical forces we experience stopping, starting, and moving in a car.

Physicists have traditionally studied forces and the response of objects to forces. One of the first questions they asked was, "What is the natural state of an object?" This problem was solved by Isaac Newton in 1687, and is expressed by his First Law of Motion that states that the natural state of an object, **in the absence of external forces**, is to remain at rest if it is at rest, and to continue to move in a straight line if it is already moving. This law leads to a rather interesting question, "Why does a car slow down and stop if I take my foot off the gas?" Newton's First Law would seem to suggest that it should keep moving, **provided there are no external forces**. The fact that the car slows and stops indicates that external forces are acting on the car, in this case friction and air-resistance. Still, this is by no means an easy question to answer. Most people's intuition suggests that the natural state of an object is at rest, so that something must be applied to an object to make it move at constant velocity. Newton's First Law states that this "common sense" reasoning is not the case.

We must now explain the relationship between a given external force and the changes in motion of an object. The rate of change of velocity is called acceleration. Thus for example, an object traveling at constant velocity has no acceleration. The relationship is described by Newton's Second Law, which states that the acceleration, a, of an object is proportional to the applied force, F. From experience, you know that a force can be a push or pull. (We will generalize this idea later.) You also know that, for a given force, a heavy object is harder to move, or accelerate, than a light object. We account for this by introducing the concept of inertia. Inertia is a measure of an object's resistance to changes in motion, and is quantified by the mass, m. An object with a large amount of inertia will have a large mass, and an object with a small amount of inertia will have a small mass.

Ignoring **for now** the directional character of both acceleration and force, Newton's Second Law of Motion may be expressed mathematically as:

$$F = ma \tag{4.1}$$

where F is the net or total force acting on a mass m, and a is the acceleration of the mass. Notice once more that an object traveling at constant velocity has no acceleration and hence has no net force acting upon it.

The object's response to a given force, which may depend on time and position, is found by solving Newton's Second Law for the acceleration. Since the acceleration can be related to the velocity, and velocity to position, we may find the position of the object as a function of time. The details of this procedure need not concern us at this point. The important point is that forces cause changes in motion and that the position of an object can be related to the external force and the initial conditions (such as the initial position and the initial velocity).

A very important property of all forces and acceleration is that they have both **magnitude** (size) and **direction**. Such quantities, sometimes called directed quantities, follow certain specific rules for addition and are usually called vectors.[1] In sketches vectors can conveniently be represented by arrows. The length of the arrow is proportional to the vector's magnitude and the vector's direction is parallel to the arrow. When necessary, we will indicate that an object is a vector by bold type. Thus the vector A, will be printed: **A**. When hand writing, a vector can be indicated by drawing an arrow over the letter: $\vec{A}$.

In general, the addition of vectors requires the use of trigonometric relations. However, this is not the case for vectors that are parallel or antiparallel, which can be added like regular numbers. (Two vectors pointing in the same direction are said to be parallel. Two vectors pointing in opposite directions are said to be antiparallel.)

So, properly considering the vector nature of forces and acceleration, the correct form of Newton's Second Law is:

$$\mathbf{F} = m\mathbf{a}. \tag{4.2}$$

Forces may be broadly divided into two types: contact and non-contact. A contact force requires contact between two or more objects. For example, a blow from a hammer onto a nail, or the push on a desk as one moves it across the floor, or the pull on a rope in a tug-of-war game. Other forces

[1]The mathematical definition of vectors is more complex than this. This distinction is unimportant in our analyses.

do not require contact between the objects, but rather reach across space. Examples of such non-contact forces include the magnetic force that aligns a compass needle. The electric forces between a blanket, fresh from a clothes dryer during the winter, and the hair on your arm is another example, as is the gravitational force between the earth and the sun, which leads to the earth's yearly orbit around the sun. In our studies we will find that non-contact forces, especially the electrical force, is critical to the operation of liquid crystal and semiconductor devices and critical to life. For this reason, we will we now focus on the electric force and the electric field.

4.2 Electric forces and fields

The electric force has been known for millennia. The essential experimental facts can be summarized as follows:

(1) Electric forces exist between charged bodies.
(2) Electric forces can be either repulsive or attractive.
(3) The magnitude of the force between two very small charged bodies separated by a distance r is proportional to the inverse square of the separation.
(4) The direction of the force between two charges is along the line joining the two charges.

The second fact lead to the realization that there are two types of charge, which we call positive and negative, and that like charges repel and unlike charges attract (the push and pull we mentioned earlier). The expression for the magnitude of the electric force is written as:

$$F = K\frac{q_1 q_2}{r^2}. \tag{4.3}$$

Equation 4.3 is called Coulomb's law,[2] where q_1 and q_2 are two charges, physically small compared to r, and r is the distance between the two charges. The constant K depends on the system of units used to measure charge and force and is an empirical (derived from experiment) constant. Since the force depends on the inverse of the distances between the objects squared, this is called an"inverse square law" In the SI system of units which we employ, $K = 9 * 10^9$ $\mathrm{Nm}^2/\mathrm{C}^2$ and has dimensions of [force *

[2]Charles Augustin Coulomb (1736-1806) measured the electric force quantitatively and confirmed this law. The unit of charge, the coulomb, is named for him.

distance^2]/[charge^2]. In the SI system of units, force is measured in Newtons, N, distance in meters, m, and charge in coulombs, C.

The experimental situation is summarized in Fig. 4.1. The open circles

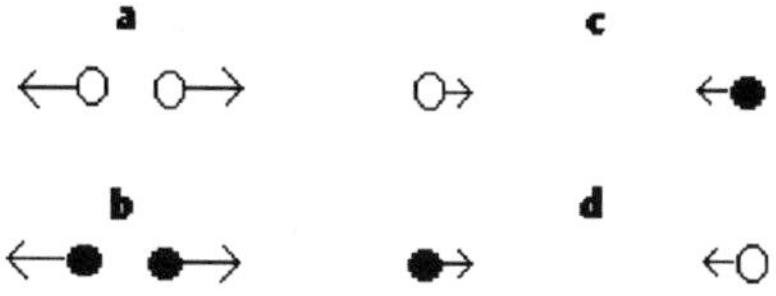

Fig. 4.1 Direction and magnitude of forces between like and unlike charges.

represent small positive charges, the filled circles small negative charges, and the arrows the forces on the charges. Assume that a given pair of charges are the only charges in the universe, and consider each case, a through d, separately. Cases a and b correspond to like charges close together, and cases c and d to unlike charges further apart. Note that unlike charges attract, while like charges repel. Further note that the force on a given charge due to the other charge is weaker when the charges are farther apart. The smaller arrows in cases c and d are meant to represent this property.

There are two fundamental properties of charge. The first is summed up by the Law of Conservation of Charge. This law states that, in an isolated system, the sum of the positive charge plus the negative charge is a constant. For an electrically neutral body, the number of positive charges equals the number of negative charges. The other fundamental property of charge is that it is quantized. This means that charge can not be continuously changed, but must be changed in discrete amounts. The fundamental charge is given the symbol e, and has a magnitude of $1.602 * 10^{-19}$ C. Any charge q can be written, as ne where n is an integer.

Matter is composed of protons, neutrons and electrons. Only the proton and the electron are charged and hence may experience a force in an electric field. The protons are generally too massive and well bound to remove them from the material. Hence, electrons cause the excess or deficient of charge in most common materials. Thus, a negative net charge corresponds to to a surplus of electrons; while a positive net charge (or just charge) corresponds to a deficit of electrons.

The quantization of charge is not apparent in our daily experience because one usually deals with a very large number of elementary charges. For example, an ordinary 120-volt, 100-watt light bulb has approximately $5*10^{18}$ electrons, each of charge $-e$, pass through it each second it is turned on.

Example 4.1 An object has a charge, q, of $4.5 * 10^{-8}$ C. How many fundamental charges does it take to create this charge on the object?

Recall that $q = ne$, where e is the fundamental charge of magnitude $1.6*10^{-19}$ C. Solve this equation for n by dividing both sides of the equation by e to get $n = q/e$. Finally, plugging in the numbers:

$$n = q/e = (4.5 * 10^{-8}\text{C})/(1.6 * 10^{-19}\text{C}) = 2.8 * 10^{11}$$

fundamental charges. The charge on an electron is $-e$ and since electrons are the mobile charge carriers, this means the object has $2.8 * 10^{11}$ (280 billion) too few electrons.

Example 4.2 In a vacuum, there is a fixed (non-moveable) particle of charge 10^{-8} C located 4 meters from a free (moveable) particle of mass 10^{-21} kg and charge 10^{-9} C. Find the force and acceleration on the free particle.

First, one must find the electric force between the two charges. Coulomb's law states:

$$F = K\frac{q_1\, q_2}{r^2}.$$

Now, $K = 9*10^9$ Nm^2/C^2, $q_1 = 10^{-8}$ C, $q_2 = 10^{-9}$ C, and $r = 4$ m. Thus,

$$F = Kq_1q_2/r^2 = (9 * 10^9)\text{Nm}^2/\text{C}^2 * (10^{-8}\text{C}) * (10^{-9}\text{C})/(4)^2\text{m}^2.$$

Canceling out the units[3] and simplifying, one obtains

$$F = 5.6 * 10^{-9}\ \text{N}.$$

Now apply Newton's second law, $F = ma$, to find the acceleration, a. Dividing both sides of Newton's second law by m yields $a = F/m$. Substituting the numbers one finds

$$a = (5.6 * 10^{-9}\ \text{N})/(10^{-21}\ \text{kg}) = 5.6 * 10^{12}\ \text{N/kg}.$$

[3]For answer to make sense the units of the final quantity must be correct. For example, a calculated length must have the dimensions of length. By including all the dimensions in the expression and operating on them like normal numbers the calculated dimension may be found. In this case coulombs and meters cancel and the calculated dimension is N, or Newtons, a force.

As an aside, the units of a Newton per kilogram are m/s^2, the same as that of acceleration.

4.3 The importance of Coulomb's law

The importance of Coulomb's law goes well beyond just describing the electrical force between two small charged balls. It has been incorporated into quantum mechanics, and correctly describes the electrical forces between charged particles in atoms. It also correctly describes the forces that bind atoms together to form molecules and the forces that bind atoms and molecules together to form solids and liquids. Most of the forces that we encounter in daily life which are not gravitational in nature are electrical in nature. Thus, Coulomb's Law is the basis of understanding a large number of interactions in our lives.

The force between two charged objects acts across the empty space between them. How is this force communicated between the two charges? The view that is presently held by physicists is that an electric charge q_1 sets up an invisible electric field in the space around itself. The second charge q_2 then interacts with this field and experiences a force. The field thus plays an intermediary role between the two charges. The existence of an electric field can be inferred by measuring the force on a small positive test charge.

The electric field is so useful that we try to visualize it by drawing field lines or lines of force. Lines of force emanate from positive charges and terminate on negative charges. Where lines are closer together the electric field is stronger, while where they are further apart the electric field is weaker. The electric field is everywhere locally parallel to a line of force. Figure 4.2 illustrates lines of force for isolated positive and negative charges.

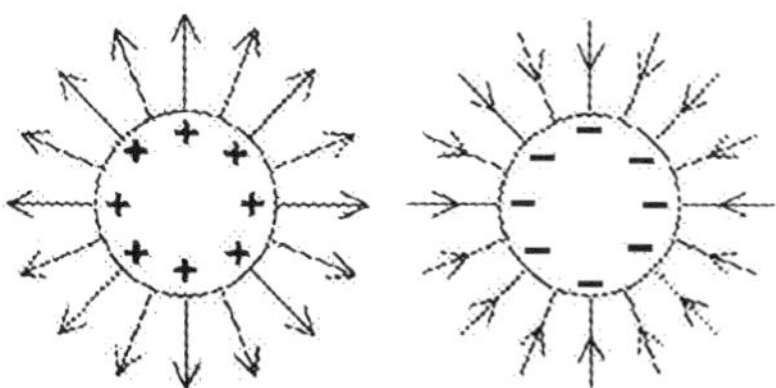

Fig. 4.2 Lines of force from positive and negative charges.

Notice that the field lines are close together near the charged objects and farther apart away from the object. This represents the fact that the electric field strength is decreasing with increasing distance from the charge.[4]

Of course, not all charged bodies are small spheres or their idealization, point charges. The idea of electric fields due to electric charges can be extended to other objects besides point charges. The two most important cases in the remainder of this book will be the parallel plate capacitor, and the electric dipole.

4.4 The parallel plate capacitor

The geometry of a parallel plate capacitor is shown in Fig. 4.3. It consists

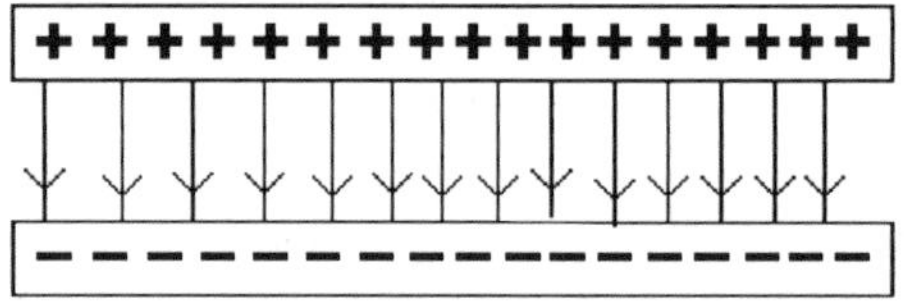

Fig. 4.3 The parallel plate capacitor. The field lines are parallel except very near the edge and emanate from the positively charged plate and terminate on the negatively charged plate.

of two uniformly charged parallel plates, one with a charge q and the other with a charge $-q$ separated by a distance d that is very small compared to the length of either side of the plate. While Fig. 4.3 does not illustrate it, the electric field in the parallel plate geometry is uniform except very close to the edges of the plates.

One can "charge" the capacitor by connecting the two plates to a battery as shown in Fig. 4.4. For any battery voltage, V, the same voltage or potential difference will occur between the two plates. The electric field between the plates is given by $E = V/d$, and points from the plate connected to the positive (+) battery terminal to the plate connected to the negative (-) battery terminal. This is equivalent to having the electric field point from the excess charge (+) to the deficit charge (-) discussed above. This geometry will have great practical applications since energy is stored in the

[4]The magnitude of the electric field due to a point charge q_1 located at the origin is $E = Kq_1/r^2$.

electric field due to the separated charges.

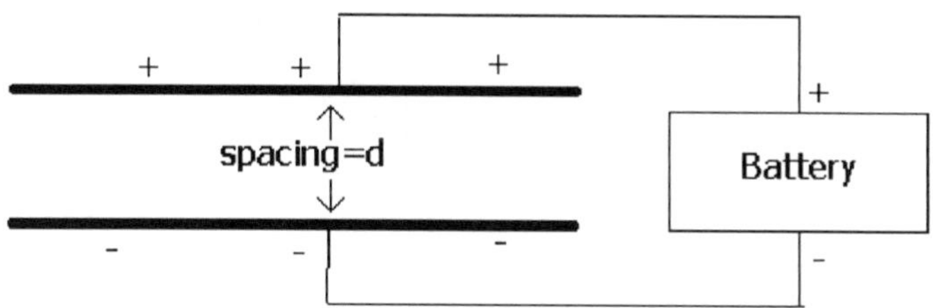

Fig. 4.4 A parallel plate capacitor connected to a battery.

Example 4.3 A parallel plate capacitor has a voltage of 5V between two plates separated by 4 microns (1 micron $= 10^{-6}$ m). Find the electric field between the capacitor plates. $E = V/d$, substituting for V and d one obtains:

$$E = V/d = 5\text{V}/(4 * 10^{-6}\text{m}) = 1.25 * 10^6\text{V/m}.$$

This is a fairly typical electric field found in a liquid crystal display.

4.5 The electric dipole

An **ideal** electric dipole consists of two equal charges, q, of opposite sign separated by a distance $2a$.[5] The electric dipole moment, $\mathbf{p}$, is a vector which has a magnitude $2aq$ and points from the negative charge to the positive charge,

$$\mathbf{p} = 2\mathbf{a}q. \tag{4.4}$$

An electric dipole may be pictured as shown in Fig. 4.5.

The electric field far away from an electric dipole decreases as $1/r^3$, where r is the distance between the field point (where the field is measured) and the center of the dipole. Many molecules have permanent electric dipole moments. A common example is water. Many liquid crystals also

[5]Why $2a$? This notation makes standard dipole calculations (that are beyond the scope of this book) notationally easier. They routinely involve dividing the charge separation by 2. So, by starting with $2a$, you only have to write the quantity a, instead of $a/2$.

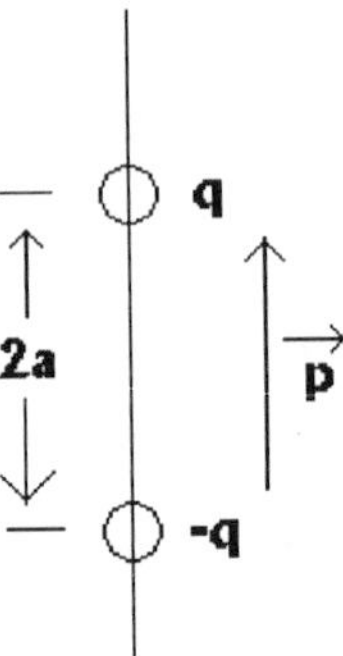

Fig. 4.5 An ideal dipole consisting of two point charges separated by a distance $2a$. Real dipoles can often be modeled as ideal dipoles.

have permanent electric dipoles that allow them to interact strongly with electric fields. Generally, one cannot independently determine q and $2a$. This is because the measured quantity is p, the dipole moment, and in real dipoles (as opposed to this model dipole) the charge is not all located at two points.

Many molecules and all atoms do not have permanent dipole moments. However, when such objects are placed in an external electric field the positive charges are pulled in the direction of the field and the negative charges are pushed in the opposite direction. This leads to a separation of charge and the formation of an electric dipole. We say that the atom or molecule has become *polarized* and has acquired an *induced electric dipole moment.* This induced dipole moment disappears when the field is removed. Since all bulk materials are made of atoms or molecules we can also talk about a material becoming polarized. The electric dipole moment per unit volume is called the *electric polarization.*

A very important situation occurs when an electric dipole is placed in a uniform external electric field. This occurs when liquid crystals are placed in an electric field produced by the electrodes in liquid crystal displays. A schematic of an electric dipole in a uniform electric field is shown in Fig. 4.6. The positively charged end of the dipole experiences a force in the direction of the applied electric field while the negatively charged end of the dipole experiences a force in the opposite direction. These two forces lead to a twisting force or torque that causes the dipole to rotate so that it aligns parallel to the field. The torque, the rotational analog to force, is a minimum when the dipole is parallel to the electric field and a maximum when it is perpendicular to the field.

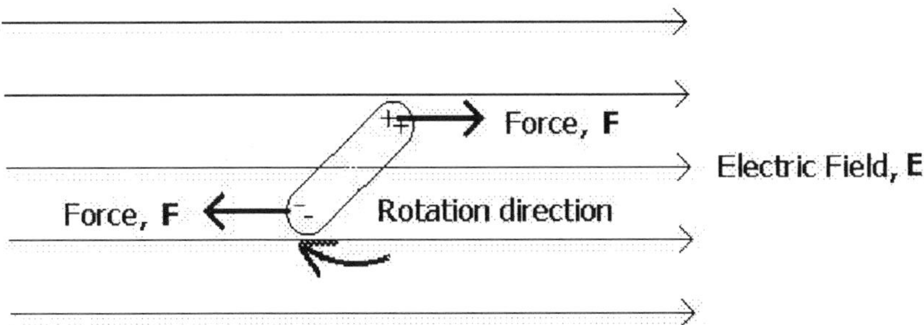

Fig. 4.6 An electric dipole in a uniform electric field.

We will find that the minimum energy configuration is the preferred state of any system. In our case, where the "system" consists of the dipole and the electric field, the energy is a minimum when the dipole moment **p** is parallel to the electric field **E**. The energy, U, of an electric dipole of magnitude p, in an electric field of magnitude E, is given by $U = -pEf(\theta)$. The function $f(\theta)$ depends only on the angle between **p** and **E**. It is 1 when **p** and **E** are parallel, 0 when they are perpendicular, and -1 when **p** and **E** point in opposite directions.

Liquid crystal molecules are anisotropic molecules, which means that their physical properties are different when they are oriented in different directions. Like all molecules, liquid crystal molecules are three-dimensional and can possess multiple permanent or induced electric dipoles both parallel and perpendicular to any symmetry axis of the molecule. When there is more than one electric dipole in a molecule, the molecule tends to orient so that the largest electric dipole is parallel to the electric field.

Example 4.4 Two point charges (charges of no spatial extent)of opposite sign and magnitude 10^{-18} C are separated by a distance of 1 nm (nanometer; 1 nanometer $= 10^{-9}$ m). Find the magnitude of the electric dipole moment. In what direction does the dipole moment **p** point?

The magnitude is calculated using the ideal dipole equation: $p = 2aq$, where q is the magnitude or absolute value of the charge, and $2a$ is the distance between the charges. Thus,

$$p = 2aq = (10^{-9}\text{ m}) * (2 * 10^{-18}\text{ C}) = 2 * 10^{-27}\text{ Cm}.$$

The dipole **p** points from the negative charge to the positive charge.

4.6 The dielectric constant

Often, the gap between the electrodes of a parallel plate capacitor is not filled with vacuum but with an electrically insulating material.[6] The material property that characterizes the electric properties of a dielectric is called the *dielectric constant.*

When a dielectric is placed in an electric field, the electric field inside the dielectric is smaller than the electric field outside of the dielectric. This effect can be understood by modeling the dielectric material as a collection of electric dipoles that all orient in the same manner as the single dipole shown in Fig. 4.6. This collection of aligned dipoles creates an excess charge on the surface of the dielectric which will reduce the electric field inside the dielectric. The internal charge cancels out since the positive charge of one dipole is adjacent to the negative charge of the next dipole parallel to the field except at the surfaces where the field enters and leave the dielectric.

The notation that is best for discussing dielectrics depends on the system of units one uses. In SI units, the dielectric constant is the ratio of the "permittivity of the media", ϵ to that of free space, ϵ_0. We use the Greek letter epsilon, ϵ, to represent permittivity and the Greek letter kappa, κ, to represent the dielectric constant. Thus, $\kappa = \epsilon/\epsilon_0$, where $\epsilon_0 = 8.85 * 10^{-12}$ F/m is the permittivity of free space (and the smallest permittivity there is) and F stands for farads, the SI unit of capacitance. The dielectric constant κ of vacuum, (and, for all practical purposes, of air) is 1. In another system of units called Gaussian units, $\epsilon_0 = 1$, and the dielectric constant is simply given by ϵ. Thus, one must be careful when reading other books on dielectrics.

Example 4.5 The capacitance of a parallel plate capacitor that is filled with a dielectric material of dielectric constant κ is given by $C = \kappa\epsilon_0 A/d$. If the plate area is 10^{-4}m^2 and the plate spacing is 10^{-5}m, calculate the capacitance when the capacitor is "filled" with vacuum, $\kappa = 1$, and when it is filled with dielectric constant, $\kappa = 10$. Recalling that $\epsilon_0 \equiv 8.85 * 10^{-12}$ F/m,

$$C_{vacuum} = \kappa * \epsilon_0 A/d = (1) * (8.85 * 10^{-12}\ \text{F/m}) * (10^{-4}\ \text{m}^2)/(10^{-5}\ \text{m}),$$

[6] An electrically insulating material or *insulator* is a material that does not conduct electricity. Electrical conductivity is due to free electrons that can move about, so insulators ideally have no free charges. Examples include glass, quartz, and most plastics.

so that $C_{vacuum} = 8.85 * 10^{-11}$ F. Furthermore,

$$C_{dielectric} = \kappa * \epsilon_0 A/d = (10) * (8.85 * 10^{-12} \text{ F/m}) * (10^{-4} \text{ m}^2)/(10^{-5} \text{ m}),$$

so that $C_{dielectric} = 8.85 * 10^{-10} \text{ F} = 10 * C_{vacuum} = \kappa * C_{vacuum}$.

Note that the dielectric has increased the capacitance of this capacitor by an amount equal to the dielectric constant. The reader may see an apparent paradox. Earlier it was stated that the electric field inside a dielectric was smaller than that outside the dielectric. However, since $E = V/d$ and V and d are unchanged in this problem, E is unchanged. The difficulty is that the potential (voltage) not the electric field is constant. The battery has provided more charge for the two plates. If the charge on the capacitor were held constant, the electric field in the dielectric would be smaller than that in vacuum, and the potential smaller by a factor of $1/\kappa$. Further discussion of this point is beyond the scope of this book.

4.7 References

This material is found in a large number of introductory physics texts. The author's book of choice is **Fundamentals of Physics** by Halliday and Resnick. However, there a great many different books at this level. The reader should use the one that they find particularly readable.

4.8 Exercises

(1) A small mass is attached to a string. A student in deep space, where there is no gravity and no air resistance, swings the string so that the mass is moving in a circular path.

   (a) What forces are acting on the mass? In particular, does a force act on the mass to cause it to move in a circle?
   (b) The student then releases the string and the mass goes flying off in some direction. What forces are acting on the mass. Describe the motion of the mass.

(2) Two charges, $q_1 = 3.2 * 10^{-9}$ C and $q_2 = -4.8 * 10^{-11}$ C, are separated by a distance of 2 m.

   (a) How many fundamental charges make up q_1?
   (b) How many fundamental charges make up q_2?

(c) Do these two charges represent a surplus of protons, or a deficit of electrons, or both? Explain.
(d) What is the electric force between these two charges?

(3) A force of 10^{-23} N acts on an electron of mass $9.1*10^{-31}$ kg. Find the acceleration of the electron.

(4) *This exercise explores the energy stored in a capacitor, and the relationship between charge, capacitance, and voltage.*
Consider a parallel plate capacitor. The expression for the energy, U stored in this capacitor is given by any of the following expressions: $U = \frac{1}{2}CV^2 = \frac{1}{2}Q^2/C = \frac{1}{2}\epsilon E^2 * Ad$. Consider a capacitor of capacitance 10^{-11}F (Farads). This capacitor has a voltage of 5 V (Volts) between its plates .

(a) What is the energy stored in this capacitor?
(b) What is the charge on one plate of this capacitor?
(Use the equality between the first two expressions for the energy to show $Q = CV$, and then calculate Q.)

(5) *This problem is a continuation of problem 3 and uses some information contained in the statement of that problem.*
A parallel plate capacitor consists of two metal plates of area $A = 0.005$ m^2 separated by a distance $d = 10^{-5}$m. The material between the plates is filled with a dielectric of permittivity $\epsilon = 10^{-9}$ F/m.

(a) Calculate C.
(b) A voltage of 5 V is placed between the two plates of this capacitor. What is the electric field between the two plates?
(c) What is the charge on each of the plates?
(d) Show that the calculated energy from all three equations given in problem 3 are the same.

(6) A water molecule has a permanent dipole moment. Let the distance between the separated charges be $0.9 * 10^{-10}$ m and the magnitude of the charges be $0.66 * 10^{-19}$ C. Calculate the dipole moment of water. Why is the charge not an integer times e (ne)?

(7) The energy of an electric dipole in an external field is $U = -pEf(\theta)$, where θ is the angle between the dipole and the electric field. Consider a molecule of dipole moment $p = 10^{-27}$ Cm, in a uniform electric field $E = 10^5$V/m. $f(\theta)$ is given in the following table.

θ *(degrees)*	*f*(θ)
0	1.000
30	0.866
45	0.707
60	0.500
90	0.000
120	-0.500
135	-0.707
150	-0.866
180	-1.000

(a) At what angle is U a maximum? Sketch the directions of $\boldsymbol{E}$ and $\boldsymbol{p}$. Explain why this is the case.
(b) At what angle is U a minimum? Sketch the directions of $\boldsymbol{E}$ and $\boldsymbol{p}$. Explain why this is the case.
(c) Graph f(θ).
(d) From your graph estimate $f(140^o)$.
(e) If the magnitude of the electric field changes from 10^5V/m to 10^3V/m, how do your answers to parts a) and b) change?

(8) Consider a rectangular prism of a dielectric material. Then imagine putting this dielectric in a uniform electric field so that the electric field is perpendicular to one of the sides of the dielectric prism. Assume the dielectric does not change the electric field.

(a) Draw the electric field lines and the prism. Draw the lines through the prism.
(b) Since the dielectric can be modeled as a collection of dipoles, draw dipoles of the correct orientation inside the dielectric.
(c) Show, using your figure and descriptions, that internal dipoles "cancel" leaving just surface charges.
(d) Draw the electric field due to these surface charges. Is it in the same direction as the applied field or the opposite direction?
(e) Is the electric field in a dielectric larger or smaller than the electric field outside the dielectric?

(9) In our everyday experience, 280 billion is a very large number. However, the number of atoms or molecules in a normal size piece of material is so large that even 280 billion is a *very small fraction* of the total number of atoms or molecules. Suppose a plastic cup has an excess charge of

$+280 * 10^9 e$ and is made of 10^{20} molecules. (Later we will see that this is not an unreasonably large number).

(a) What fraction of the plastic molecules have gained or lost an electron? (Assume each molecule gains or loses only one electron.)
(b) Does the plastic have too few or too many electrons?

Chapter 5

An Introduction to Light Waves

5.1 Overview

A wave is a traveling disturbance. Waves are important because all information that we receive via hearing and seeing is transmitted via waves. While light has both particle-like and wave-like properties, the present discussion will focus on the wave-like character of light. The focus in this chapter is on the interaction of these light waves with "polarizers," such as are found in some sunglasses, and material media. Some related topics will be addressed in later chapters, those needed to understand displays will be discussed in this chapter.

We will address the following questions.

(1) What is a wave? How is a wave represented?
(2) What is light? What makes light waves different than waves on water or sound waves in air?
(3) What is a polarizer, and how does it work?
(4) How is the interaction of light with essentially transparent materials explained?

5.2 Waves

Almost all of the applications of liquid crystals that we shall discuss concern how light and liquid crystals interact. To understand this phenomenon, one must understand some of the basic properties of light.

We begin by noting that light has wave-like properties, and in elementary discussions may be treated as a wave. A wave is characterized by a traveling disturbance in which the disturbance propagates but the media, as a whole, does not move. Broadly, there are two different types of waves: transverse waves in which the displacement, the movement caused by the disturbance, is perpendicular to the direction the wave propagates, and longitudinal waves where the displacement is parallel to the direction of propagation. Mathematically, the simplest waves are "periodic waves" which continue unchanged for very long periods of time, ideally for all time. The simplest type of periodic wave is the *sine wave.* A sine wave has a displacement that is periodic in space at constant time with the functional form of a sine function. Its temporal evolution at constant position is also described by a sine function.[1] A sine wave can be characterized by knowing its frequency and the speed of the disturbance, the *phase velocity* v, in the media. A snapshot of a sinusoidal traveling wave is shown in Fig. 5.1.

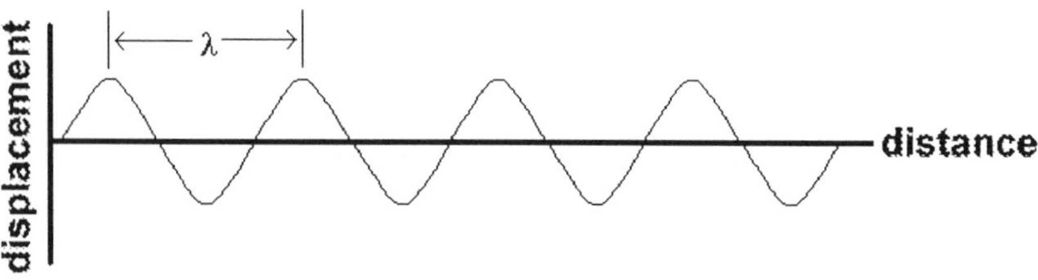

Fig. 5.1 A snapshot, or constant time sketch of a sinusoidal wave as function of position. The Greek letter lambda, λ is used to indicate the wavelength of the wave.

This figure also illustrates one important property of waves, its *wavelength.* Following the custom used in most of the scientific literature, the Greek letter lambda, λ will be used to designate the wavelength. The wavelength is the distance between two identical points, also called points of corresponding phase, on consecutive cycles of a waveform.

The frequency, f is the number of times per second the displacement returns to a given position, and is determined by the source. A consequence of the conservation of energy is that frequency does not change as the wave travels from medium to medium. Since the frequency does not change, **the wavelength of a given wave is different in different media**. The relationship between wavelength, velocity of propagation (phase velocity),

[1]You have undoubtably seen the sine function in a math course. The wave could also be a *cosine function*, the only difference being the value of the displacement at zero time. The generalization to include cosines, sines and intermediate cases is indicated by using the term *sinusoid.*

v, and frequency is:

$$\lambda = \frac{v}{f}. \tag{5.1}$$

This equation tells us that, for a given velocity of propagation, the higher the frequency, the smaller the wavelength. Long wavelengths correspond to low frequencies and short wavelengths correspond to high frequencies.

This discussion has been general. There are many different types of waves. For example, surface waves on the water of a pond, mechanical waves on a string or slinky, sound waves in the air that let us hear, radio and TV waves, and x-rays. The last three types of waves are examples of *electromagnetic* waves. Only a very small region of all electromagnetic radiation is visible to the human eye, and is what we normally call light. In this context, you may have seen a version of the last equation that is valid for light *in a vacuum*, $\lambda f = c$, where c is the speed of light in vacuum (such as exists in outer space). The speed of light in air is so near that of vacuum that we take $v_{light\ in\ air} = c$, except in high precision measurements.

Electromagnetic waves are produced by electrical and magnetic disturbances and are transverse waves. Thus, in Fig 6.1 the displacement represents either the electric or the magnetic field of the wave. An important characteristic of electromagnetic waves is that in free space (vacuum) all electromagnetic waves travel at the same speed, the speed of light, c. The speed of light in vacuum is essentially $3 * 10^8$ m/s. The speed of electromagnetic waves in all media is less than c.[2] There is no upper or lower limit to the wavelength of electromagnetic waves, and the electromagnetic spectrum covers at least 24 orders of magnitude. This corresponds to frequencies as low as 1 vibration per second, or 1 Hertz $=$ 1 Hz. to 10^{24} Hz. The wavelength range is from at least $3 * 10^8$ m to less than 1 fm (10^{-15}m). A very long wavelength could be the waves produced by power lines (which in the US have a frequency of 60 Hz) that have a wavelength of $5 * 10^6$ m. A radio wave from WTAM, 1100 AM, in Cleveland, Ohio, has a frequency of 1100 kHz, and a wavelength of 273 m. WKSU, 89.7(MHz) FM, in Kent, Ohio corresponds to $\lambda = 3.34$ m. A typical microwave wavelength is 1 cm. Visible light has a frequency of approximately 10^{15} Hz, and a wavelength of approximately 0.5 μm. A typical x-ray wavelength is 10^{-10} m. Gamma rays and high energy x-rays have even shorter wavelengths, 10^{-12} m and

[2] There is one know exception. Recently, a structure with a negative index of refraction for microwaves has been made, and the speed of light in this media is greater than c. "Experimental verification of a negative index of refraction," Shelby RA, Smith DR and Schultz S, Science, 292:(5514) 77-79, April 6, 2001.

10^{-15} m respectively.

5.3 Visible light

Visible light occupies a very small part of the electromagnetic spectrum. Light is defined as radiation that can be discerned by the human eye. It extends from wavelengths of about 400 to 700 nm. The instrument used to measure the intensities and wavelengths of the various components of a beam of light is called a spectrophotometer, and allows one to discern the spectral components of the light. Our everyday experience is often with *nonspectral hues* that can be obtained from mixing these spectral colors. For waves of a *single* wavelength (or a small range of wavelengths), a correspondence may be made between the wavelength and the perceived color. However, it is important to note that the situation when color addition and subtraction such as occurs in color televisions or color printing is more complex. This will be discussed in chapter 17. The following table gives "ballpark" values for the colors that correspond to various spectral wavelengths.

Color	**Wavelength (nm)**
violet	425
blue	475
green	525
yellow	550
orange	600
red	625

The transverse nature of electromagnetic waves is somewhat more complex than that of mechanical waves because both the electric field and the magnetic field of the wave are transverse to the direction of propagation. They are also perpendicular to each other in isotropic materials. This relationship is further complicated because, generally, waves do not last forever, but only for a finite period of time. In this case, the periodic wave looks like the transverse wave in the previous picture, but only for a short time, corresponding to many cycles. A periodic wave which lasts for a finite time is called a *wavetrain*. Since in nature there are no periodic waves that last forever, observed waves are made up of a series of wavetrains. Light from a normal light bulb or the sun consists of a large number of wavetrains.

5.4 Polarization and polarizers

Wavetrains have an important property that will be critical to our later study of liquid crystal devices. The electric fields of electromagnetic waves and wavetrains have a direction. This direction may be varied and even controlled. A special case of this condition occurs when the wavetrains are *linearly-polarized* or *plane-polarized.* These two terms have the same meaning, and are used interchangeably in the scientific literature. When a wave is plane-polarized the electric field of **all** wavetrains are parallel to each other and to a common plane called the *plane of polarization.* The plane of polarization is defined as the plane that contains the electric field and a vector parallel to the direction of propagation of the wave. Unfortunately, this term is somewhat ambiguous. For that reason, we will use the term linearly polarized light and then specify the direction of the electric field. The direction of the electric field of a wave for polarized and unpolarized light is shown in Fig. 5.2. This picture of unpolarized light is **not** totally correct.[3] However, there is no completely satisfactory picture of unpolarized light, and this type of picture is commonly used.

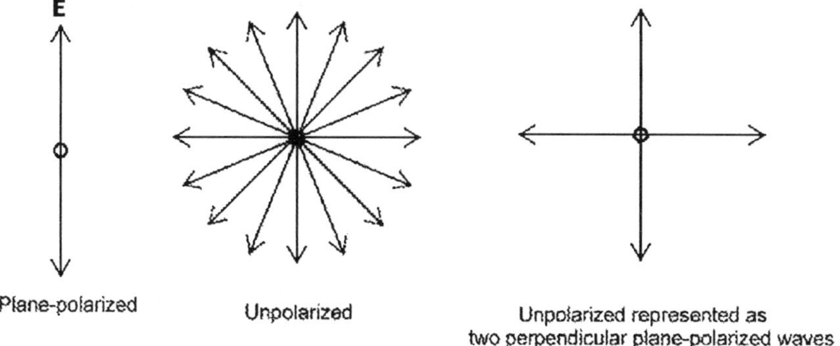

Fig. 5.2 An pictorial representation of linearly polarized and unpolarized light. When unpolarized light is incident on an object, the representation on the far right is often useful.

This figure describes light propagating out of the page. The arrows represent the electric field vectors of *some* of the wavetrains. The left most part of this figure illustrates linearly-polarized light such as comes out of a polarized laser. The center and right images in this figure both illustrate

[3]This is discussed by Shurcliff (See references at the end of this chapter).

unpolarized or natural light. This figure illustrates the following important points. Natural light is the superposition of wavetrains with planes of vibration in every possible direction. From the point of view of manipulation with polarizers or by reflection, this is equivalent to assuming that half of the light intensity is linearly polarized in the vertical direction and half of the light intensity linearly polarized in the horizontal direction. However, while these two directions must be be perpendicular, their absolute orientation is completely arbitrary. Thus, these two perpendicular directions can be rotated by any angle in the plane of the page.

The modeling of unpolarized light as two perpendicular linearly-polarized wavetrains is not unique. There are several variations. First, the two perpendicular directions are arbitrary. Second, the decomposition need not be into linearly-polarized waves. It is sometimes convenient to decompose randomly polarized light into equal amounts of right- and left-hand circularly polarized light. In circularly polarized light, the plane of polarization continuously rotates. The handedness of the light depends on whether the rotation is clockwise or counterclockwise. This representation will be of great utility when the optical properties of *chiral nematic* liquid crystals are discussed later in the book.

There are commercial devices that convert unpolarized light into polarized light. These devices are called *Polaroids*[4] or *polarizers*. A polarizer only allows the component of the electric field parallel to a certain characteristic direction of the polarizer to be transmitted. Light in the perpendicular polarization is absorbed. The action of a polarizer is illustrated in Fig. 5.3.

In the process of passing through an ideal polarizer, natural light losses 1/2 of its total intensity. This follows from and verifies the previous assertion that unpolarized light can be modeled as two perpendicular polarizations each with 1/2 of the intensity. Moreover, one obtains 1/2 of the intensity through the polarizer for any orientation of the polarizer because the two mutually perpendicular directions are arbitrary. (This may seem strange, however, the direction of the electric field in unpolarized light can be represented by a vector going from the origin to anywhere on the circumference of the circle. All points on the circumference are equally likely; thus, there can be no preferred polarization.)

In many applications, a medium other than vacuum follows the first polarizer. After this medium, a second polarizer is inserted into the opti-

[4]Polaroids are commercially available polarizing materials that where originally invented by Edwin Land. The Polaroid Corporation was formed in part to manufacture these devices.

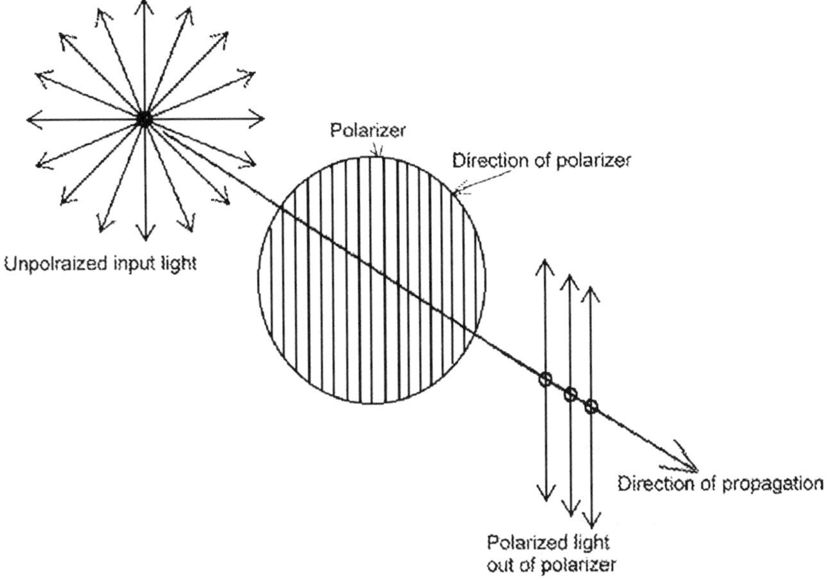

Fig. 5.3 The effect of a polarizer on unpolarized light.

cal path. In the absence of such media, the amount of light transmitted through the second polarizer, also called the analyzer, depends on the difference in the angle of the polarizer and the analyzer. This angle will be called phi, "ϕ," pronounced fie. The Greek letter phi, ϕ is often used to represent angles in mathematics and physics. The amount of light transmitted through the analyzer is given by the Law of Malus:

$$I = I_{in} \cos^2(\phi) .$$

Where I_{in} is the intensity just before the analyzer, and *cos* is the cosine function. The value of $\cos^2(\phi)$ for several important values of ϕ is in the following table.

ϕ in degrees	$\cos^2(\phi)$
0	1
45	$1/2$
90	0
135	$1/2$
180	1

The use of two polarizers is essential to understanding some liquid crystal displays.

5.5 Interaction of electromagnetic waves with matter

A wave that never interacts with matter would be impossible to detect. We are aware of waves and observe their behaviors because waves interact with matter. There are several effects that we will discuss. First, one of the most basic ramifications of the interaction of waves with matter, the speed of electromagnetic waves, will be discussed. The behavior of light at the interface between two different materials will be considered next. Birefringent materials will then be briefly explored. Lastly, scattering of light by media will be investigated.

Suppose that a light wave of frequency f is propagating in a medium other than vacuum. What are the properties of this wave? First, the **frequency** of the wave or wavetrain **does not** change when it propagates in a media other than vacuum. However, the speed at which the electromagnetic wave propagates through the media is smaller than in vacuum. The speed of propagation also depends on the wavelength of the wave traveling through the media. Because the speed of light is so large, and the change in speed is typically no more than a factor of two to five, the reduction in the speed is characterized by a material property called the *index of refraction*, $n(\lambda)$. The index of refraction in general depends on the wavelength of the light.[5] For this reason, it has been written as $n(\lambda)$ rather than just n. The relationship between the index of refraction,the speed of propagation, v, and speed of light in vacuum, c, is $v(\lambda) = c/n(\lambda)$. Since $v < c$, n can not be less than 1.

Typical values of n are 1.0003 for air (however, we will use $n_{air} = 1.00$), 1.33 for water, and 1.5 for glass. The index of refraction is a sensitive indi-

[5]The index of refraction can also vary with temperature, composition and other material properties. For now we will focus on the wavelength dependent behavior.

cator of purity and the presence of additives. The above indices for water and glass are only "ballpark" values. The variation of index of refraction with wavelength is called *dispersion.* This is usually not a large effect for light. A typical variation in index of refraction between 400 and 700 nm is 0.05 or less. In spite of the smallness of this variation, spectacular effects can result. A simple example is the rainbow, which has colors because the index of refraction of water depends on wavelength.

An effect related to the index of refraction is the *absorption* of electromagnetic media. Very generally, dispersive media is absorptive media. Thus, matter absorbs some of the energy in an electromagnetic wave and changes this energy into heat. You are very familiar with this effect in a microwave oven. In this appliance, the wavelength of the microwaves corresponds to absorption regions of water and fats. In the visible region, absorption is most obvious in colored glasses and liquids. A bottle appears to be green when illuminated by white light (light of a broad range of wavelengths such as produced by the sun or room lights). This is because it absorbs the red and violet light more strongly than green light.

What happens when light is incident on the interface between two different media? Our experience with the sun reflecting off windows suggests that some of the light is reflected and some of the light is transmitted. This is indeed the case, as illustrated in Fig 5.4.

Several features of the behavior of light at the interface between two different media are illustrated in this figure. All of these properties are experimentally verified. The first is that when the light is incident at an angle θ_i (θ is the Greek letter theta, and is very commonly used to represent angles) with respect to a line perpendicular to the interface, the reflected light is at an angle θ_r with respect to the same perpendicular, **and** $\theta_i = \theta_r$. This may be read as, "theta i equals theta r" or "theta incident equals theta reflected;" that is, the angle of reflection equals the angle of incidence. The line that is perpendicular to the interface where the incident, transmitted, and reflected rays hit the interface is called the *normal* to the interface. The angle the transmitted light makes with the same perpendicular line is called the transmitted angle, θ_t (theta t). The relationship between the incident angle and the transmitted angle is given by *Snell's Law:* [6]

[6]The law of reflection was known to the Greeks. The law of refraction (Snell's law) was experimentally discovered in 1621 by Willebrord Snell, but had not been made public by the time he died in 1626. This law was published by René Descartes (of philosophy fame) in 1637 in the book **Dioptrique, Météores**, without acknowledging Snell. However, it is believed that Descartes had seen Snell's manuscript. This is further discussed in the historical introduction to **Principles of Optics**, by Born and Wolf. This introduction

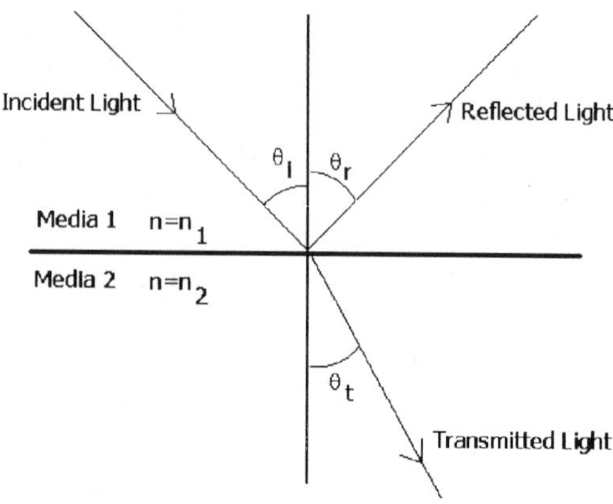

Fig. 5.4 Light at an interface. The incident, transmitted, and reflected beams are all shown. The vertical line where all three meet is the "normal" to the interface.

$$n_1 \sin(\theta_i) = n_2 \sin(\theta_t)\,. \tag{5.2}$$

Thus, the index of refraction is also important in understanding the direction of propagation of the transmitted light. *Sin* is the sine function. A few representative values are in the following table.

θ in degrees	$\sin(\theta)$
0	0
30	0.5
45	0.707
60	0.866
90	1

It is important to note that when $n_2 > n_1$, $\theta_t < \theta_i$, and the light is bent towards the "*normal,*" the line perpendicular to the interface. The amount

refers the reader to E. T. Wittaker's **A History of the Theories of Aether and Electricity** for a more extensive account.

of light that is reflected depends on the indices of refraction, the angle of incidence and the polarization of the incident light in a complicated way, and are described by the Fresnel equations.[7]

So far, we have assumed that material media can be characterized by a single index of refraction. This is only true for "isotropic" materials. (An isotropic material has the same physical properties in all directions within the material.) Many of the materials that we will discuss, especially liquid crystals, are *anisotropic*. This means that their physical properties are **direction dependent**. This anisotropy will often manifest itself in the optical properties of the material. In particular, the index of refraction will depend on the direction the light propagates and the light's polarization. This is called *birefringence*. Since different polarizations of light "see" different indices of refraction, the waves of different polarization travel in different directions at different speeds. This can lead to a number of interesting and useful optical devices.

The liquid crystal devices that we shall study are optically *uniaxial*. This means that they are optically anisotropic, have one unique direction or axis where the index of refraction does not depend on the polarization of the incident light, and may be describe by two different indices of refraction. These indices can be visualized by considering a right rectangular prism, shaped like a shoe box with square ends, made of the uniaxial material. Assume that the material is aligned so that the unique axis is parallel to the long axis of the box. The two other directions, perpendicular to the long axis are equivalent. In fact, light incident normal (perpendicular) to the square ends will see the same refractive index regardless of its polarization. The index of refraction for light incident perpendicular to the long axis of the box but polarized parallel to the long axes of the box will be called $n_{\parallel}$. The index of refraction for light incident perpendicular to the long axis of the box and polarized perpendicular to the long axis of the box will be called $n_{\perp}$. To obtain some insights into the propagation of light in this material consider Fig. 5.5.

Suppose that unpolarized light is normally incident to a side of this prism of uniaxial material (the left half of Fig. 5.5). This light can be described by two mutually perpendicular polarized light rays. Since the direction of these two perpendicular directions is arbitrary, we will pick one parallel to the long axis of the prism and the other perpendicular to the

[7]Augustin Jean Fresnel (1788-1827), helped establish the wave theory of light, devised the first model describing dispersion, and developed the laws that describe the polarization and intensity of light produced by reflection and refraction.

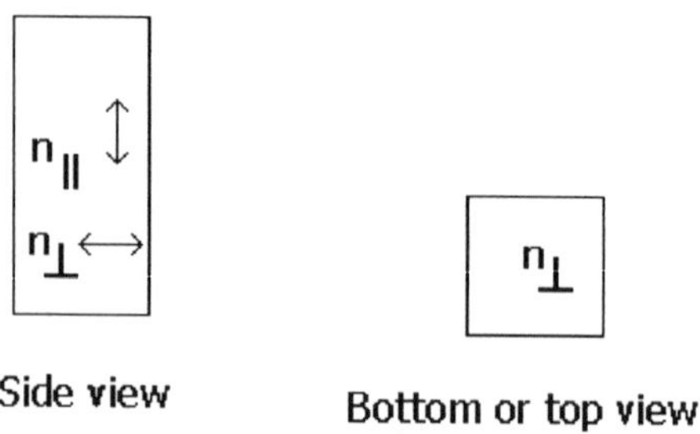

Fig. 5.5 Side and bottom views of a piece of uniaxial material. The indices of refraction are as shown.

long axis as seen on the left in the figure. The wavetrains in the medium that are polarized parallel to the long axis of the prism will "see" an index of refraction $n_{||}$. Similarly, within the media the wavetrains polarized perpendicular to the long axis of the prism will "see" an index of refraction $n_{\perp}$. These two wavetrains will propagate at different speeds because the two indices of refraction are different. Furthermore, when the incident light is not normal to the interface the light can still be modeled as two sets of perpendicularly polarized wavetrains that propagate in different directions and at different speeds.

Now suppose we turn the block so that unpolarized light is normally incident on the square bottom of the prism. We would find that all polarizations of light incident in this direction propagate with the same speed because they all "see" a single index of refraction, $n_{\perp}$. The unique axis of the uniaxial prism is therefore perpendicular to this square end.

Thus, an optically uniaxial material behaves optically like an isotropic material for unpolarized light incident in a single special direction and like a material with two different indices for unpolarized light in all other directions. The index of refraction of one polarization, the so-called "ordinary ray" is independent of the direction of propagation, while the index of refraction of the perpendicular polarization, or "extraordinary ray" depends on the direction of propagation. In the above discussion, the ordinary index of refraction is $n_{\perp}$. The extraordinary index of refraction depends on the angle θ from the unique axis and is given by:

$$1/n^2_{extraordinary} = cos^2(\theta)/n^2_{\perp} + sin^2(\theta)/n^2_{||}.$$

This behavior will be important in later discussions of liquid crystal devices.

Since the indices of refraction are important material properties, they are commonly measured. In liquid crystals, the index of refraction depends on the state of the material, which, in turn, depends on temperature and other physical variables. A typical thermotropic liquid crystal, a liquid crystal where the phase depends strongly on temperature has a high temperature *isotropic* phase, characterized by a single index of refraction n_{iso}, and uniaxial optical behavior characterized by $n_{\parallel}$ and $n_{\perp}$ in the lower temperature *nematic* phase. The change from one index of refraction to two occurs at the isotropic to nematic phase transition temperature. A schematic graph of this type of behavior is shown in Fig. 5.6. An explicit example will be considered in the exercises.

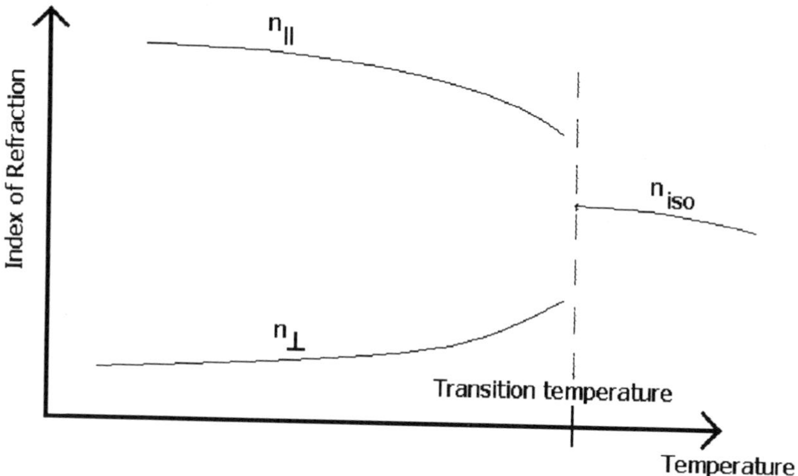

Fig. 5.6 The index of refraction as a function of temperature for a typical thermotropic liquid crystal that changes from an isotropic phase with a single index of refraction to a nematic phase which is uniaxial and has two indices of refraction.

When light interacts with material, it can also be scattered. Light will travel in a straight line only in regions were the optical properties of the medium are uniform. Due to thermal motion of the molecules that constitute matter, there will always be very small variations in the index of refraction in all matter. Thus, there will always be some scattering of light. In the case of a mixture of small particles in a liquid, there will be scattering because of the difference in refractive index between the particles and the liquid. As an example, milk appears white because of the strong light scattering that results from the large index of refraction difference between the liquid and the protein and fat particles in milk.

The equations that describe scattering are complex. However, they predict that for weak enough scattering, or thin enough slices of the material media, some of the incident light will be transmitted and some of the light will be scattered in almost all directions. The important feature that leads to scattering is that the medium must be optically non-uniform[8] on length scales similar to the wavelength of the light that is scattered. If the inhomogeneities of the media vary with time, then the scattered light intensity will also vary with time. The amount of light scattered depends on the properties of the media and the wavelength of the incident light. Rayleigh[9] was one of the first scientist to study light scattering. He found that there is an overall prefactor of 1 divided by the fourth power of the wavelength of the light in the expression for the scattered intensity,

$$I_{scat} \propto \frac{1}{\lambda^4} I_{incident}.$$

This factor, when coupled with the variations in the density of gas molecules in the air, explains why the sky is blue and sunsets are red.

5.6 References

Most introductory physics texts have a good discussion of light and waves. A particularly good text is **Light and Color,** David L. Wagner and R. Daniel Overheim, John Wiley and Sons, 1982. Shurcliff's text **Polarized Light: Production and Use**, Harvard Univ. Press, 1962 is a tremendous reference on light, polarized light and anisotropic optical materials. In spite of its age, it is still one of the finest references on polarized light. At a somewhat higher level, **The Physics and Chemistry of Color**, K. Nassau, Wiley, 1983 has many interesting and careful discussions.

---

[8]When describing the uniformity of a media, many terms are interchangeable: uniform = isotropic = homogeneous and non-uniform = anisotropic = inhomogeneous.

[9]Lord Rayleigh (John William Strutt), 1842-1919. Lord Rayleigh's was the second Cavendish Professor of Physics at Cambridge, and performed research in nearly all of the areas of 19th century physics including optics, sound and vibration, wave theory, color vision, electricity and magnetism, light scattering, hydrodynamics, elasticity, and photography.

5.7 Exercises

(1) Calculate the frequency of light in vacuum that has a wavelength of 400 nm. Repeat for $\lambda = 700$ nm.

(2) A neon sign downtown is red. Estimate the wavelength of this light. (If you feel the need to calculate, (you need not), take $n = 1$.)

(3) A TV station broadcasts at a frequency of 454 MHz. What is the wavelength of this electromagnetic wave in air, which is essentially the same as free space (vacuum)?

(4) Two polarizers are arranged in a line so that light propagates first through one then the other. This is the standard polarizer-analyzer setup. Use a calculator or a spreadsheet to obtain representative values of the transmitted light intensity as the angle between the direction of the polarizer and the analyzer varies from 0 to 180°. Assume that $I_{in} = 1$. Graph your results. Describe where the transmitted intensity is most sensitive to small variations in the angle between the polarizers.

(5) An electromagnetic wave of frequency $6*10^{14}$ Hz passes from air ($n = 1$) to glass with an index of refraction of $n = 1.52$.

   (a) What is the frequency of the electromagnetic wave in the glass?
   (b) What is the wavelength of the wave in air and in glass?
   (c) Generalize and make a statement about the relationship between the wavelength and frequency of electromagnetic waves in materials with index of refraction greater than one compared to when $n = 1$.

(6) A beam of light in air is incident on glass of index of refraction 1.60 at an angle of incidence of 30°. What is the angle of reflection? What is the angle of the transmitted beam?

(7) A liquid crystal is characterized by two indices of refraction: $n_{\parallel} = 1.65$ and $n_{\perp} = 1.45$. Light of wavelength 565 nm in air is incident on this material. Let $n_{air} = 1$.

   (a) What is the frequency of the light?
   (b) Approximately what color is the light?
   (c) What is the speed of light propagating with $n_{\parallel}$?
   (d) What is the speed of light propagating with index $n_{\perp}$?

(8) The following data was obtained for the index of refraction of a liquid crystal as a function of temperature.

   (a) Graph this data, n vs. T.
   (b) Estimate the transition temperature of this material.

T (^oC)	$n_\perp$	$n_\parallel$	$n_{average}$
40			1.633
34			1.624
33			1.623
32			1.621
31			1.620
29.9	1.557	1.670	
28.6	1.550	1.689	
26.5	1.544	1.703	
24	1.542	1.714	
21.8	1.540	1.721	
19.7	1.539	1.727	
17.8	1.538	1.733	
15.5	1.536	1.738	

(9) Assume that the scattering of light by density fluctuations of air molecules can be modeled by the following equation:

$$I_{scattered} = \frac{2.401 * 10^{11}}{\lambda^4}.$$

In this expression $I_{scattered}$ has units of watts, and wavelength is measured in nm. Complete the following table:

λ(nm)	$I_{scattered}$ (watts)
400	
450	
500	
550	
600	
650	
700	

(a) Graph $I_{scattered}$ versus wavelength.
(b) Is bluish light or reddish light more strongly scattered? Explain.
(c) Based on your answers to the last two parts, why do you think the sky is blue? Explain your reasoning.
(d) During a sunset, the sunlight passes through a much longer path length of air than during the middle of the day. Based on your answers to parts a) and b), do you expect more reddish light to be

transmitted through a long path length or more bluish light? Explain your reasoning. Explain why you think a sunset is dominated by orange and red colors.

(10) *Total internal reflection* occurs when light is incident on an interface of a material that has a lower index of refraction than the media in which it is propagating. To see how this might occur, assume that light propagating in a piece of glass of refractive index 1.53, and is incident onto an interface with air on the other side.

(a) Let the angle of incidence in the glass be 20°. Find the angle of the transmitted beam. Repeat this for several other angles less than 40°. Is the transmitted angle larger or smaller than the incident angle?

(b) Now, at some incident angle the transmitted angle will be 90°. Find this incident angle and draw a sketch of the incident and transmitted rays.

(c) For larger angles of incident, no transmitted wave can be found. Why? (Hint: Can the *sine* function become larger than 1?) Thus, larger angles than that found in b) the incident light is totally reflected.

5.8 Research questions

(1) Calcite and quartz are uniaxial materials. Find some optical devices that use these materials.

(2) These exercises have focused on electromagnetic waves in which the speed of propagation v is only weakly dependent on the wavelength through the index of refraction. This is not the case for surface waves on water. Find the relationship between wavelength and propagation speed for surface waves on water, and graph the speed as a function of wavelength.

(3) What is an optically biaxial material? Give an example. (Try searcing for "optics" and "biaxial" on the Web.)

(4) What are the Fresnel equations? If you can graph them using a spread sheet, do so. Are there any special angles where, for certain polarizations, there is no reflected or transmitted light?

(5) What is the CIE chromaticity diagram or chart? What are metameric colors? How many ways are there to obtain the same metameric color?

Chapter 6

Atoms and Molecules

6.1 Overview

Since early in your education you have heard that matter is made of atoms, and that atoms combine to form molecules. This chapter will treat this assertion as fact and not try to convince you of the atomic theory of matter. Instead, we will look at some of its consequences, and discuss, using simple models, the structure of atoms and molecules. By using your knowledge of electric forces, you will understand that the forces between the nucleus and electrons in single atoms and between different atoms and molecules are electrostatic in origin. With this in mind, the structure of the atom will be reviewed. The ways in which atoms combine to form molecules and solids, and the reasons they do so will be the primary focus of most of this chapter.

The following questions will be addressed:

(1) What is an atom? What are an atom's constituent parts?
(2) Is there a relationship between force and energy? If so, what is it, and why is it useful?
(3) What is a molecule? Why do molecules form?
(4) What distinct types of chemical bonds occur between atoms and groups of atoms?
(5) What are the characteristic energies of these chemical bonds?

6.2 Atoms

In order to understand the material in this book at more than a cursory level, some understanding of molecules and atoms is necessary. Once you understand atoms you will be better prepared to understand molecules, which are simply groups of atoms bonded together. It is then straightforward, although not trivial, to understand how large numbers of molecules behave.

We will begin describing why atoms are important, then discuss some of the properties of atoms and conclude with a simple model of atoms. First, you should remember that everything we physically experience depends on the material universe, which is made of atoms. The diversity of the material universe follows from the wide range of chemical compounds that exist.

Experiments have confirmed that all matter is made of atoms. However, this begs the question: What is an atom? An atom is the smallest fragment of an element which (a) takes part in a chemical reaction, and (b) has all the properties of the element. This further begs the question by introducing the word element. Fortunately, this "begging" dance ends here because an element is defined as a substance that can not be separated into simpler substances by *chemical* means. [1] Thus, elements are the basic building blocks of nature.

Here are a few facts about elements. Most readers will be familiar with all of them. First, there are approximately 90 naturally occurring elements and about 20 artificial or man made elements. Second, the properties of the elements are well known and are often presented in summary form on the periodic table of the elements. A copy of the periodic table is included at the end of this chapter. The elements are assigned one or two letter symbols; for example, H for hydrogen, He for helium, Na for sodium, Fe for iron, and so forth. These symbols may be thought of as the letters of the chemical alphabet. By using this analogy, we can think of chemical compounds as being equivalent to words. (Don't push this analogy too far! Words exist along a line, while compounds exist in three-dimensions.) Note that we have already gone to a very abstract model of atoms. We have proceeded from the actual substance to a symbolic representation of

[1]The term chemical is important. *Physical properties* are those properties that can be detected or measured using our five senses or their extensions through scientific instruments. Examples include shininess, hardness, and melting and boiling temperatures. The *chemical properties* of an element describe how it may interact with other atoms and molecules. In the current context, a *chemical change* means that the chemical properties of a molecule change; while a *physical change* is a change in physical form without a change in the chemical properties of the molecule. Is must also be noted that elements can be changed into other elements by radioactivity and/or bombardment by high energy particles. Neither radioactivity or particle bombardment are considered *chemical* reactions.

the actual element in less than one page!

Our model of the atom will consist of a diffuse cloud of electron(s) about a very massive nucleus. It is not important for us to go into great detail about the nucleus because we are more concerned with groups of atoms or molecules and how large number of molecules behave. While the nucleus is interesting, there is simply not enough time to study it in this book. The following outline about atoms and the nucleus will have to suffice:

(1) Atoms are made of electrons, protons, and neutrons.
(2) The radius of an atom is roughly 0.1 to 0.5 nm.[2]
(3) The nucleus of the atom contains protons and (generally) neutrons and has it a radius of roughly 10^{-5} to 10^{-6} nm. Thus, by comparing atomic radii to nuclear radii it is clear that there is lots of empty space in an atom. By way of comparison, if a nucleus were to have a radius of 1 ft., then a typical atomic radius would be about 19 miles.
(4) In spite of its small size, almost all of the mass of an atom is in the nucleus.

The following table lists some properties of protons, neutrons and electrons:

Particle	**Charge**	**Mass in kg**	**Mass in u**
Electron	-e $\approx -1.6 * 10^{-19}$ C	$9.1094 * 10^{-31}$	$5.49 * 10^{-4}$
Neutron	0	$1.6749 * 10^{-27}$	1.0087
Proton	e $\approx 1.6 * 10^{-19}$ C	$1.6726 * 10^{-27}$	1.007276

From this table you should note the following:

(1) The mass of a proton is very nearly the same as the mass of a neutron.
(2) An electron is much less massive than either a proton or a neutron. In fact, an electron's mass is approximately 1/1836 that of a proton or neutron.
(3) Neutrons are uncharged, protons have a positive charge of one fundamental charge, *e* and electrons have a negative charge of one fundamental charge, -*e*.
(4) We have introduced a secondary mass standard, *the unified atomic mass unit*, u, where u $= 1.6605 * 10^{-27}$ kg. The advantage of this system

[2]Since the location of electrons in an atom may be described mathematically by a probability "cloud" that extends to ∞, this radius corresponds to a surface that will enclose where the electron is 90% of the time.

of mass is that the mass of an atom of an element in these units is approximately an integer and ranges from 1 to several hundred.

An essential point that this table suggests (and that the periodic table illustrates) is that **the number of protons in the nucleus characterizes the element.** The number of protons is called the *atomic number* and is given the symbol Z (capital z). The number of neutrons is called the *neutron number* and is given the symbol N (capital n). The *mass number*, given the symbol A (capital a), is equal to the number of protons plus the number of neutrons, *i.e.* A=Z+N.

Why does this last distinction matter? Nature provides atoms with the same number of protons (making them the same element) but with different numbers of neutrons and thus with different atomic masses. Atoms of the same element, but with different atomic masses are called *isotopes.* Several hundred isotopes exist. Furthermore, **all atoms of a given isotope of a given element are identical**, but different from all other isotopes of the same element as well as all other elements. This last statement means that all iron atoms of mass number 57 are indistinguishable from all other iron atoms of mass number 57. The origin of a piece of pure iron 57 cannot be determined. Interestingly, this is not true of iron ores. The impurities and trace materials in an ore sample can be used to trace its origin.

Consider the following examples of isotopes:

(1) Hydrogen (H): 1 proton, 0 neutrons; Z=1, N=0, A=1 (this is the only isotope with no neutrons)
(2) Deuterium (D): 1 proton, 1 neutron; Z=1, N=1, A=2
(3) Chlorine (Cl): 17 protons, 18 neutrons, A=35, (75.77%)
(4) Chlorine (Cl): 17 protons, 20 neutrons, A=37, (24.23%)

We see again that isotopes of a given element are atoms that have the same number of protons (the same value of Z) but different number of neutrons, hence different values of the neutron number, N. If this were not the case, then the isotope atoms would not be of the same element. Lastly, since A=Z+N, isotopes of a given element have different values of A. The percentages included with the examples of chlorine indicate that 75.77% of naturally occurring chlorine is the isotope of mass number 35. The balance is of mass number 37.

We will now discuss electrons in atoms. In their lowest energy state **atoms are neutral.**[3] This means that they have no **net** electrical charge.

[3]This is sometimes further clarified by stating they are electrically neutral. An atom

Since the protons have a quantized positive charge and the neutrons have no charge, there must be an equal amount of negative charge elsewhere in the atoms. In our model of the atom, the only other particles that can carry charge are the electrons, which indeed are negatively charged. Furthermore, since there is no net charge on a neutral atom, the number of electrons must equal the number of protons, $n_{electrons} = \mathrm{Z}$.

It is important to realize that while this discussion has treated electrons as particles, and that electron behavior in computer terminals, TVs and the like are accurately described by this particle picture, electrons in atoms cannot be thought of as particles or as waves. They exhibit both types of behavior at the same time. In fact, electrons in atoms exhibit very complex behavior.

Exploring the details of electron behavior in atoms will take us too far afield. For now, it is sufficient to recognize that scientists have a very good understanding of electrons in atoms and, that the electrons in atoms can be thought of as being a negative charge cloud that surrounds the nucleus. You should also be aware of a few important details. **In an atom, electrons can only have certain allowed energies**. Electrons of a given energy are said to be in a specific *atomic orbital.* An orbital can be related to the probability function describing where the electron can be found. With this understanding, we can start to discuss chemical bonds and molecules.

6.3 Molecules

Most substances are combinations of two or more atoms, often of different elements, with a fixed composition, *i.e.* a fixed ratio of constituent elements. A molecule is the smallest particle of a substance that exists in the free state and has all the characteristics of the substance. Molecules are chemically stable—they will not spontaneous change to (a) different chemical substance(s) without an external driving force. Thus, TNT is chemically stable, but not stable to excessive physical vibration (*e.g.* a shockwave from an exploding blasting cap). Examples of common molecules and their chemical formulas are:

Water: H_2O. (Read as two hydrogen atoms and one oxygen atom. The one is always implied.)

Hydrogen peroxide: H_2O_2. A different compound than water, with different properties.

---

can be in a higher (internal) energy state and still be neutral.

Sodium chloride, table salt: NaCl. The molecular formula is not Na_2Cl_5 or Na_4Cl_4.

Molecular oxygen, oxygen in the air we breath: O_2.

The existence of specific compounds (molecules) with fixed numbers of different kinds of elements implies there must be certain rules of combination. The discovery of these rules was a long process that might well be summarized as the early history of chemistry. The study of chemical reactions, whereby atoms or molecules combine to form other molecules, helped reveal these rules.

We will illustrate chemical reactions using the following type of notation: atoms $\rightleftharpoons$ molecules. The double arrow indicates that atoms can combine to form molecules and the molecules can dissociate to form atoms. More generally, the left side of the $\rightleftharpoons$ corresponds to reactants and the right side to products. Of course, molecules can combine to form other molecules so we could write molecules $\rightleftharpoons$ molecules, or even atoms + molecules $\rightleftharpoons$ molecules. Once more, the details are not critical; molecules can be formed. We will go into more detail later. Finally, while the reaction can go both ways (forwards, to the right, and backwards, to the left, in the above examples) there are occasions in which the backwards or reverse reaction takes place very slowly or hardly at all.

6.4 Forces and energy

We are now led to ask questions like: "Why does nature work this way?" and "What holds molecules together?" To answer questions like these we must carefully look at the force between two atoms. Recall that a force may be thought of as a push or a pull. It has both magnitude and direction. In the case of two spherical atoms the direction of the force is along the line joining the centers of the two atoms. The magnitude of the force is a more complex function. It depends on the distance between the atoms. A schematic figure of the magnitude of the force as a function of distance, r is shown in Figure 6.1.

This figure is not as foreboding as it first appears. First, observe that forces can be attractive, $F < 0$, or repulsive, $F > 0$. This is indicated along the vertical axis. For two objects separated by a distance r, the separation cannot be negative. Thus, r varies from 0 to ∞. For large separations, the force is negative and small. This means that there is a small attractive force between the atoms. At some separation the force

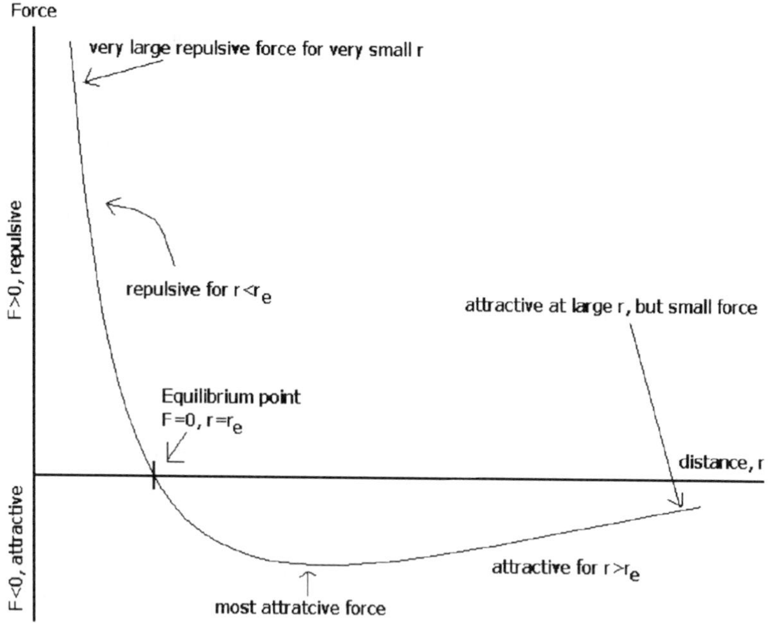

Fig. 6.1 A generic force versus distance curve for two spherical atoms.

becomes zero which means there is no net force between the two atoms. This separation is the equilibrium separation. If the atoms approach more closely than the equilibrium separation, a positive repulsive force results. This force becomes very large for very close separation between the atoms.

While this force versus distance curve is useful, it is often easier to discuss atomic behavior in terms of energy. The importance of energy cannot be overstated. Unlike force, energy is a scalar; thus, it can be mathematically manipulated just like normal numbers. Moreover, under many conditions, energy is conserved. The conditions needed for energy conservation to occur are not too difficult to find, so this is a very useful general principal that we will have occasion to use later.

Systems of all sorts will be found in their lowest energy state consistent with external constraints, provided the system can move to this state. The existence of stable, lowest energy states explains the occupied electron orbitals in atoms and is crucial to the formation of molecules. Atoms form molecules because a molecule has a lower total energy than the separate atoms that combine to form the molecule.

To use energy to discuss atomic behavior, we would like to change from a force versus distance curve to an energy versus distance curve. Such a change is made possible by introducing the concept of *potential energy*. In physics, energy is defined as the ability to do work. *Potential energy*, PE, is the ability (potential) to do work (expend energy) that an object possesses because of its location or configuration. A common example is a mass, m a distance h above a reference height in the gravitation field of the earth. This mass has a potential energy mgh, where g is the acceleration due to Earth's gravity.

To see how this mass might have potential energy, consider, as an example, dropping a mass onto a nail. If there is energy due to position, one should be able to use this energy to pound a nail into a piece of wood by dropping a large block of metal onto the nail. From your everyday experience, you know that dropping a heavy metal block onto a nail will pound the nail into the wood. Thus, a metal block held above the ground possesses energy.

Not surprisingly, for one drop of the block, the higher one holds the block above the nail (the larger h, where the top of the nail can be the reference point) before it is released, the further the nail goes into the wood. Since the block is initially at rest, the block is said to possess only potential energy. The metal block is doing work on the nail by pushing it into the wood. Furthermore, it takes more energy (requires more work) to pound a nail 2 cm into the wood than to pound it in 1 cm. This extra work is accomplished either by repeated blows from the block released from the same low height, or by one blow from the block released from a greater height.

Springs may also be used to supply potential energy. In this case, the potential energy is not due to the location of the spring, but to its configuration (stretched or compressed). Other forms of potential energy are also possible. In particular, we will use the potential energy that is a result of the electric charges of electrons and nuclei.

There is another common form of mechanical energy. This is energy of motion or *kinetic energy*. For a point object,[4] the kinetic energy depends on the mass of the object and its velocity. In fact, we write kinetic energy as $KE = 1/2mv^2$, where m is the object's mass and v is the object's speed.

[4] A point object has no physical extent (size) so it cannot rotate. This is an idealization that allows us to ignore rotational motion. More complete analysis shows that motion can always be described as motion of a point mass concentrated at the center of mass of the object and motion about the center of mass.

There should generally be little confusion between v as the speed of a wave and v as the speed of a particle. The meaning of v will be clear from the context.

The total mechanical energy of a system, E is given by the sum of the kinetic energy plus the potential energy,

$$E = KE + PE. \tag{6.1}$$

The total mechanical energy is conserved in an isolated system *that has no dissipation due to friction, viscosity, and other loss mechanisms.* This is the principal of **conservation of energy**. It is one of the basic principles of science. Furthermore, for a system of two or more objects, it is the total energy that will be a minimum value at equilibrium. For objects at rest, the minimum in the PE corresponds to the minimum in E.

We can now relate potential energy to force. This cannot always be done, but for the forces that we are considering, we may make such a correlation. Mathematically, we find that the force depends on the negative rate of change of the potential energy with center-to-center distance between the two objects. Thus, we may sketch curves of the force, F and the potential energy, PE as a function of center-to-center distance as shown in Fig. 6.2. Note that the zero in force, which occurs at r_e corresponds to a minimum in the potential energy. The symbol r_e (pronounced "are e") is the equilibrium **r**adius or separation between the two objects.

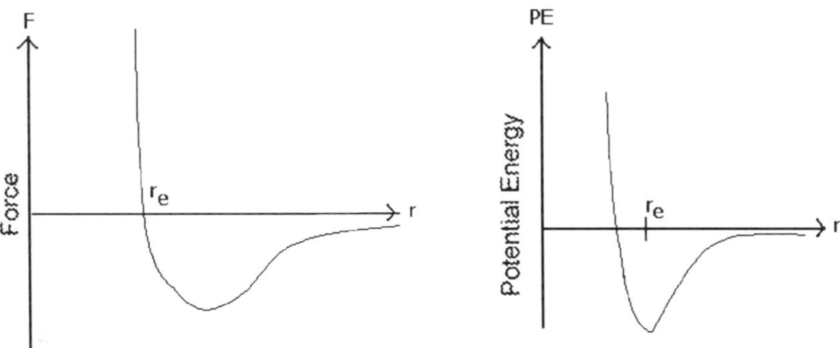

Fig. 6.2 The force vs. distance curve and the corresponding potential energy curve for a hypothetical interaction.

Now consider the potential energy curve that we have just constructed as it applies to two separated atoms. Notice that the system can achieve its lowest total energy by having the atoms separated by a distance of r_e. Further note that if the two atoms come too close together, the potential energy becomes very large. It is this large increase in potential energy as two atoms approach one another that keeps matter from collapsing. However, as the first figure of force verses distance illustrates, the force is attractive at distances greater than r_e. This means the atoms experience a force "pulling them together" when they are separated by a distance greater than r_e.

In fact, we know that the form of the force shown above must be essentially correct. First, an attractive force must exist; otherwise, chemical bonds between atoms would not form, and we know that compounds exist. That compounds exist tells us that the interatomic force is attractive at large enough distances (although it does not address the issue of its mathematical equation or "functional form"). The fact that matter does not collapse means that there must be a force preventing this collapse to a lower energy state. Thus, the force must be repulsive at very small separations. The repulsive part of the interaction is largely due to the positively charged nuclei repelling each other and a quantum mechanical principle known as the Pauli exclusion principle. Experiments also indicate that there is an equilibrium separation or bond length between chemically bonded atoms. This equilibrium bond length must correspond to zero force.

To understand chemical bonds, it helps to understand a little more about electrons in atomic orbitals, orbital electrons for short. A careful examination of the periodic table (see Fig. 6.7) indicates that as one moves in a row (or period) from left to right, the orbitals become filled with more and more electrons. A special class of atoms forms the last column (VIIIA) of the periodic table. The atoms in this column have totally filled orbitals. These atoms are called the *noble* or *inert* gases, because they have very low reactivity or "low chemical affinity." There are very few compounds formed from inert gases, and for a long time there were no known compounds formed with this class of atoms. Because of this experimental fact and detailed study, it is known that completely filled orbitals are (a) spherical, (b) particularly stable and (c) associated primarily with repulsion. Thus, we may state that **partially filled orbitals govern chemical bonding.**

To summarize, we may state the following. Molecules form because the molecule has a lower total energy than the separate atoms that form it, **and** there is a mechanism for the molecule to form.[5] Bonding of atoms

[5]This is important. Molecular oxygen and molecular hydrogen in the gas form are stable unless a major change, such as a spark of electricity, occurs. Then, liquid water

is governed by the electrons in the partially filled atomic orbitals of the atoms that form the molecule. Filled orbitals are particularly stable and desirable. Lastly, since molecules can react to form new molecules, the points that apply to constituent atoms also apply to constituent molecules.

At this stage we will need to define an ion. **An ion is an atom or molecule with a net charge due to too many or too few electrons.** The notation we use to indicate this will be the atomic or molecular symbol and a superscript + or − or several of them to indicate the number of excess or deficient electrons. Thus, Cl^- is a chlorine atom with one excess electron, Na^+ is a sodium atom with one too few electrons, and Mg^{++} is a Magnesium atom with a deficiency of two electrons.

6.5 Types of chemical bonds

We will discuss five **idealized** types of chemical bonds: **ionic, covalent, metallic, van der Waals**, and **hydrogen** bonds. These types of bonds **rarely** occur as pure forms; however in this introduction this idealization will provide the necessary insight into the bonds that form molecules. In all cases these bonds may be associated with attractive forces between the bonding species.

6.5.1 *Ionic bonds*

The ionic bond is the simplest bond we will discuss. The essence of this bond is the transfer of an electron from one atom to another. In this type of bond, two atoms, one with one too many electrons to have an **inert gas electron configuration** and one with one too few electrons to have an inert gas electron configuration, combine, transfer an electron between themselves so that positive and negative ions are formed, and then reduce their overall energy by forming a molecule. In essence, the electrostatic attraction between oppositely charged entities discussed earlier can explain this type of chemical bond. The atoms that form ionic bonds are usually at opposite ends of the periodic table such as the alkali metals in Group IA (Li, Na, K, Rb, Cs, Fr) and the halogens in Group VIIA (F, Cl, Br, I, At). Common ionic bonded compounds include NaCl, LiBr, etc. It is important to notice one more feature of this bond: it is nondirectional. A positive ion attracts a negatively charged ion equally no matter what their

will form explosively.

relative position or orientation. This has important structural implications. Additionally, every ion interacts with every other ion, so the ionic bond is more complex than just two atoms interacting, even though that is the simple language often used.

To reiterate the basic idea: In ionic bonding an electron is transferred between two atoms so that the atomic orbitals of both atoms achieve inert gas electron configurations. The common examples are the Alkali-Halides, such as NaCl.

6.5.2 *Covalent bonds*

Covalent bonds occur when the outermost electrons of atoms are shared more or less equally between atoms. We must start by defining *valence electrons*. To a first approximation, the electrons in the outermost orbitals that participate in bonding are called valence electrons. A covalent bond is formed when valence electrons are shared more or less equally by neighboring atoms so that, on the average, each atom obtains an inert gas electron configuration. This is a highly cooperative sharing of electrons. Covalent bonds always involve pairs of electrons, and we may identify the following types of covalent bonds:

Type of Covalent Bond	# of Shared Electrons	Symbol	Example
Single	2	- or :	C-H or C:H
Double	4	= or ::	O=O or O::O
Triple	6	≡ or :::	N≡ N or N:::N

A covalent bond (a pair of shared electrons) is typically indicated by a short line (-) or a pair of dots (:).

It is not at all obvious that such a sharing of electrons should lead to a lowering of the total energy of the system. However, both experience and calculations show that this is indeed the case. The explanation is ultimately a quantum mechanics problem. Ideal covalent bonding with equal sharing of electrons occurs when the two bonding atoms are identical. When two different molecules bond as for instance in water, H_2O, the electrons are not equally shared, causing an imbalance in the electron charge distribution, and the bond takes on a partial ionic character. This is discussed in many chemistry texts.

Covalent bonding explains diatomic molecules such as O_2, N_2, H_2, and most organic molecules. Furthermore, a covalent bond is directional. There are well-defined angles and distances associated with the various types of covalent bonds. A deep understanding of covalent bonding allows one to predict the three dimensional structure of many molecules.

6.5.3 *Metallic bonds*

To a first approximation, metals may be thought of as an array of ions surrounded by a sea of free valence electrons. It is these free valence electrons that determine many of the physical properties of solids, such as their high electrical and thermal conductivities and their shininess. Thus, metallic bonds involve electron sharing, but are non-directional. The valence electrons are delocalized and are not attached or associated with a given atom.

How do we understand metallic bonding? The simplest picture regards solids as very large molecules. This fairly accurate picture allows us to extend the ideas we just used for simple molecules to very large molecules. The argument proceeds as follows. Consider two atoms that form a molecule. The particular type of atom we will pick is lithium, Li, because lithium has only three electrons, and is known to form one of the "simpler" metals. When the two lithium atoms form a lithium molecule, the atomic orbitals combine and form molecular orbitals. This is shown in Figure 6.3.

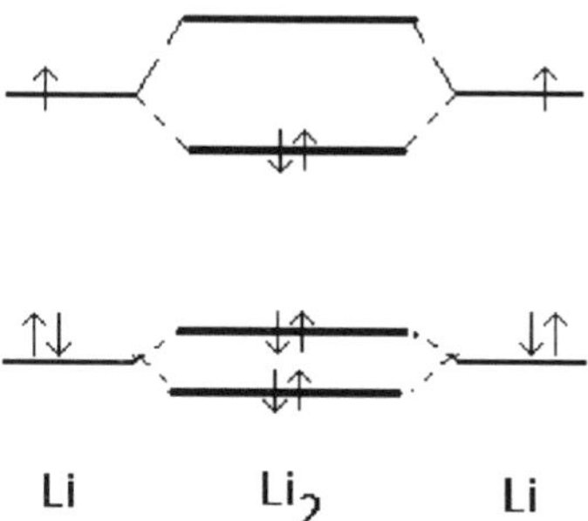

Fig. 6.3 The energy levels of two separated lithium atoms and those of a hypothetical di-lithium molecule.

In Figure 6.3, energy is plotted vertically and the horizontal direction illustrates the status of the atoms. The two horizontal lines on the left and right hand sides of the figure represent the two lowest energy levels of lithium atoms. The lower energy is at the bottom, and the vertical arrows represent electrons. Note that the individual lithium atoms have the same energy levels, and each has three electrons. Furthermore, electrons fill the energy levels so the total energy is as low as possible.[6] Their combination to form Li_2 is shown in the center of the figure. When two lithium atoms combine to form the lithium molecule, Li_2, the atomic energy levels combine and split as shown. The stability of Li_2 is demonstrated by the fact that its six electrons have a lower total energy than the six electrons in the two separate Li atoms.

Now envision bringing a large number of lithium atoms together. It is found experimentally, and verified by detailed calculations, that the spacing between the highest and the lowest molecular energy levels in Li_n (the subscript indicates a molecule with n atoms) remains finite. Interestingly and importantly, the region between the two molecular energy levels fills in with more and more energy levels. These energy levels become so close together that they form an essentially continuous *energy band.* This very general phenomenon is schematically shown in Figure 6.4.

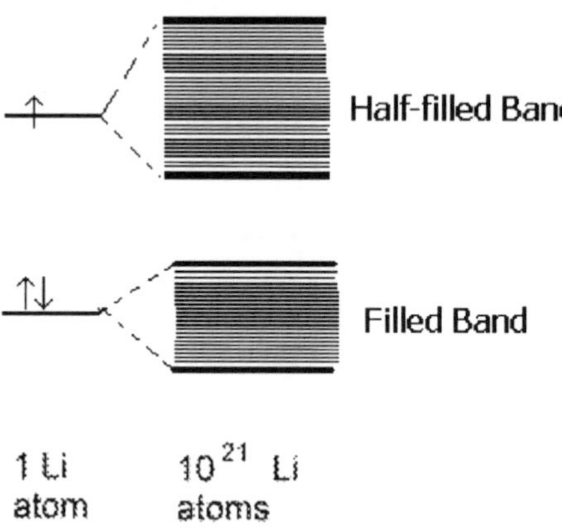

Fig. 6.4 The formation of energy bands by bringing a large number of atoms together to form a metal.

The bands which form when a large number of atoms come together may be totally filled with electrons, so they can hold no more electrons, empty, containing no electrons, or partially filled. In order to deal with

[6] The electrons are drawn as arrows to indicate that they have a property called *spin*. The rules of quantum mechanics indicate that the spins should be drawn as shown.

these situations scientist refer to full bands as *valence bands.* This can be confusing because the electrons in many valence bands were not originally valence electrons in the sense valence electron was used in the discussion of covalent bonding. Once more, this somewhat obtuse language will generally be clear from the context. (You might also correctly surmise that these two fields were largely explained by scientists who did not communicate enough.) Partially filled bands are called *conduction bands.* This allows us to develop the *band theory of solids*, which we will find useful later.

6.5.4 *Band theory of solids*

The simplest version of the band theory of solids is based totally on how full or empty the energy bands are **and** the energy spacing between the top of the highest energy filled band and the bottom of the next partially filled band. This theory explains the electron contribution to electrical conduction in conductors, insulators, and semiconductors. In many practical applications, this is the prime concern.

First, what does one mean when one states that a solid is a conductor, an insulator, or a semiconductor? In a conductor, there are electrons that are free to move to other energy levels within an energy band, and within the solid. Most metals are good conductors of electricity. They are also good conductors of heat. This is because the electrons are free to move in the solid making it is easy for charges to move through the material. Hence, it is easy for a current, a flow of electrons, to move or be conducted through the solid. This is the basis of how copper wiring conducts electricity. An insulator is characterized by a very small number of electrons (ideally zero) being free to move about. Because of the lack of free charges, insulators do not conductor electricity. Intermediate between conductors and insulators are semiconductors. Among the elements, the best known examples of semiconductors are silicon, Si, and germanium, Ge. The conductivity of pure semiconductors is a very strong function of temperature. At very low temperatures, they act like insulators, while at high temperatures they act more like conductors. Furthermore, the electrical properties of semiconductors can be greatly influenced by the controlled addition of small amounts of other elements, called dopants. The three cases, conductor, insulator, and semiconductor, are illustrated in Figure 6.5.

Notice that a typical conductor has (a) filled valence band(s) (only one is shown) and an energy gap and then a conduction band that is approximately 1/2 full of electrons. The existence of many very closely spaced

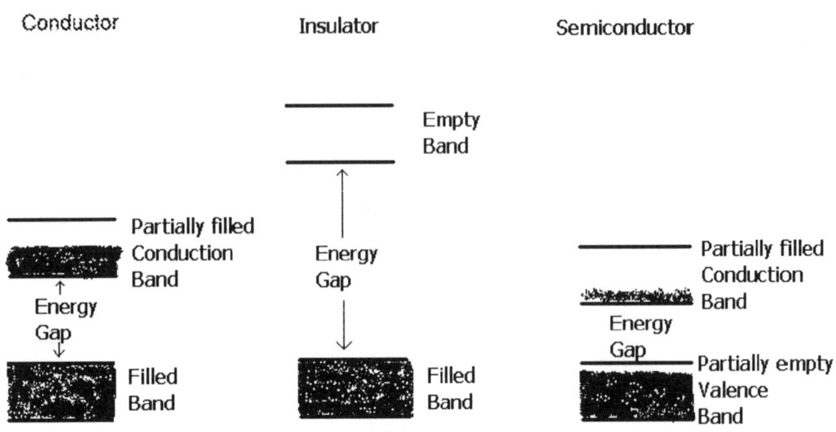

Fig. 6.5 Simplified band structures of conductors, insulators, and semiconductors.

energy levels within the conduction band allows the electrons to gain the small amount of extra energy needed to move, and hence make the metal a good conductor. The insulator has (a) filled valence band(s), a rather large energy gap, and a totally empty (or nearly totally empty) conduction band. Finally, a pure semiconductor has a partially empty valence band (the emptiness depends on the temperature, at absolute zero the valence band is totally filled), a small energy gap and a conduction band with a relatively small number of electrons promoted from the valence band. The number of electrons in the conduction band and the matching number of "holes" in the valence band are very temperature dependent in semiconductors. This is not the case for metals or insulators.

There are certain rules of thumb that allow one to determine if a solid will be a metal, insulator, or semiconductor. By counting the number of valence electrons (in the covalent bonding sense), we may often get a good first guess. For example, solids with one free valence electron per atom are always metals. This includes the alkali metals (Li, Na, K, Rb, Cs) and the "noble metals" (Cu, Ag, Au).[7] A solid formed from atoms with an odd number of valence electrons is almost always a metal. For example, Al, Ga, In, and Ti are all metals. However, this does not work that well for atoms with five valence electrons: As, Sb, and Bi are called "semimetals." A solid

[7] The noble metals are somewhat more complex since they actually have eleven valence electrons; however, ten are in a full atomic orbital and do not participate in chemical bonding. Silver halides (AgCl, etc.) are common in photographic film, and copper chloride (CuCl) are common ionicly bonded noble metal solids.

with an even number of valence electrons is not necessarily an insulator. In this case, detailed calculations of the band structure must be undertaken in order to determine if the bands overlap.

A very interesting case, both technologically and scientifically, are the tetravalent elements in Group IVA of the periodic table which can be insulators, semiconductors, or metals. Carbon, C, is a familiar insulator as diamond. It may also be thought of as a semiconductor with a very large energy gap. As one moves down this column of the periodic table one encounters silicon, Si, a semiconductor, germanium, Ge, a semiconductor, tin, Sn, either a semiconductor (in one structure) or a metal (in a different structure), and finally lead, Pb, a metal.

Clearly, the properties of atoms, as well as the structure into which the ions arrange themselves determine in very subtle ways the electronic properties of a solid, and this simple band theory is only a first step towards a more complete understanding.

6.5.5 *van der Waals bonds*

van der Waals bonds are weak bonds. A component of this type of bonding is always present, but often much weaker than the other bonds in a system. In order to understand van der Waals bonds we must first reconsider the electric dipoles that we discussed previously. An electric dipole may be modeled as two charges of equal magnitude and opposite sign separated by a distance d. Since the electrons and positive ions in atoms or molecules are not at rest, at any given instant of time one part of the atom (or molecule) may be more positively charged than another part. This results in the formation of a time varying electric dipole. Since the dipole moment of these dipoles does not average to zero, there is always a dipole associated with the atom or molecule. There will be a resulting energy of attraction between the dipoles, called the van der Waals bond. There are other types of dipole-dipole interactions, that we will not consider, that also lead to similar bond energies. The energy, U, of a van der Waals bond has the following functional form with distance, r, between the two dipoles: $U_{van\ der\ Waals} \propto 1/r^6$. The corresponding force varies as $1/r^7$.

It is important to realize that van der Waals forces and bonds exist between all types of atoms and molecules, but they are usually masked by other types of bonding. van der Waals bonding predominates in solids formed from inert gas molecules. A different type of everyday example of van der Waals bonding dominating is the clinginess of Saran Wrap^{TM}which seems to cling to nearly everything, especially itself.

6.5.6 *Hydrogen bonds*

There is one more type of bond that is extremely important to life, polymers, and many liquid crystalline systems. This is the hydrogen bond. This bond is much weaker than some of the other bonds discussed, yet its existence is critical to the binding of proteins in our bodies and the properties of water. This bond is essentially electrostatic and occurs between hydrogen and atoms in other molecules that typically have a permanent dipole moment and consists of the sharing of a hydrogen atom between two different entities that may be on the same or different molecules. Most often this bond is between hydrogen, H, and oxygen, O, nitrogen, N, or fluorine, F. The dipole weakly attracts the hydrogen. In spite of its weakness there are usually so many hydrogen bonds that can lead to a large effect.

The weakness of the hydrogen bond compared to other bonds masks the considerable importance of the existence of the hydrogen bond to life. One way to see this is to compare the freezing and boiling temperature (in degrees Celsius) of various molecules formed from two hydrogen atoms and one atom from the sixth column of the periodic table. These are shown in the following table:

Substance	Boiling Temperature (^oC)	Melting Temperature (^oC)
H_2Te	-2	-51
H_2Se	-42	-65.7
H_2S	-60.7	-85.5
H_2O	100	0

The following observations are in order. As the atomic mass of the atom forming the molecule with hydrogen decreases, the boiling temperature steadily falls until we reach H_2O, where it suddenly and very dramatically increases. Similarly, the freezing temperature is decreasing until water is encountered. The temperature range of the liquid state also increases significantly for water. If all we had were the first three substances in this table, we would predict that water would boil at about -100 ^oC and freeze at about -125 ^oC. Since this is not the case, there must be some additional force that binds the molecules of water so that it does not boil until the much higher temperature of +100 ^oC. This additional force comes from the hydrogen bond. Hydrogen bonding is responsible for many of the most important properties of water.

The hydrogen bond is sufficiently weak and of a transient nature that it is often indicated in pictures by a dashed line rather than a solid line. A diagram showing a hydrogen bond is shown in Figure. 6.6

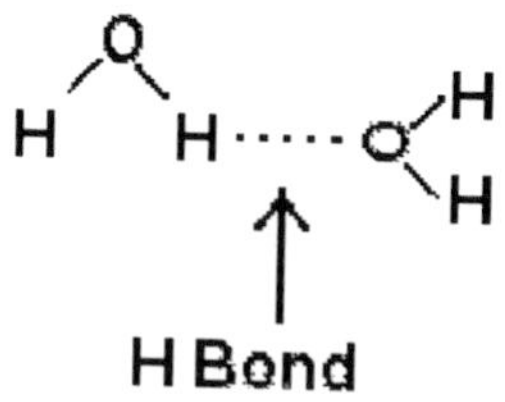

Fig. 6.6 An example of a hydrogen bond in water.

6.6 Summary

This chapter will end with a short summary of the energies involved in these various types of bonds. The amount of energy, U, needed to destroy or form various types of bonds can be quantified. On the basis this work:

$$U_{covalent} \approx U_{ionic} \approx 10 \ to \ 20 \ U_{hydrogen \ bond} \approx 10 \ to \ 100 \ U_{van \ der \ Waals}.$$

Typical equilibrium spacing between atoms, r_e in a molecule can also be measured. This distance, as well as the energy needed to break such bonds, depends on the bonded atoms and the type of bond. The following table gives a few examples.

Type of Bond	Bond Energy (kJ/mole)	Equilibrium separation (nm)
NaCl, ionic	764	0.282
KI, ionic	627	0.353
C-C, single covalent	347	0.154
C=C, double covalent	611	0.133
C≡C, triple covalent	837	0.120
C-N, covalent	293	0.151
Xenon, crystal, van der Waals	16	0.435

In this table, the bond energies are expressed in kJ/mole. You might recall that 1 mole is $6.02 * 10^{23}$ particles (atoms or molecules). In this case, it refers to the number of bonds. The table shows some of the basis for the energy relationship discussed above.

6.7 References

Most chemistry texts cover this material in more detail. The following specialized texts are especially useful. **Introduction to Solid State Physics** by C. Kittel, Wiley, exists in many editions, though the 4th edition is perhaps the best. The material in this chapter is expanded on in all editions. **The Nature of the Chemical Bond** by L. Pauling, (Cornell University, 1940) is still a very useful text. **Coulson's Valence**, 3rd ed. by R. McWeeny (Oxford, 1990) is a good introduction to the quantum mechanical aspects of these problems. **Electronic Structure and Chemical Bonding** by D. Sebera, (Blaisdell, 1964) is written to augment an introductory course and is very readable. **Elementary Electronic Structure** by Walter Harrison (World Scientific, 1999) goes into this material in great detail. **Basic Biophysics for Biology** by E. K. Yeargers (CRC Press, 1992) has a good introduction to the various chemical bonds. Finally, at about the same level of the present discussion is a delightful book, **Molecules** by P. W. Atkins (W. H. Freeman and Company, 1987).

6.8 The periodic table of the elements

The elements are organized in The periodic table of the elements. An example is seen in Fig. 6.7.[8] The columns in this table are called groups and elements of the same group have similar chemical properties. The rows of the periodic table are called periods. The details of how the periodic table was developed will take us too far afield.

There are several good versions of the periodic table and many web sites with information about the elements, and the reader is urged to check these for more information. In the table on the next page, the average atomic mass of all stable isotopes is shown in unified atomic mass units. When this number is in parenthesis it indicates that the element is unstable and the mass of an isotope is given.

[8]The author has used this particular version for a number of years and has lost the source. There are a number of versions available in the public domain.

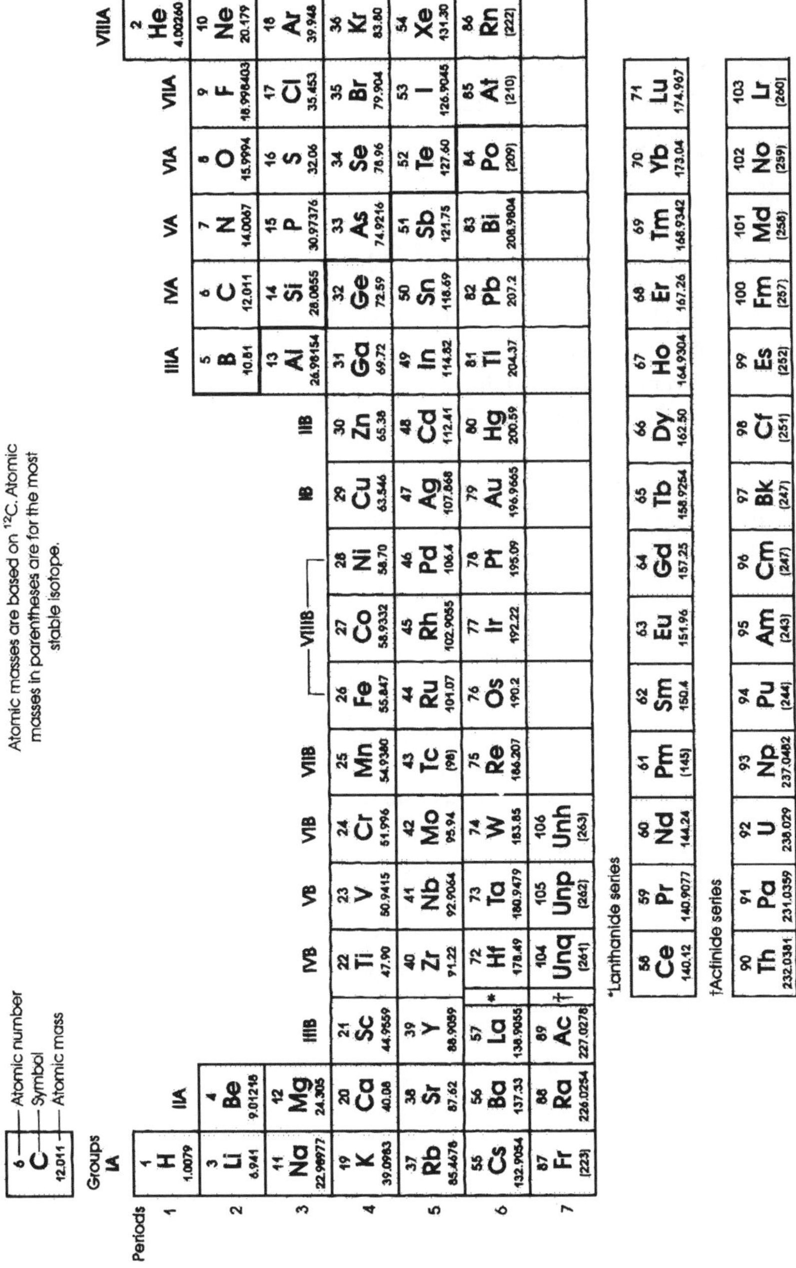

PERIODIC TABLE OF THE ELEMENTS

Atomic masses are based on ^{12}C. Atomic masses in parentheses are for the most stable isotope.

6 — Atomic number
C — Symbol
12.011 — Atomic mass

Periods	IA	IIA	IIIB	IVB	VB	VIB	VIIB	VIIIB	VIIIB	VIIIB	IB	IIB	IIIA	IVA	VA	VIA	VIIA	VIIIA
1	1 H 1.0079																	2 He 4.00260
2	3 Li 6.941	4 Be 9.01218											5 B 10.81	6 C 12.011	7 N 14.0067	8 O 15.9994	9 F 18.998403	10 Ne 20.179
3	11 Na 22.98977	12 Mg 24.305											13 Al 26.98154	14 Si 28.0855	15 P 30.97376	16 S 32.06	17 Cl 35.453	18 Ar 39.948
4	19 K 39.0983	20 Ca 40.08	21 Sc 44.9559	22 Ti 47.90	23 V 50.9415	24 Cr 51.996	25 Mn 54.9380	26 Fe 55.847	27 Co 58.9332	28 Ni 58.70	29 Cu 63.546	30 Zn 65.38	31 Ga 69.72	32 Ge 72.59	33 As 74.9216	34 Se 78.96	35 Br 79.904	36 Kr 83.80
5	37 Rb 85.4678	38 Sr 87.62	39 Y 88.9059	40 Zr 91.22	41 Nb 92.9064	42 Mo 95.94	43 Tc (98)	44 Ru 101.07	45 Rh 102.9055	46 Pd 106.4	47 Ag 107.868	48 Cd 112.41	49 In 114.82	50 Sn 118.69	51 Sb 121.75	52 Te 127.60	53 I 126.9045	54 Xe 131.30
6	55 Cs 132.9054	56 Ba 137.33	57 La * 138.9055	72 Hf 178.49	73 Ta 180.9479	74 W 183.85	75 Re 186.207	76 Os 190.2	77 Ir 192.22	78 Pt 195.09	79 Au 196.9665	80 Hg 200.59	81 Tl 204.37	82 Pb 207.2	83 Bi 208.9804	84 Po (209)	85 At (210)	86 Rn (222)
7	87 Fr (223)	88 Ra 226.0254	89 Ac † 227.0278	104 Unq (261)	105 Unp (262)	106 Unh (263)												

*Lanthanide series

58 Ce 140.12	59 Pr 140.9077	60 Nd 144.24	61 Pm (145)	62 Sm 150.4	63 Eu 151.96	64 Gd 157.25	65 Tb 158.9254	66 Dy 162.50	67 Ho 164.9304	68 Er 167.26	69 Tm 168.9342	70 Yb 173.04	71 Lu 174.967

†Actinide series

90 Th 232.0381	91 Pa 231.0359	92 U 238.029	93 Np 237.0482	94 Pu (244)	95 Am (243)	96 Cm (247)	97 Bk (247)	98 Cf (251)	99 Es (252)	100 Fm (257)	101 Md (258)	102 No (259)	103 Lr (260)

Fig. 6.7 The periodic table of the elements

6.9 Exercises

(1) Consider two types of atoms. One type has Z=19 and N=21, and the other has Z=18 and N=22.

   (a) For both types of atoms, find A.
   (b) For both types of atoms, find the number of electrons.
   (c) Are these atoms isotopes of the same element? Explain.
   (d) Identify the two atoms. (Use the periodic table)

(2) Summarize the properties of protons, neutrons, and electrons.
(3) The force between two atoms is greater than zero. Is this force attractive or repulsive? Explain.
(4) A spring in an automobile is compressed. Is energy stored in the spring? Explain.
(5) A person of mass 50 kg is running at a speed of 5 m/s. Does this person possess any mechanical energy? Explain. If the answer is yes, calculate the energy.
(6) The equilibrium force between two atoms is zero. What do you know about the PE when two atoms are at their equilibrium separation?
(7) Sketch a generic force versus distance curve. Explain the physical reasons for this shape.
(8) Suppose that forces where repulsive for large distances and attractive for small distances, so that a force vs. distance curve is the negative of the curve in the text. Explain the behavior of matter and the universe if this force curve described nature.
(9) The van der Waals force is considered short range compared to the electrostatic force. The magnitude of the electrostatic force is proportional to $1/r^2$ where r is the distance between the two atoms. The magnitude of the van der Waals force is proportional to $1/r^7$. Graph $1/r^2$ and $1/r^7$ versus r for r between 0.8 and 4. Why is the van der Waals force considered short range? Explain.
(10) Ionic bonding occurs in KCl. Discuss how and why this occurs.
(11) The common gases nitrogen and oxygen occur in air as molecules, N_2, and O_2. Is the bonding in these gases predominately ionic or covalent? Explain your reasoning.
(12) A solid has a partially empty valence band and a partially filled conduction band. Is this solid a metal, insulator, or semiconductor? Explain.
(13) The attractive force and repulsive force (in appropriate units) between two atoms as a function of distance in nm is shown in the following

table.

Distance (nm)	Repulsive Force (pN)	Attractive Force (pN)	Total Force (pN)
0.1	100	-80	
0.2	50	-20	
0.3	25	-10	
0.4	12.5	-5	
0.5	6.7	-3	
0.6	3.3	-2.5	
0.7	1.67	-1.63	
0.8	0.83	-1.25	
0.9	0	-1	
1.0	0	-0.8	
1.2	0	-0.55	

(a) Graph the repulsive force, the attractive force and the total force (the sum of these two forces) versus distance.

(b) At what distance does the total force, $F_{total} = 0$? What does this distance correspond to physically? What can you state about the potential energy at this value of r?

(c) Estimate the total force at a distance of 1.5 nm.

(14) Which of the following elements do you expect to be metals? Explain your reasoning. Francium, Fr (Z=87). Tantalum, Ta (Z=73). Cobalt, Co (Z=27). Platinum, Pt (Z=78). Nickel, Ni (Z=28). Sulfur, S (Z=16).

(15) A student states that the chemical bond between two different atoms must always be polar. Do you agree or disagree? Critically discuss your answer. Be detailed.

(16) In terms of interatomic forces, why do ionic solids have a very high melting and boiling temperature? Verify that ionic solids have a very high melting temperature by finding and listing the melting temperatures of several ionic solids.

(17) Consider the table of covalent bonds in the text. Explain why you think stronger binding energies and smaller equilibrium separations are correlated.

6.10 Research questions

(1) Explain covalent bonding more completely than in the text. In this explanation discuss Lewis dot diagrams.
(2) Discuss the atomic nucleus and nuclear forces.
(3) What is a semimetal? Why are elements such as arsenic and antimony called semimetals?
(4) Hydrogen bonding does not just occur between water molecules. Discuss other cases where it is important.
(5) Radioactivity is useful in medicine and archeology. Discuss these uses including what isotopes are used, why they are used, and the techniques of measurement.
(6) Discuss the development of the periodic table. Pay special detail to the work of Mendeleev.

Chapter 7

Molecules and Matter

7.1 Overview of molecules and matter

At this juncture, the reader should be comfortable with a range of basic ideas such as symmetry, light, the electrostatic force, and atoms and molecules that are essential to understanding the laptop computer and life. However, the discussion has not dealt with macroscopic objects that contain large numbers of molecules. We are finally in a position to rectify this omission. This chapter will begin by reminding you that in your experiences with matter you almost always deal with a very large number of molecules. Thus, our discussion must evolve and begin to focus on the behavior of a large number of molecules acting as a group. The present discussion will be limited to (essentially) standard solids, liquids and gases. In later chapters the discussion will focus on the specific types of materials that are important in laptops and life.

This chapter will answer the following questions.

(1) What physical variables determine how molecules interact and the properties of the resulting *phase* of the material?
(2) What kinds of arrangements of atoms are observed?
(3) A set of related questions: What is heat? What is thermal energy? What is temperature? How do these quantities affect the arrangement of molecules?
(4) How do we characterize the properties of the various phases of matter? Are there any general rules that describe this behavior?
(5) What are the characteristics of the common phases of matter?

7.2 Introduction

So far we have discussed individual atoms or a relatively few atoms that combined to form molecules. Everyday we see or experience a very large number of atoms or molecules interacting with a large number of other atoms or molecules. **In order to avoid the continual use of the term "atoms or molecules," the term "molecules" will henceforth be used to mean atoms and/or molecules. If the discussion pertains only to atoms it will be clear from the context of the discussion.**

How large are the numbers we typically experience? One cm^3 of water contains about $3*10^{22}$ molecules, one cm^3 of steel contains about $9*10^{22}$ molecules, and 1 cm^3 of air contains about 10^{19} molecules. Thus, even a small drop of water contains 10^{19} to 10^{20} water molecules.[1] By comparison, the population of the world is approximately $6*10^9$ people, and the population of at a typical state university is roughly $2*10^4$ students. Even a good vacuum (achieved with standard laboratory equipment) contains approximately 10^{11} air molecules per cm^3. Thus, it is only under very special conditions that one can experience and study individual molecules.

We may now ask our first key questions. **When so many molecules are in such close proximity, what determines their arrangement? How does this arrangement depend on the prevailing physical conditions?**

To begin, we ask what physical conditions might be important, and why are they important? At the familiar large scale (macroscopic) level, an answer would be temperature, pressure, and purity or composition (Can you think of others?). At the microscopic, atomic level, one would say potential and kinetic energy. Keeping these two perspectives in mind, we will now start to explore the arrangements of atoms.

How do we model atoms and molecules when discussing their spatial arrangements? We should use the simplest model that incorporates the essential properties and features that we want to represent. For this purpose, a *space-filling model* will be adequate. In a space filling model, an atom may be represented by a sphere, and hence, when drawn on a page, it will appear to be a circle. Molecules can then be formed from groups of different sized spheres, where the radius of the sphere roughly corresponds to the size of the atom it represents. We can even go a step further and represent molecules by rectangles, triangles and other approximations to

[1] This is in part why scientists discuss *moles* of a substance. A mole corresponds to approximately $6.02*10^{23}$ molecules.

their overall shape.

In space filling models, there are two important lengths which characterize the system being modeled. The first characteristic length is the size of the molecule. This is roughly the diameter of the molecule, which we will designate by d. Note that for very unsymmetric molecules, this can be a rather crude approximation since one length may be many times longer than the other. Fortunately, the exact details are not going to be critical in our simple models. The second characteristic length is the average distance between molecules. We will designate this distance by l.

Now, just using our imagination, we can draw pictures that represent many kinds of materials and types of arrangements. It turns out that it is quite difficult to think of an arrangement of molecules that is not observed in nature or has not been developed in the laboratory. Let's construct a number of examples. These will be referred to as "cartoons" not because they are amusing but because they are rather crude approximations to nature.

7.3 Gases

Gases: Gases are characterized by $d \ll l$, a molecular diameter much less than the average molecular spacing. You might recall that the PE of two atoms that are far apart is very small. Also, the gas molecules are moving about. Hence, at room temperature, $KE \gg PE$[2] in a gas.

Atomic Gas: An atomic gas, a gas composed of only one type of atom (*e.g.* argon) is illustrated in Fig. 7.1.
The atoms are represented by circles, and the lines are meant to indicate that the atoms are moving about in all directions. Note that the typical distance between the atoms is much greater than the diameter of an atom.

Molecular Gas (one type of diatomic molecule): A gas that contains one type of diatomic molecule could be modeled as in Fig. 7.2, where the molecules are represented by two small circles near one another. The straight lines indicate that the molecules are moving. The short curved lines indicate that the atoms that make up the molecule are not rigidly connected to each other, but can vibrate about their equilibrium separation and rotate about various axes.

[2]The symbol $\gg$ means"much greater than." It may be taken to mean several times larger in this text. Similarly, $\ll$ means "much less than," which should be taken to mean several times smaller.

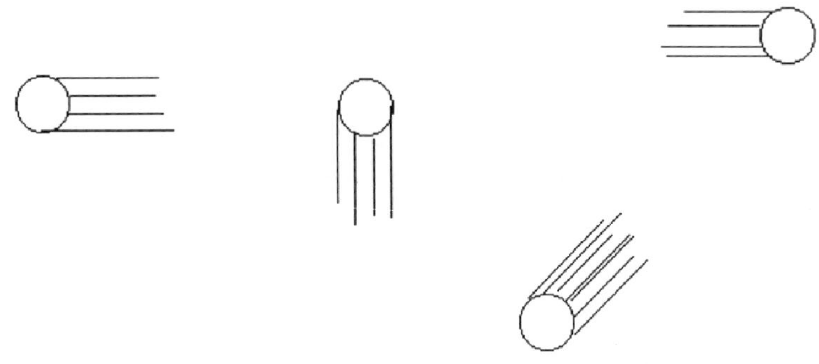

Fig. 7.1 Cartoon of an atomic gas.

Fig. 7.2 A cartoon of a gas containing one type of diatomic molecule.

Molecular Gas (two different types of diatomic molecules): Figure 7.2 is not a good model for a gas that is composed of two types of diatomic molecules. To better model this material, we need a different symbol to represent each of the gases, as shown in Fig. 7.3.
Here the vibration and rotation have not been indicated. To the extent that air is mainly composed of molecular oxygen, O_2 (21%), and molecular nitrogen, N_2 (78%), this is a fair cartoon of air.

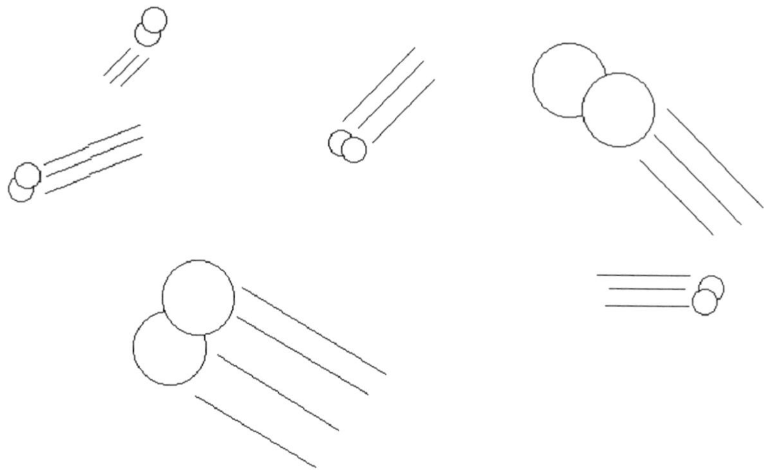

Fig. 7.3 A cartoon of a gaseous mixture of two diatomic molecules. Rotational and vibrational motion are not indicated.

7.4 Liquids

Liquids: Liquids are in many ways the most complex phase of matter. They are characterized by $l \approx d$, a molecular diameter on the order of the average molecular spacing, and a random arrangement of the molecules. The term $l \approx d$ may cause some confusion. Since $l > d$ in a liquid, $l \approx d$ means that while l and d are very similar in size, l is a just a little bit larger than d. Since $l \approx d$ and the atoms of a liquid are quite free to move about, we can correctly conclude that in a liquid $KE \approx PE$.

Simple Liquid: A simple liquid is composed of atoms of a single element. An example is liquid argon. Figure 7.4 illustrates a cartoon of such a material. It is important for you to realize that in a real liquid the atoms are not fixed as shown in this figure, but able to move, vibrate and rotate. Thus this cartoon is more correctly a snapshot of a liquid.

Molecular Liquid: A molecular liquid is composed of one type of molecule. An example is liquid nitrogen. Figure 7.5 is a snapshot of a molecular liquid where the overlapping circles represent diatomic molecules. Note that the cartoon does not show rotation or vibration. In an actual liquid there are no large holes or gaps between the molecules as shown in this cartoon.

Liquid Mixture: A liquid mixture is a liquid containing more than

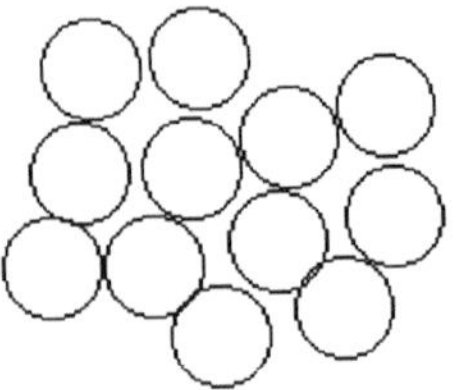

Fig. 7.4 A cartoon of a simple (atomic) liquid. This cartoon is a snapshot. At different times, the arrangement of the atoms will be different.

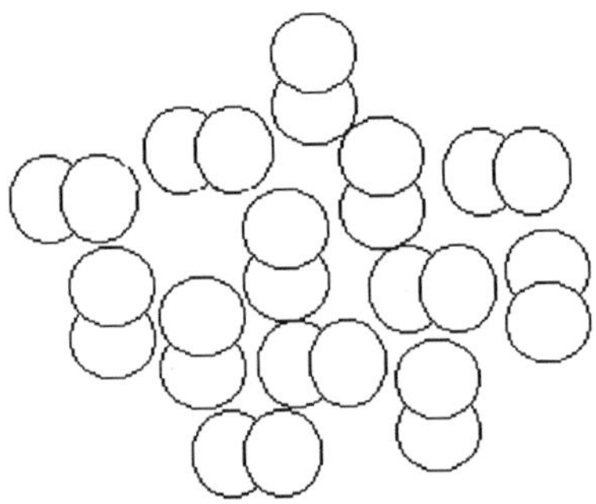

Fig. 7.5 Cartoon snapshot of a molecular liquid.

one type of molecule. An example would be windshield washer fluid that contains both alcohol, water, and a blue dye. A cartoon of such a liquid is shown in Fig. 7.6. Once more, there are no large voids in actual liquids. This figure uses shapes other than just circles to represent molecules. The triangle a fair representation of water, H_2O, and the rectangle with rounded corners a fair representation of a linear molecule such as CO_2.

7.5 Solids

Crystalline Solids: Crystalline solids, what we commonly refer to as solids, are characterized by $l \approx d$, as are liquids, but have the added constraint that there must be a **periodic arrangement** of the molecules. In this arrangement of matter, the atoms are not free to move, and can only

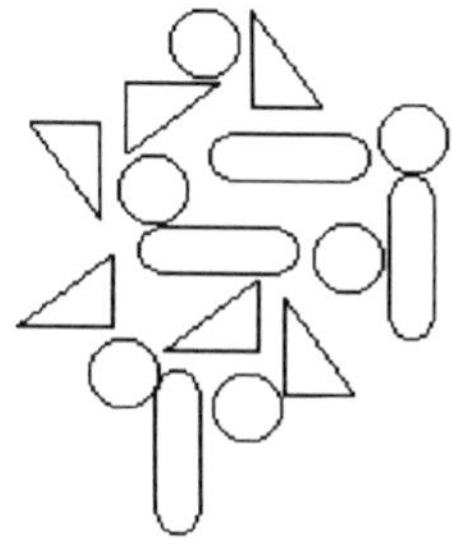

Fig. 7.6 A cartoon snapshot of a molecular liquid mixture.

vibrate about their equilibrium positions, and $PE \gg KE$. It is important to realize that the cartoons of solids are two-dimensional slices of a repeating (periodic or crystalline) three-dimensional pattern. Figure 7.7 is a cartoon of a crystalline solid composed of a single element.

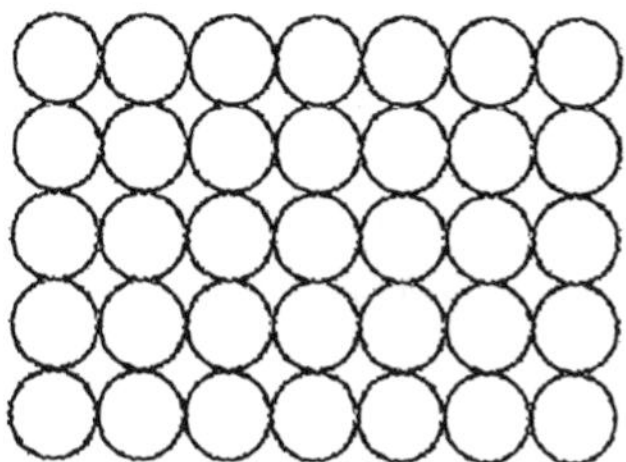

Fig. 7.7 A two-dimensional slice of a crystalline solid composed of atoms of a single element.

Metal Alloy: A metal alloy is a homogeneous solid mixture of two or more molecules.[3] A cartoon of an metal alloy composed of two types of atoms is shown in Fig. 7.8.
Notice that the repeating pattern is a large circle with small circles on its circumference in the northeast and southeast directions as shown in Fig. 7.9. Once more, a real solid does not have the voids like our cartoon.

Almost all metals you use are composed of many tiny *crystalites* (small crystals), that *individually* exhibit this periodic behavior. However, the

[3] An alloy is defined in the dictionary as "any of a variety of materials having metallic properties and composed of two or more intimately mixed chemical elements, of which at least one is a metal; *e.g.*, brass is an alloy of copper and zinc. Alloys are produced to obtain some desirable quality such as greater hardness, strength, lightness, or durability."

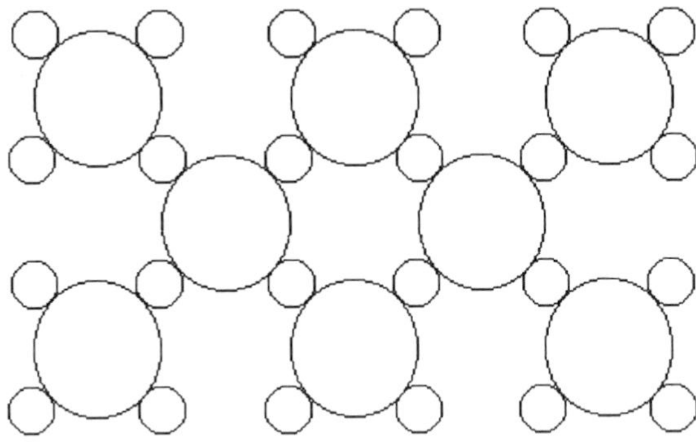

Fig. 7.8 A cartoon of a metal alloy.

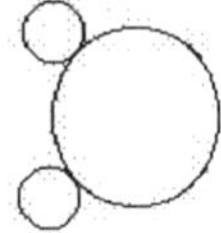

Fig. 7.9 The basic repeating structure in the alloy shown in Figure 7.8.

crystalites are randomly oriented, so the solid as a whole does not exhibit this periodic behavior. Interestingly, some modern turbines use components that, instead of being composed of crystalites, are grown from a single crystal.

Amorphous Solid: An amorphous solid has structure very similar to that of the liquid, so $l \approx d$. However, the molecules **cannot** move about *as easily* as in a liquid. Thus, for all intents and purposes, it appears to be a solid. The correct distinction between an amorphous solid and a liquid ultimately hinges on the careful definition of a liquid and a solid.

You are already familiar with the most common example of an amorphous solid – glass. Glass is made of many components, including SiO_2 and Na_2O, that contribute to its amorphous structure. The cartoon of glass in Fig. 7.10 illustrates how the amorphous structure arises. In this figure, the lines indicate local covalent bonds between Si^{4+} and O^{2-} atoms . Note that the sodium ions (Na^+) locally break the Si^{4+} - O^{2-} bonds.

At this point, we have introduced a variety of different materials demon-

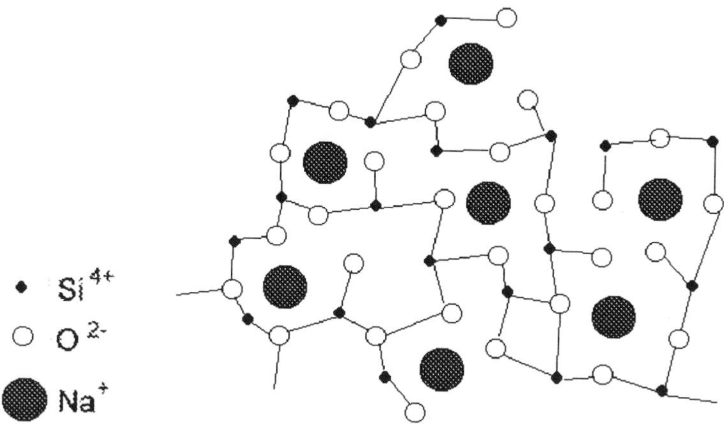

Fig. 7.10 Cartoon of a normal silicon based glass. An example of an amorphous solid.

strating many phases of matter. High technology devices use these "simple" materials in many components, especially the metals. However, many of the materials used in a laptop computer are not simple solids, liquids, or gases. The case is made of *polymers*, and the display contains *liquid crystals*, and *semiconductors* are the electronic heart and soul of the machine. We will discuss each of these materials in separate chapters.

What ultimately limits the structures of matter observed in nature? While this is a complex topic, at this level we may state that structure of matter is determined by the interplay between intermolecular forces and the prevailing physical conditions. In order to understand this interplay, we need to examine the phases of matter in more detail.

7.6 The common phases of matter

Everyday experience tells us that the highly detailed picture of the molecular and atomic gases is sometimes too complex. Our experience leads us to think that there are only three *phases*[4] or states of matter: solid, liquid, and gas. These "common" phases are by no means the only possibilities. By looking more carefully, we see that other phases of matter such as liquid crystals, plasmas, polymers and surface phases are all possible. Furthermore, these "other phases" are becoming technologically more important. However, in this chapter we will focus our attention on the three common

[4] A substance that is physically and chemically uniform is considered a single phase.

phases of matter.

We will begin by summarizing the macroscopic and microscopic properties of solids, liquids and gasses.

(Crystalline) Solids

Macroscopic: Rigid, retain their shape unless distorted by a large external force. Solids are difficult to compress (squeeze into a smaller volume). Solids have densities (mass/unit volume) very similar to the liquids from which they solidify. They may have "any" desired shape, and have defined boundaries.

Microscopic: Attractive forces between molecules are very strong. The atoms do not readily move, but vibrate about their equilibrium positions. The molecules are very close together, $l \approx d$, where l is a typical separation between molecules and d is a typical molecular size. The molecules have a regular periodic arrangement, and the potential energy of the solid is much greater than the kinetic energy of the solid. An ideal solid consists of an array of molecules at the absolute zero of temperature, and corresponds to a system with no kinetic energy and all potential energy.

Liquids

Macroscopic: Flows readily (pours). Conforms to the bottom of the container that holds the liquid. Liquids have well defined surfaces, and are difficult to compress. The density of a liquid is similar to that of its solid form, and much greater than that of its gaseous form.

Microscopic: The molecules of a liquid are very close together. The molecules in a liquid can move about, although they remain in contact, so that $l \approx d$. The molecules have a random arrangement. There are strong attractive forces between molecules in a liquid. The average potential energy and average kinetic energy of a molecule are comparable in a liquid.

Gases

Macroscopic: Gases flow very readily. Gases completely fill the container that holds them, **do not** have well defined surfaces, and are very easy to compress. The density of a gas is much less than that of a solid or liquid, except under unusual conditions.

Microscopic: The atoms of a gas move freely; the only strong interactions with a neighbor occur during collisions. On the average, the molecules are very widely separated, $d \ll l$. The attractive forces between molecules are very weak. The average kinetic energy of a molecule is much greater than the average potential energy.

7.7 A digression on temperature and heat

We must introduce temperature, heat, and internal energy in order to better understand the phase a material adopts under a set of conditions including temperature and pressure, and to start to understand the relationships between phases and the minimum energy principle we discussed earlier.

Temperature is related to a fundamental physical property of an object - how warm or cold the object is. Temperature is independent of the amount or type of material present. A drop of water at 75°F has the same temperature and "feels" the same to a bacterium as a swimming pool at 75°F. When a physical quantity is independent of the amount of material present, it is said to be an *intensive* quantity. Other common examples of intensive quantities include pressure and density. Macroscopically, temperature is one of the dominant factors that determine the phase of a material. Microscopically, temperature is a measure of the average kinetic energy of the molecules.

For non-quantum liquids and gases (liquids and gases at all but very low temperatures) we find:

$$\langle KE_1 \rangle = \left\langle \frac{1}{2} m v^2 \right\rangle = \frac{3}{2} kT.$$

This equation has the following meaning. KE once more means kinetic energy, while the subscript 1 indicates that this expression is for 1 molecule. The angular brackets, $\langle\rangle$,[5] indicate an average value. The kinetic energy was earlier defined as $(1/2)mv^2$, so the second expression reminds you of this equality. The third expression is the key. It states that the average kinetic energy of one molecule is a constant (3/2) times k, the Boltzmann constant, times the absolute temperature, T;[6] ($k \approx 1.38 * 10^{-23}$J/K, and has dimensions of energy measured in Joules, J, per Kelvin, K.) Thus, we

[5] In any system, all molecules do not have the same speed. Rather, they are described by a distribution of speeds. If all molecules possessed $\langle KE_1 \rangle$, the average kinetic energy of the system would be the same as the average kinetic energy of the system with a distribution of speeds. So, using the average kinetic energy of one molecule is a justifiable convenience.

[6] Absolute temperature will be explained in the next section.

may state that $\langle KE \rangle \propto T$, the average kinetic energy is proportional to the absolute temperature, T.

Kinetic energy is an example of a quantity that depends on the number of molecules in the system, or, more generally, on the amount of material present. Quantities with this dependence are said to be *extensive.* Other common examples of extensive properties include volume, V, number of particles, N and, as we shall soon see, total energy.

Because energy is an extensive quantity, we expect that the (average) energy of N molecules to be N times the (average) energy of one molecule. Thus, for N molecules in the liquid or gaseous phase,

$$\langle KE_N \rangle = \frac{3}{2} NkT,$$

which may also be written as

$$\langle KE_N \rangle = \frac{3}{2} nRT,$$

where n is the number of moles of substance, and R *is the universal gas constant*, 8.314 J/mol-K.

It is important to keep in mind that in these expressions for kinetic energy, T is the absolute temperature. This is the temperature scale that is most useful in scientific work. There are a number of common temperature scales. The most common, ones are discussed in the following section.

7.7.1 *Temperature scales*

Fahrenheit: The common English temperature scale. The freezing point temperature of water is 32°F. The boiling point temperature of water is 212°F.

Centigrade: The common metric temperature scale. Also called the Celsius temperature scale. The freezing point temperature of water is 0°C. The boiling point temperature of water is 100°C.

Kelvin or Absolute: The temperature scale used in the study of materials. The freezing point temperature of water is 273.15 K. The boiling point temperature of water is 373.15 K. Note that there is no degree symbol (°) associated with absolute temperatures.

We will use the Kelvin or Absolute temperatures in most of our analysis. The reason is that when temperature is measured on this scale the average kinetic energy of the constituent molecules is proportional to the temperature. There are straightforward conversions between these different temperature scales. These are shown below:

Conversions between different temperature scales

$$^{\circ}F = \left(\tfrac{9}{5}\right) {}^{\circ}C + 32$$

$$^{\circ}C = \tfrac{5}{9}\left({}^{\circ}F - 32\right)$$

$$K = {}^{\circ}C - 273.15$$

The following table lists representative temperatures in these three temperature scales.

Description	°**F**	°**C**	K
Absolute zero	-459.67	-273.15	0
Boiling temp. of N_2 (STP)	-320.4	-195.8	77.35
Lowest measured terrestrial temp.	-125	-87	184.15
H_2O freezing temp. (STP)	32	0	273.15
"Normal" body temperature	98.6	37	310.15
Highest measured terrestrial temp.	136.4	58	331.15
"Red" hot (approx.)	800	427	700
Aluminum melting temp.	1220	660	933
Surface of the sun (approx.)	10,000	5550	5800

Two entries in this table require further explanation. The first entry, "Absolute zero," is the temperature at which all thermal motion ceases. This temperature cannot be reached experimentally. However, scientists have gotten to within a few millionths of a Kelvin of this temperature. The second entry that needs explanation is "STP." In this context, STP means "standard temperature and pressure," and corresponds to a pressure of one atmosphere and a temperature of 0°C.

7.7.2 *Internal energy and heat*

We will now discuss internal energy and heat. The internal energy of a system of molecules is the sum of the potential energy and kinetic energy of all of the molecules, *excluding* both the kinetic energy of the system as a whole and the potential energy due to external sources. As an example, consider a gas filled ball that is thrown into the air. Its total energy includes the energy the ball had at rest (which is the internal energy of the gas in the ball) plus its kinetic energy (KE) due to its motion plus its potential energy (PE) due to the earth's gravitational field. In order to focus on the energy of the system of interest, the gas in the ball, we exclude from the internal energy the kinetic and potential energy added to the ball by throwing it. The internal energy of the gas depends on the amount of material present, so it is an extensive quantity.

We may now make the following definitions. Temperature is a measure of the **average kinetic energy** of the system of particles. Internal energy is the **total energy** of the system of particles, excluding any energy of the container that houses the particles. Lastly, heat is the **energy transferred** between two systems due to a difference in temperature between the systems.

This set of definitions helps us to properly understand the following situation. Suppose a scientist places a 100 gram piece of aluminum at 50°C on a 1000 gram block of ice at 0°C. There is much more ice than aluminum, so it has a much larger internal energy than the aluminum, even given the aluminum's higher temperature. However, since heat flows from hot to cold, heat will flow from the warm aluminum block to the cold ice. This may seem a bit obvious; however, in science one must be very careful about definitions in order to properly describe and understand physical phenomena. During the summer we might say, "The heat and humidity will wear you out" when we really mean "The temperature and humidity will wear you out."

The internal energy is an example of a more general class of energies called *free energies* that are used to describe materials. In systems with thermal energy, all of the energy cannot be used to do work. The free energy is defined as the energy available to perform work for a given a set of constraining parameters such as constant temperature and pressure or constant temperature and volume. **Free energy is important because the equilibrium state of a material is the state of minimum free energy.**

7.8 An example - water at atmospheric pressure

To illustrate these ideas consider water, H_2O, at atmospheric pressure. At low temperatures, below 273.15 K (0°C, 32°F), water is a solid in the form of ice. Detailed structural studies of ice indicate that it is made of many small solid crystals with a periodic structure such as discussed above. At 273.15 K (0°C, 32°F), the ice and water coexist (given sufficient time and a temperature slightly above 273.15 K, ice melts and becomes liquid water). What causes this abrupt phase change from solid to liquid at a specific temperature? The details are complex; however, basically, at this temperature the conditions are such that the thermal energy of the molecules can overcome the intermolecular potential energy due in part to the hydrogen bonding that causes the ice to remain in the crystalline phase.

A further increase in the temperature of the water does not lead to a change in phase until the sample reaches a temperature of 373.15 K, (100°C, 212°F) and the water boils and becomes a gas. Studies of gases including water vapor and steam (gaseous water) indicate that the macroscopic and microscopic models discussed earlier work well for these substances. Why does this phase transition occur so abruptly? Once more the details are complex, but we may state that at this temperature, the thermal energy of the molecules is so great compared to their potential energy that the molecules literally break free from the intermolecular forces that contained them in the liquid state and escape into the atmosphere. However this is not an all or nothing situation; there are always water molecules leaving and arriving at the surface. A water molecule may escape and return to the surface many times before becoming part of the bulk liquid or the gaseous state.

Nature does not always present systems at atmospheric pressure. For example, there is presently great interest in water on the Moon (no atmosphere) and Mars (a very thin atmosphere). For this reason, and countless others, we also care about substances at pressures other than atmospheric. From studies of materials over a wide range of temperatures, volumes, and pressures, scientists can construct three-dimensional plots of the data to form *PVT surfaces.* These surfaces represent the **equilibrium** (lowest free energy) phase (or state) of the substance in terms of pressure, P, volume, V, and temperature, T for a fixed amount of material. These PVT surfaces are much too complex for our present needs. We will focus on the PT projection of the PVT surface, which is called the *phase diagram* of the material. Almost all of the important observations about the equilibrium phase at a given T and P (the two easiest variables to control) are on this diagram. Later, we will consider an even simpler diagram that gives the phase behavior at a pressure of one atmosphere.

Figure 7.11 is a schematic phase or PT diagram for a substance like water. The scale of temperatures and pressures is very non-linear, equal distances do not indicate the same change in pressure or temperature. For example, assuming this is the phase diagram for water, the "triple point"[7] occurs at a temperature of 273.16 K and a pressure of 611.2 Pa. The "critical point" occurs at a temperature of 647.14 K and a pressure of $22.064 * 10^6$ Pa.

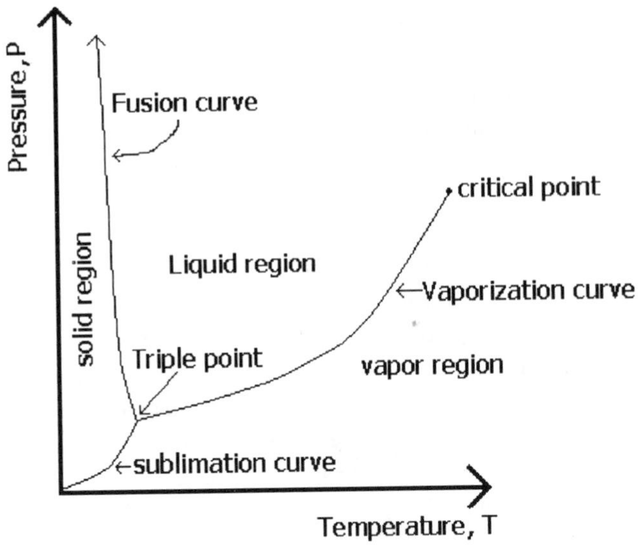

Fig. 7.11 Schematic phase diagram of water.

This figure may look rather foreboding. It contains many terms that have not yet been defined. We will begin by defining them.

Triple Point: Though a point on a PT diagram, this is not a point on the PVT surface, but rather a line. It corresponds to the temperature and pressure at which the solid, liquid and gas all coexist at equilibrium. Keep in mind that these three phases have different volumes, making the triple point a very special occurrence. Except under unusual conditions, the triple point temperature is the **highest** temperature at which the solid

[7]This term and critical point will be defined shortly.

can exist as an equilibrium phase and the **lowest** temperature at which the liquid can exist as an equilibrium phase.

Critical Point: This is a point on the PVT surface. The coordinates of this point are T_c, the critical temperature, P_c, the critical pressure, and V_c, the critical volume. Physically, the critical point corresponds to the highest temperature at which a liquid and gas an coexist. It is the **highest** temperature at which a liquid can exist as an equilibrium phase.

Vapor: A vapor is the gas phase at temperatures **below** the critical temperature.

Fusion Curve: The fusion curve is is the curve (line) on the phase diagram that separates the liquid region from the solid region. Along this curve the solid and the liquid are in equilibrium. When one crosses this line the phase of the material changes from solid to liquid or vice versa depending on the path taken. This could also be called the freezing or melting curve.

Vaporization Curve: This curve or line separates the liquid region from the vapor (gas) region. Along this curve the vapor and liquid coexist in equilibrium. The phase changes from liquid to vapor or vice versa when crossing this line.

Sublimation Curve: The curve along which solid and vapor coexist. As one crosses this curve, the equilibrium state of the material changes from solid to vapor **without** going through the liquid phase. A common example of sublimation is dry ice at room temperature. It transforms from a solid to a vapor without becoming liquid CO_2.

The following generalizations are useful in understanding what phase of matter to expect, given a material's critical point and triple point temperatures. At high temperatures ($T > T_c$) the gas will be the equilibrium phase. At low pressures ($P < P_{triple\ point}$) the gas will also be the equilibrium phase. At low temperatures ($T < T_{triplepoint}$) and intermediate pressures, the solid will be the equilibrium phase. The liquid phase can exist at temperatures between the critical temperature and the triple point temperature.

Often, the details of the phase diagram are not too important. In such cases one may consider the projection of this phase diagram at a pressure of one atmosphere. This is shown in Fig. 7.12.

The one-dimensional simplified phase diagram at the bottom indicates the temperatures at which phase transitions occur and the equilibrium phase at intermediate temperatures. Except in systems with very strong pressure dependencies, or in unusual cases, this is the information that is

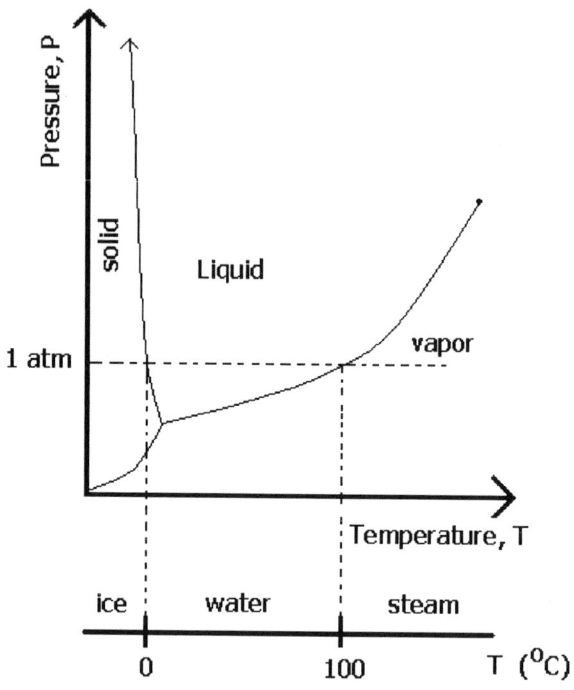

Fig. 7.12 The phase diagram and simplified phase diagram of water.

of most interest since most of our lives are spent at atmospheric pressure.

7.9 References

Most introductory chemistry and physics texts cover the material in this chapter. Various specialized books can be very useful when searching for more information. Most physical chemistry textbooks cover this material in greater detail. **Physical Chemistry**, Gordon M. Barrow, (6th edition, McGraw-Hill, 1996) has been through many editions and is becoming a classic. **A Textbook of Physical Chemistry** by Arthur Adamson (Academic Press, 1986) is one of the author's favorite physical chemistry texts as is **Physical Chemistry** by Berry, Rice, and Ross (John Wiley and Sons, 1980). Introductory material science texts also cover these topics. **Introduction to Materials Science for Engineers**, J. F. Shackelford (Macmillian, 1988) is very readable. While the focus is a bit different,

Engineering Materials: Properties and Selection, K. G. Budinski (Prentice-Hall, 1996) contains a wealth of information.

7.10 Exercises

(1) Why should we use the simplest model that incorporates and describes the essential properties and features that we want to represent when discussing molecules and matter?

(2) CsCl exists as a crystalline solid. Represent Cl by a large circle and Cs by a small square and draw a possible "cartoon" of this crystal. This cartoon should contain at least 20 atoms. Check your answer by finding a picture of this structure.

(3) A "gin and tonic" is essentially water, sugar and alcohol. Assume that a given gin and tonic is, by number of molecules, 60% water, 20% alcohol, and 20% sugar. Draw a cartoon of at least 20 molecules representing this gin and tonic.

(4) Sketch a cartoon of a gas mixture that consists of two different types diatomic molecules and one kind of atom.

(5) Using a square, a triangle, and a circle to represent three different types of atoms, sketch a hypothetical solid alloy that contains all three types of atoms.

(6) Sketch the structure of a molecular liquid mixture that contains two types of molecules. Represent the first type of molecule by a triangle, and the second by a rectangle with rounded vertices (corners).

(7) Calculate the average kinetic energy of a molecule of mass 10^{-26} kg at a temperature of 300 K. What is the average kinetic energy of these molecules at 600 K? What is the kinetic energy of 10^{20} of these molecules at both temperatures?

(8) Convert the following temperatures from the given temperature to temperature in the other two units discussed in the notes.

    (a) 38°C
    (b) 100°F
    (c) -25°F
    (d) 200 K
    (e) 435 K
    (f) -25°C

(9) Why are the critical point and triple points of materials considered so

important?

(10) The *ideal gas law* can be written as $PV = nRT$, where P is the pressure, V is the volume, n is the number of moles of gas, R the gas constant and T the absolute temperature. Show that this equation can be rewritten as $PV = \frac{2}{3}\langle KE\rangle$. Explain what this equation means in words. Why do you think the pressure only depends on the volume and the average kinetic energy $\langle KE\rangle$?

(11) A particular material is studied and the following data obtained.

Temperature (°C)	**Pressure (atmospheres)**	**Observed Phase(s)**
25	0.001	solid and gas
34	0.005	solid, liquid and gas
35	1	solid and liquid
36	100	solid and liquid
300	1	liquid and gas
360	45	liquid and vapor
361	45	vapor
400	50	vapor

(a) Where is the critical point?
(b) Where is the triple point?
(c) Is any point taken along the melting line? If yes which one(s)?
(d) Draw the one-dimensional phase diagram at 1 atmosphere.

(12) Based on Fig. 7.11, does the freezing temperature of water increase or decrease with increasing pressure? Why is this important to ice skating?

(13) Consider the phase diagram on the next page.
Describe the phases at the points labeled a through d.

7.11 Research questions

(1) What is the difference between a super cooled liquid and an amorphous solid or glass.

(2) Discuss the use of single crystal materials in modern technology. (A place to start is searching for single crystal casting on the web)

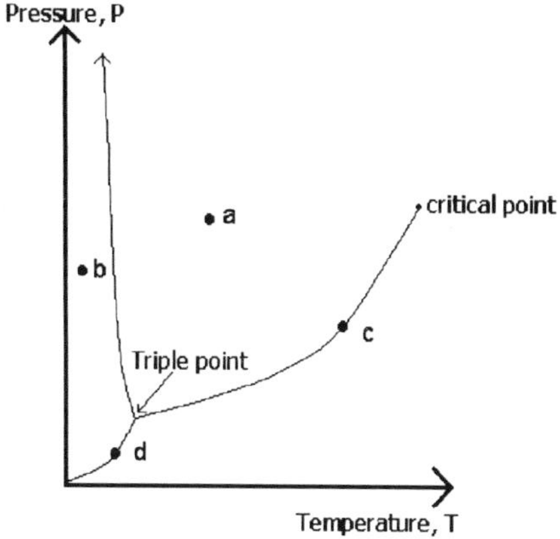

Fig. 7.13 Exercise 13.

(3) Stainless steel is a very common alloy. What elements are found in various types of stainless steel? Why is stainless steel "stainless?"
(4) Problem(10) states that there is an ideal gas law. What is an ideal gas? What are its experimental and theoretical underpinnings? Is the air in a room an ideal gas? Explain.

Chapter 8

Historical Interlude - A Brief History of Polymers, Liquid Crystals, and Semiconductors

8.1 Introduction

In the next three chapters of this book we will study the properties of three unique classes of materials that are essential to the portable laptop computer: polymers, liquid crystals, and semiconductors. These materials have properties that in many ways are similar to, yet in other ways are significantly different from, the more familiar solids, liquids, and gases that have already been discussed. They possess unique properties that allowed new processing techniques and enabled components unheard of thirty years ago to be developed and ultimately incorporated into a device that was not conceived of at the time. Polymers are important to computers in a number of "behind the scene" ways. Commonly known as plastics, they are used in connectors, circuit boards, displays and memory media, as well as "on stage" in the case and keyboard. Liquid crystals are used in many displays, especially those of the laptop computers that we will be studying. Transistors are the electrical heart of the computer, and transistors are manufactured from semiconductors. Individually, each of these materials is so important to businesses and technology that there are often whole academic departments and industrial laboratories dedicated to their study and development. Their combination in a single device that is altering our world was unexpected by the pioneers of these three fields.

This chapter will briefly survey the history and current economic importance of polymers, liquid crystals, and semiconductors. One purpose of this historical interlude is to add an essential human element to the science and technology. It is also important for you to see that the scientific and technological progress that came together in the laptop computer was not always the result of design, but, rather often, of accident. Furthermore, the growth of large markets for some of these materials arose from totally unexpected areas. The synergistic relationship between scientific progress and the growth in the use of these materials has lead to large, important financial enterprises. These enterprises, the changes they have brought to society, and their emergence in other scientific areas have lead to a great improvement in our standard of living.

8.2 Polymers and plastics

The distinguishing feature of a polymer molecule is that it is a very long organic molecule. A typical polymer is polyethylene, the material used in dry-cleaning bags. In this application, the molecules are typically 50,000 atoms long, and have a length of 2500 nm. Later we shall see there is a profound relationship between molecular structure and the properties of these materials.

Almost any classification of polymers is somewhat arbitrary. Still, in order to get a handle on the topic, we will use the following classification that reflects the similar underlying molecular structure and physical properties of each group. This classification scheme separates polymers into three types: *thermoplastics*, *rubbers*, and *thermosets*.

Thermoplastics are often just called "*plastics*." They represent the bulk of the polymers used in everyday applications. These polymers melt upon the application of heat, and they are easily molded into a variety of shapes. Rubbers are materials that display large elastic extensions. They are easily stretched when an external force is applied, yet they rapidly return to the unstretched state when the force is released. These materials do not melt upon heating; instead, they decompose. Thermosets are polymers that often form when heated or liberate heat when they form. They are usually rigid and also decompose rather than melt. Later we will relate these behaviors to the underlying molecular structure.

Plastics may well be the defining materials of the twentieth century. In 1816, a Danish museum curator, Christian Thomsen, recognized that human development was strongly correlated to humankind's control and use of materials. He coined the terms Stone, Iron, Bronze, Copper, and Steel Ages. The Industrial Revolution was the first step into the Steel Age. Presently, it has become common to say that we are on the cusp of the Information Age. Nevertheless, progress has always been related to the development of knowledge and the exchange of information.

The two materials that are ubiquitous in this information revolution are polymers and semiconductors. Polymers are essential to almost all of the common electronic information storage media: video and audio tapes, computer disks and CD-ROMS. Polymers in many cases are used to imitate natural materials. For example, nylon, and rayon are used to replace as replacements for silk and formica for wood. At the same time, they are used to supplement and fill niches that no natural material could; for example, teflon and Bakelite. The global volume of plastic production has been greater than that of steel since 1979. One could place that date as the start of the Polymer Age. However, polymers have been part of our lives and bodies for significantly longer than this.

The economic impact of plastics alone is also quite large. The United States is the largest consumer and producer of plastics in the world. In 1996, the plastics industry accounted for 1.3 million jobs and $274.5 billion in shipments in the United States alone. 1.3 million jobs correspond to about one worker per 100 in non-agricultural fields being employed directly by the plastics industry. If upstream suppliers are included the number rises to 2.3 million workers. Employment in the plastic industry has increased steadily for the past several decades at an annual rate of 3%. The shipment of plastics worth $274.5 billion in 1996 represents a 55% increase since 1991. This amount has increased by over 4% annually since 1974. When the upstream suppliers are included, the shipments totaled $366.4 billion in 1997, which made plastics the fourth largest industrial group behind motor vehicles, petroleum refining, and electronic components and accessories.

The impact on the state of Ohio is also large. Roughly 113,000 nonagricultural workers are employed in the plastics industry, their annual payroll is approximately $3.1 billion, and the shipments total approximately $20.5 billion. This makes Ohio one of the largest states in terms of the polymer industry.

How did plastics grow to be such a large industry? Plastics owe their existence to the industrial revolution and, of all people, James Watt of steam engine fame. In the 1780s, Watt and his partner, Matthew Boulton, hired William Murdock to act as an agent and manager for some of their interests. In 1791, Murdock obtained a patent for extracting from coal "a composition for preserving ships' bottoms" — coal tar. A year later he reported experiments on using coal in essentially the same apparatus to produce coal gas that would burn in lamps. By 1805, gas lamps had been installed in some cotton mills, where they proved to be safer and less expensive than candles and whale-oil lamps. Gas lamps eventually became very popular. In 1823, there were 300 miles of piping for gas street lamps in England. This number increased to 2000 miles in 1850. The effects of gas lights on society were great. However, what of the original "composition," the coal tar? For many years, with the profitability of coal gas, it was simply disposed of in the nearest river.

In 1819, a Scotsman, Charles Macintosh, in the clothes dyeing business, and ever on the lookout for a bargain, purchased some of this coal tar (which was being thrown out anyway) so he could extract the naphtha in it. He hoped that naphtha could be used to clean his dyeing equipment. It appears that this did not work, and he was stuck with a large quantity of sharp-smelling waste product. His natural thriftiness drove him to find a use for the naphtha. He eventually discovered that naphtha could dissolve raw rubber. To that point, rubber was generally a curiosity that had found its primary use in pencil erasers. Macintosh then painted this mixture between sheets of canvas and formed a waterproof product modestly called the macintosh. He received a patent for this in 1823. About the same time Thomas Hancock, a London coach maker, had a similar idea for making waterproof clothing. They got together in 1825 and began producing the macintosh raincoat. This rubberized cloth still had problems. Rubber would become a sticky, tacky mess on hot summer days and a brittle, hard, unwieldy mass on cold winter days.

The man who solved this temperature problem was Charles Goodyear. After several financial setbacks, some so severe that he referred to debtor's prisons as his "hotels," Charles Goodyear, in 1840, finally succeeded in his quixotic quest to make improved rubber. This discovery was made when he accidentally spilled some raw latex rubber mixed with sulfur onto a hot stove. When Goodyear heated the sulfurized rubber to a high temperature and rather than melting into a gooey ball, it charred. When he exposed it to cold, it remained pliable. In short, he had succeeded in vulcanizing rubber; crosslinking the polymer molecules so that an elastic material resulted. After filing for an American patent, the financially hard-pressed Goodyear approached the flourishing Charles Macintosh and Company about licensing his cured rubber. He sent samples to London where they landed on the desk of Thomas Hancock. Hancock was intrigued by these samples and reversed engineered it to discover how this rubber was cured. He filed a British patent two months (the Goodyear corporation web site, http://www.goodyear.com/corporate/strange.html, states a few weeks) before Goodyear. Hancock offered one half his patent to settle the suit out of court, and Goodyear turned him down and then lost the case. A set of patent infringement cases were taken to the Supreme Court of the United States. In 1852 Goodyear paid then Secretary of State, Daniel Webster, $15,000 (at the time the highest fee paid to an American lawyer) for two days work in representing before this court. While Goodyear obtained a permanent injunction, the piracy did not stop. Interestingly, neither Goodyear or his family was ever connected to the Goodyear Tire and Rubber Company.

1846 was a banner year for polymers; not only was the Goodyear-Hancock case settled, but a new polymer was discovered. The French chemist Théophile Pelouze accepted into his lab a chemist-revolutionary named Louis Ménard. They developed a new substance, collodion, while working on some aspects of smokeless gunpowder. This material consisted of nitrated cellulose dissolved in a solvent. When the solvent evaporated a tough, clear, transparent film resulted. This was later used in medicine as a "band-aid," and in photography in the Archer Wet Plate Collodion Process that dominated photography until the 1880s.

The Great International Exhibition of 1862 saw Alexander Parkes exhibit a new substance, Parkesine. He described it as "hard as horn, but flexible as leather, capable of being cast or stamped, painted, dyed or carved." Parkesine was the first man-made plastic. Parkes saw that the great advantage of this material was that it could be cast or molded so objects did not have to be hand carved or tooled. A major disadvantage of Parkesine was that the raw materials, especially the solvents, were too expensive.

The next big push in the development of polymers came from, of all places, the game of billiards. After the American Civil war, billiards became very popular. In those days, the balls where made of ivory and only select elephant tusks were of sufficient quality to warrant use as billiard balls. Furthermore, since so many elephants were being killed, a shortage of ivory was developing. In 1869, American inventor John Wesley Hyatt received a patent for an improved coating for billiard balls made of collodion. Unfortunately, collodion's kinship to smokeless powder soon became apparent. A lighted cigar could set one of these balls on fire and collisions between the balls could result in an explosion, which in those days could lead to every man in the room pulling a gun. Hyatt never did solve the billiard ball problem. However, a few years later he modified Parkes' old formula by adding camphor and created the first thermoplastic: celluloid. Celluloid was easily shaped by the application of heat and pressure. Furthermore, it retained its shape after the heat and pressure were removed. This was the ideal material for the industrial age because it could be molded by mechanical means using unskilled labor in factories. Unfortunately in 1885, a judge ruled that Hyatt's patent was not valid and opened the celluloid market to everyone.

Celluloid had several applications, including photographic film. The use of celluloid laid the foundation of the Eastman Kodak Company. (As an aside, Eastman claimed that Kodak was a "purely arbitrary combination of letters ... [that] ... displayed a vigorous and distinctive personality.") Interestingly, the first Kodak cameras came preloaded with film. When the one hundred or so pictures were exposed, the camera and exposed film were both sent to Eastman for processing. After the film was developed and printed, the pictures and camera were sent back (presumably reloaded with film) to the customer. A chemical modification to make celluloid more flexible was developed a few years later.

Interestingly, celluloid film was actually invented by an Episcopal priest, in search of a better way to illustrate his lectures. He filed a patent in 1887, two years earlier than Eastman. Somehow, his application languished until 1898, when the patent was granted. Since Eastman was far richer, he tied the patent up in court for as long as possible. The suit was finally settled in 1913.

The next big step forward in polymers came when Leo "Doc" Baekeland developed a new photographic paper. Eastman bought the rights from him for $750,000. Baekeland used some of this money to become a self-employed scientist. The burgeoning electrical industry of the early twentieth century needed shellac, which is derived from the resinous secretions of a beetle, for use as an insulator. Shellac was also used as a wood finish and was costing over a dollar a pound (roughly $30/ lb. in today's dollars). In 1907, Baekeland created a liquid resin called Bakelite. When Bakelite was hardened under appropriate temperature and pressure, it took the shape of its container. It did not burn, melt, dissolve or change when acted on by common chemical and physical agents. Furthermore, once set, Bakelite could not be re-melted like celluloid. Bakelite was the first thermoset plastic and had excellent electrical insulating properties. It was of great use to the military during WWII.

Cellophane, another common polymer, was discovered by Jacques Brandenberger, a Swiss chemist employed by a French textile company. Cellophane is a clear polymer that is chemically the same as rayon; it is cellulose based and can be made into clear sheets. The use of cellophane really took off after Du Pont purchased the rights to it in 1920 from the French firm and spent seven years making it moisture-proof (so water vapor could not escape through it) instead of waterproof. The inability of cellophane to act as a vapor barrier had prevented cellophane from entering the foodstuff and tobacco businesses. By making a four-part composite that increased cellophane's moisture resistance one hundred times, and still left the material very thin, Du Pont chemist, William H. Church solved this problem. This opened new markets for cellophane including Clarence Birdseye's frozen foods, the 3M company's cellophane tape, and cellophane wrappers on cigars and cigarettes.

The pace of development of polymers continued to quicken. The 1930s were a particularly fruitful time. In 1938 Dr. Wallace Carothers, of Du Pont, developed nylon. The fiber almost immediately replaced animal hair in toothbrushes and was a great success when introduced as a replacement for silk stockings in 1939, and was used in parachutes during WWII. Other polymers were developed at about the same time. For example, Polyvinyl chloride, PVC, or "vinyl" was developed by a B. F. Goodrich chemist, Waldo Semon, while trying to bind rubber to metal. Vinyl is easily molded, inexpensive, durable, and fire-resistant, almost did not make it to the market. There was no obvious demand for this product at the time. Semon had to find one or lose future funding. Fortunately, with help from his wife, he saw this as the perfect waterproof fabric. Not since Charles Macintosh had a new foul-weather material come along, and vinyl easily beat rubber.

In 1933, Ralph Wiley, a Dow Chemical chemist, accidentally discovered polyvinylidene chloride, better know as Saran and sold today as Saran Wrap. This polymer has been keeping food fresh, military equipment dry and clinging to everything in sight ever since.

A few years later in 1938, a Du Pont chemist named Roy Plunkett discovered teflon. The uses of teflon are, of course, widespread today. Its electrical properties make it an excellent choice for wire insulation, and it is one of the slipperiest substances known. Two Imperial Chemical Industries chemists, Reginald Fawcett and Eric Gibson discovered polyethylene, a very common polymer, in 1933. You are familiar with this polymer as it is used to make milk jugs and similar items.

Progress in polymers did not stop with the beginning of WWII. Since then, new polymers and ways to make polymers have continued to be developed. Synthetic rubber has been formulated and new types of fibers that nature could never have made are now in production. The details of this modern era will be left for another day and time.

8.3 Liquid crystals and liquid crystal displays

The name liquid crystal is a bit of a misnomer. Liquid crystalline materials have phases that are liquid, yet have some orientational or positional order, and often both. They are **not** crystals. The small organic molecules that form these phases are usually either disk shaped or rod shaped. Furthermore, liquid crystals assume different states or phases at various temperatures and pressures. Because, under normal circumstances, the pressure is one atmosphere, temperature is the main physical variable that determines a liquid crystal's phase. Thus, these liquid crystals are called *thermotropic liquid crystals*. Another class of liquid crystalline materials is formed from anisotropic objects (typically large molecules of aggregates of molecules) immersed in a fluid. The phases of these materials are largely determined by the composition of the mixture. These are called *lyotropic liquid crystals* and are extremely important to life.

Unlike polymers, liquid crystals are the focus of a much younger science and an even younger commercial enterprise. The fundamental science of liquid crystals is only a little over one hundred years old, and the commercial development of liquid crystal displays only 30 to 40 years old. Thus, the liquid crystal display industry is still in its youth and is nowhere near as large as the polymer industry. Like polymer science, liquid crystals remain an active scientific field. Furthermore, the display industry is strong and growing.

Since the early 1990s, there has been extensive growth in the liquid crystal display market. The world market for liquid crystal displays (LCDs) was approximately \$12.5 billion in 1997. It is expected to grow to \$24.4 billion by 2003, a growth rate of approximately 10% per year. The demand for graphic LCDs has been growing at a rate of approximately 14% per year. The more mature market for segmented and character displays, such as used in cellular phones, calculators, and watches constitutes a 1 billion unit, \$1.3 billion market that continues to grow. The growth in shipments is roughly 4% per year, and the value of these shipments is increasing at roughly 5% per year. Interestingly, this industry, which was developed in the US and Europe, is dominated by western Pacific Rim countries such as Japan, Taiwan, China, and the province of Hong Kong.

The history of liquid crystals, like polymers, has some "pre-discovery" events that set the stage for later work. Several European scientists observed liquid crystals and recorded their observations, but did not realize what they had found. This early work took place between 1850 and 1888, and consisted, in broad terms, of three types of observations.

First, early investigations were made of biological specimens using polarized light. These experiments by Rudolf Virchow, C. Mettenheimer, and G. Valentin all presented unusual phenomena. All three observed effects under the polarizing microscope that were consistent with effects observed in certain solids, yet their samples were liquid. Since *isotropic* liquids do not produce any unusual polarization effects, this was indeed unusual. The inherent complexity of biological materials undoubtedly made understanding these observations even more complex.

The second pre-discovery experiment was performed by the German chemical-physicist Otto Lehmann. Lehmann constructed a heating stage for his microscope and used a polarizing microscope to study the melting and freezing of various materials. He observed that some materials did not crystallize directly from an isotropic liquid to a crystalline solid, but rather had an intermediate phase as well. He knew that it is common for two-phase regions, regions of temperatures and concentrations where two phases are in equilibrium, to exist. A common example of a system with a two-phase region is an insulated glass containing a mixture of ice and water. The addition of impurities to such a two-phase system can lead to "strange" effects on freezing and melting. Based on this knowledge, he thought this phenomenon was nothing unusual.

The third set of pre-discovery observations were made by a group of individuals, P. Planer, W. Lobisch, and B. Raymann who studied compounds synthesized from cholesterol. They all observed striking colors when these compounds cooled. However, they merely noted the phenomenon and moved on.

The discoverer of liquid crystals is commonly said to be the Austrian botanist Friedrich Reinitzer. Reinitzer's primary interest was in cholesterol in plants. At the time, he was studying the melting behavior of a compound related to cholesterol. In 1888, he described this behavior by saying it had **two melting points**. At 145.5 ^oC it melted and formed a cloudy liquid and at 178.5 ^oC this cloudy liquid turned into a clear liquid. He also described some of the same color phenomena reported by the earlier scientists. The breakthrough was not in what he observed, but, as is often the case, his interpretation of what he observed. He interpreted the result in terms of two-phase transitions rather than an impurity effect. This new insight lead to the idea of liquid crystals, but not directly. Reinitzer knew of Lehmann's work and sent him some samples to study. Lehmann described this material and other similar materials by a range of names. He finally concluded that these materials flow like liquids and have optical properties like crystalline solids. Thus, he called them liquid crystals. The name has stuck in spite of research that demonstrates that the optical properties arise from the liquid crystal molecules themselves and not from any crystalline structure.

The post-discovery era was not blessed with large commercial markets, but but did result in an increase in the basic scientific knowledge in the field. Lehmann was the dominant figure in liquid crystal research at the end of the nineteenth century. He performed the first experiments with nematic liquid crystals and discovered what are now called nematic and smectic liquid crystals. He also performed some of the first experiments on surface induced alignment of liquid crystals.

The actual nature of liquid crystals was in doubt during this time. Some scientists felt that liquid crystals represented a mixture of two or more phases, while others, lead by Lehmann, correctly argued that liquid crystals are a single phase. The German chemist, Daniel Vorlander, and his coworkers synthesized many new liquid crystalline materials and studied their properties. The magnitude of this work is best appreciated when one realizes that over 80 doctoral dissertations resulted from work in his group in the years 1901 through 1934.

Interestingly, the reason there were so many German chemists at this time goes back to the coal tar mentioned earlier. All the coal gas that was produced had as a "left over" a large amount of coal tar. By the 1850s, many scientists had worked on finding a use for it, and that lead to the development of aniline dyes.[1] These dyes became the basis for the German chemical industry and such firms as BASF, Hoechst, Bayer, and Agfa. This chemical expertise also led to many spin-offs including aspirin, synthetic and structural chemistry, and, of course, liquid crystals. In many ways, the birth of the pharmaceutical industry and much of today's chemical industry is directly tied to coal tar.

The French have also made many significant contributions to liquid crystal science. Lehmann introduced Georges Friedel to liquid crystals in 1909. In 1922, Freidel published a seminal paper, (*Ann. Phys.* **18**, 273 (1922)), that proposed the current classification of liquid crystalline phases using the words nematic, cholesteric and smectic. He also described liquid crystalline phases in terms of the orientational and positional ordering of

[1]Until about 150 years ago all dyes came from natural sources. For this reason they were expensive and required great skill to make and use. This changed around 1857 when William Perkin accidently discovered the first synthetic dye, mauve (mauveine) or aniline purple.

the molecules, and explained many of the defects that are observed in liquid crystalline phases. As if this were not enough, he also understood the effects of magnetic and electric fields on liquid crystals. There was renewed French interest in liquid crystals beginning in the 1960s that continues to this day. A primary scientist in this growth was Pierre DeGennes, who was awarded the 1991 Nobel Prize in Physics for his work in liquid crystals and other related areas.

The inter-war period from 1922 to 1939 continued to see advances in understanding liquid crystals. Theoretical work was performed by Carl Oseen and Dans (H) Zocher and experimental advances were made by Vsevolod K. Freedericksz and coworkers. However, this was a quiet era when liquid crystal science grew slowly. In fact, liquid crystals all but disappeared in the decade after WWII. There were many reasons for this, but the two primary ones were undoubtedly that scientists felt most of the basic problems had been solved (they hadn't!) and there was no obvious industrial/commercial application for these materials. This started to change in the late 1950s.

Around this time, it was hoped that a better understanding of these materials would lead to some applications. In 1957, Glenn Brown, a chemistry professor at Kent State University, published a review article that led to an increase in interest in liquid crystals. A year later, a Faraday Society conference on liquid crystals was held in London. At the same time, F. C. Frank in England published an influential paper on the continuum elastic properties of liquid crystals. Also in 1958, two Germans, Wilhelm Maier and Alfred Saupe, formulated a successful microscopic theory describing the phase transition from the isotropic phase to the nematic liquid crystalline phase. The basic theory and models needed to understand liquid crystals were finally in place. This achievement, along with a 1962 book by George Gray, (**Molecular Structure and the Properties of Liquid Crystals**), set the stage for spectacular growth that would follow.

Interest in liquid crystals has continued to grow through the last three and one half decades. Scientific understanding has reached the stage where there are tens of books on liquid crystals, and new ones are being published all the time, including this one, of course. This spectacular growth has been the result of the combination of work by scientists from many disciplines including chemistry, physics and engineering. The importance of chemistry and the synthesis of new compounds have been critical to the growth of this field. This is truly an interdisciplinary effort.

By the late 1960s, it seemed that the field might be ripe for industrial applications, and indeed, it was. In 1968, two RCA scientists developed the first liquid crystal display (LCD). Its contrast was low and it consumed too much power. Nevertheless, it was the first product of the LCD industry. In 1971, James Fergason in the United States and Martin Schadt and Wolfgang Helfrich in Switzerland invented the twisted nematic display. The twisted nematic display has been the workhorse display of the LCD industry ever since.

Of course, as soon as money starts to become an important issue, lawyers are soon to follow. In this case, the invention of the twisted nematic display was the issue. The question of who would have the rights to license the device, and possibly collect significant royalties became a significant issue. The dynamics of the situation were rather murky and even today the details are not discussed publicly.

The growth of liquid crystal applications has continued since 1968 and expanded to encompass new areas such as fiber optic communications. In recent years, several new types of displays have been commercialized. These include new versions of the twisted nematic device, super twisted nematic devices, polymer dispersed liquid crystal devices and cholesteric devices to name a few. Many of these were pioneered or invented at the Liquid Crystal Institute at Kent State University in Kent, Ohio.

8.4 Semiconductors

Semiconductors are materials that have electrical properties that are midway between those of good conductors such as metals and good insulators like teflon, bakelite, and glass. Semiconductors can be composed of either single elements or compounds. The most commonly used semiconductor is the element silicon. However, compounds such as gallium arsenide are also being used in commercial devices. Semiconductors are useful because their electrical properties depend very strongly upon the type and (generally very low) concentration of impurities added to a very pure material. By using this "doping effect," scientists and engineers have developed many different types of semiconductor devices.

Presently, almost every electronic device uses semiconductors. Of course, there are several applications where semiconductors are not used. A common example is the *cathode ray tube, or CRT*. This is an evacuated glass vessel with a phosphor coating on the front, a source of electrons, an "electron gun," in the back, and deflection coils or plates in between. It is used in televisions and in most desktop computers monitors. Many high power, high frequency applications, such as radio and television stations, as well as radar installations also use tubes. Lastly, some audiophiles hear a difference between tube and semiconductor amplifiers and assert that tubes make better amplifiers.[2]

The semiconductor and electronics industries are gigantic. Here we will focus on the former. In 1998, the information technology sector of the economy accounted for 11% of the US gross domestic product and 25% of the manufacturing output. US semiconductor manufacturers supply approximately 40% of the world's output. Furthermore, US manufacturers had approximately a 42% market share from 1991 through 1996. In 1997 and 1998 it rose to nearly 52%. The value of shipments in constant 1992 dollars rose from $23.2 billion in 1989 to an estimated $89.4 billion in 1997. This corresponded to an average increase of approximately 18% per year. The number of jobs in the semiconductor industry increased from roughly 281,000 in 1989 to 292,000 in 1995. This is an average of increase of approximately 0.6% per year. Since the value of the shipments has risen significantly faster than employment, there have been significant increases in productivity in this industry. At this time, the five largest corporations, in terms of sales revenue, are IBM, Motorola, Intel, Texas Instruments, and Siemens.

How did such a large vibrant industry start? The history of semiconductors is conveniently broken into four parts: the years before WWII, the war years, the post-war boom, and the post integrated circuit era. The story of semiconductors goes back to England and Michael Faraday. In 1833, Faraday observed several substances whose ability to conduct electricity increased with temperature (unlike metals that decreased) and was much smaller than metals. These observations remained isolated, unexplained facts for many years. Other experiments showed that these same materials could be used to change the character of electrical signals and used as light detectors. Their unbound charge type (positive or negative) was determined as well. Thus, by 1880, the basic properties of materials, that would only later be called semiconductors, had been established.

The first few decades of the twentieth century saw three significant technological advances. The first, though hardly technical, was the coining of the name semiconductor. The second was the discovery that silicon was a good detector of radio waves, which meant that there was some practical

[2]This is somewhat complicated since there is subjective human perception, artistic preference and engineering data. The interested reader should consult the references at the end of this chapter.

reason to study and understand this material. Finally, the role of chemical composition in controlling the properties of semiconductors was established, although not necessarily appreciated.

Silicon and a wire "cats whisker" was the only good detector in the early days of radio. This detector consisted of a rough piece of silicon and a wire sharpened to a fine point. In order to tune the radio, the contact between this fine point and the silicon was varied until the best signal was found. In the mid 1920s, semiconductor rectifiers (electronic components that allow electricity to pass in only one direction) were invented and their manufacture became commonplace. However, the manufacturing of these devices was largely an empirical art. This soon started to change. By the 1930s, theoreticians started to try to understand these materials. Alan H. Wilson and (Sir)Nevill Mott in England, (Sir)Rudolf Peierls in Germany and later England, Walter Schottky in Germany, Jacov Frenkel in the Soviet Union and others worked on these problems. In light of their later leadership role, American scientists were conspicuous by their absence.

Common wisdom suggests that WWII considerably enhanced our knowledge of semiconductors and greatly accelerated their technological applications. However, several historians believe that this point is debatable. What is clear is that semiconductors were very important as radar detectors, and radar was very important to the war effort. Therefore, semiconductors received a great deal of attention. In 1940, the semiconductor rectifier problem became important and a new research group to study the elemental semiconductor silicon was established as part of the war effort at the University of Pennsylvania. Later, a group at Purdue University that studied another elemental semiconductor, germanium, was established. Not by accident, Bell Laboratories was also included in this work on radar detectors. Several Bell Laboratory scientists were critical to the effort. Some of these same scientists continued to provide a large impetus towards the development of the transistor after WWII.

After the war, Bell Labs decided to make a major push in semiconductor research with the long-term goal of replacing their tube amplifiers with smaller semiconductor amplifiers. At that time, silicon and germanium were the only known semiconductors. This was a basic scientific program that was to build on the pre-war work at Bell Labs and the knowledge gained during the war. In late 1947 John Bardeen, Walter Brattain, and William Shockley invented the transistor, a semiconductor device that could act as a switch or as an amplifier. The transistor was announced to the world in June of 1948. Bell labs made the information about their work available to a wide audience. This effort was unclassified by the military and was not regarded as a great proprietary industrial secret. In spite of this, the military funded Bell Labs transistor research between 1949 and 1958. The total amount was $8.5 million, and represented approximately 25% of the total Bell Labs expenditures on device and material development during that time.

It soon became apparent that in order to fabricate good reliable transistors, the basic properties of silicon and germanium had to be understood. This led to the development of techniques for producing extremely pure silicon and germanium; a heroic effort produced truly amazing results. By 1957, silicon could be purified to the point where the impurity atoms represented two atoms in 10^{10} atoms of silicon. This is equivalent to adding one crystal of table salt, 1 mm. on a side, to 50,000 pounds of sugar. Furthermore, certain contaminants, especially boron and phosphorus, were very difficult to remove. The detective work needed to discover that phosphorus was a contaminant and how to remove it was tremendous. The entire semiconductor industry benefited from this improvement in material purity. More pure material lead to better devices, which suggested the need for purer material, which lead to even purer materials in an ever repeating dance.

The first compound semiconductor that was seriously considered for use as a transistor was indium antimonide, developed by Heinrich Welker in 1951. Further progress in silicon processing prevented compound semiconductors from dominating the market. However, there are certain niche markets including light-emitting diodes and very-high-speed integrated circuits where compound semiconductors have established a sizable presence.

In 1955, Shockley left Bell Labs and went west to what is now Silicon Valley to start his own semiconductor manufacturing company (Shockley Semiconductor Company). Although brilliant, Shockley was a difficult individual with which to work. Eventually seven of his team members left the organization and founded Fairchild Semiconductor. This company is the "parent of all semiconductor companies." Twenty-three semiconductor and related corporations trace their origins back to Fairchild, including Intel, Advanced Micro Devices, and National Semiconductor.

The most important breakthrough of this period occurred almost simultaneously in two places in 1958. This was the development of the integrated circuit. Robert Noyce of Fairchild and Jack Kilby of Texas Instruments developed slightly different versions of the "chip" It turned out that Fairchild's method was simpler to manufacture, however, both Fairchild and Texas Instruments grew into a leading semiconductor companies. In 1968 Noyce left Fairchild, and with a colleague, formed Intel. They set out to manufacture a computer memory chip and came to dominate the industry by developing the first microprocessor.

Since the early 1970s, the semiconductor industry has grown and flourished, but not without some pain. In the 1980s foreign competition eroded the US' worldwide market share. In fact, the US semiconductor industry almost died. However, in the early 1990s the industry started to capitalize on its strengths in design and innovation and tried to minimize its exposure in the commodity market (common RAM chips and other generic devices). The US government also passed laws to help the industry during this time. However, the semiconductor industry remains tied to three very economically sensitive industries: the automobile, computer, and the consumer electronics industries. Thus, in spite of its size the semiconductor industry remains volatile.

8.5 References

Plastic: The Making of a Synthetic Century, Stephen Fenichell, Harper Collins (1997) is a very readable history of polymers and plastics.

Connections, James Burke, Little, Brown and Company (Boston, 1978). This book describes how various inventions and discoveries are linked.

Liquid Crystals, Peter Collings, Princeton University Press (Princeton, 1990). This is an understandable introduction to liquid crystals.

The Physics of Liquid Crystals, P.G. De Gennes, and J. Prost, Oxford University Press (Oxford, 1993). This is a classic research monograph; however, some parts are rather descriptive.

Liquid Crystal Displays 7th Edition, D. Mentley, *et al*, Stanford Resources, Inc., Stanford California. This report describes the size and economic importance of liquid crystal displays.

Out of the Crystal Maze, L. Hoddeson, Ernest Braun, Jurgen Teichmann, and Spencer Weart, Editors, Oxford University Press (New York, 1992) is a history of the rise of the solid state physics and semiconductors.

Crystal Fire, M. Riordan and L. Hoddeson, W. W. Norton & Company (New York, 1997) is a good history of the rise of the semiconductor industry. The IEEE Spectrum, April and August, 1998. "Tubes: Still Vital After All These Years," and "The Cool Sound of Tubes."

Chapter 9

Polymers

9.1 Overview

Polymers are one of the most common materials encountered in our daily lives. They have replaced naturally occurring materials in many products, and in other cases they have enabled the production of new products that could not be manufactured before the advent of manmade polymers. While most people can identify a plastic or a polymeric bottle, they are less certain of a polymer's chemical composition, why certain polymers cannot be recycled together, or how to describe the chemical and physical structure of a polymer. This chapter will introduce the reader to polymers. The following questions will be addressed.

(1) What is a polymer?
(2) What are some of the unique properties of a polymer?
(3) What are some common polymers and how are they used?
(4) What are commonly observed polymer structures?
(5) How do structure and physical properties relate?
(6) Why are materials added to polymers? What is their affect on the physical properties of polymers?

9.2 Introduction

Polymers are an essential structural material for modern life. Commonly referred to as plastics, these substances are available in a wide variety of forms: fibers, sheets, thin films, foam and bulk to name a few. The term plastic is used because polymeric materials are easily shaped and deformed during fabrication.[1] Polymers are differentiated from metals, another common engineering material, because they are most often organic molecules. Unlike the popular use of the term 'organic,' which generally means pesticide free, organic in this sense means carbon based, and organic chemistry is the chemistry of carbon. When we discuss polymers in more detail, you will see the role carbon plays in polymers.

We will begin by discussing the basic chemistry of polymers. An important difference between polymers and materials such as fluids or metals is that polymers are long-chain or network type molecules. We will illustrate this using cartoons. We will then discuss the major types of polymers. From the materials viewpoint, there are two main types of polymers: *thermoplastic* polymers that become less rigid when heated, and *thermosetting* polymers that become more rigid when heated. Both types of polymers are often used with *additives* that improve color, strength, and other physical and chemical properties.

9.3 What is a polymer?

The term polymer literally means "many parts." It is derived from Greek where *poly* means many and *meros* means unit or part; that is the simple, repeated building block of the chain or network. Thus, a polymer is a large molecule made up of many smaller and simpler chemical units covalently bonded together. For example, polyethylene (CH_3–$(CH_2)_n$–CH_3) is a long chain molecule composed of ethylene molecules ($CH_2{=}CH_2$), Notice however that the terminal groups are CH_3. This is a fairly standard occurrence – the terminal moieties[2] are frequently different from the central moieties that make up the polymer. Moreover, the two terminal groups need not be the same. Molecules with these general properties, that is, long chain molecules consisting of three different moities, a distinct one at each end and many repetitions of the central moiety, are ubiquitous. Such molecules occur in nature, and are also synthetically produced.

This chapter will primarily focus on artificially produced polymers. Synthetic polymers often have a *central structure* of the form:

$$-A-A-A-A-A-A- \text{ or } -A-B-A-B-A-B-.$$

[1] A *plastic deformation* is a permanent deformation that does not change upon removal of the source of the deformation.

[2] A moiety is a sub-section of a molecule that has characteristic properties. Thus, in the above example both CH_3 and CH_2 are different moieties.

The basic repeat unit in this long sequence, –A– or –A–B–, is called the 'structural unit' or 'monomer unit,' or simply the 'monomer.' The number of monomers in the molecule is called the *degree of polymerization*. A molecule is generally called a polymer when the degree of polymerization exceeds about 100. Typical polymers have degrees of polymerization between 10^3 and 10^5. The normal processes of synthetic polymerization do not yield polymers with only one degree of polymerization. Rather, there are polymers with a variety of degrees of polymerization, some greater and some less than the average. To describe this situation, one says that the polymers are *polydisperse*, which means there is a distribution of polymer lengths. Some naturally occurring polymers such as DNA have degrees of polymerization as high as 10^9 to 10^{10} and are *monodisperse*, having just one degree of polymerization or molecular weight. Because they contain so many atoms, polymers are also sometimes called macromolecules.

Materials composed of very long molecules and networks of molecules have physical properties that are very different from materials composed of small molecules. Generally, polymeric materials are very flexible and easily formed into a wide variety of shapes. One of our goals is to understand how these properties depend on the structure and interactions of these molecules.

9.4 The structure of polymers

A polymer is a molecule that is produced through the process of polymerization in which monomers react chemically to form three-dimensional linear chains or networks. Often polymers are named by adding the prefix "poly" to the name of the monomer. (*i.e.* poly(monomer)) We will begin with the chemical formula model of a polymer, in this case poly(vinyl chloride) (PVC). We can represent the center portion of such a polymer by the formula shown in Fig. 9.1. Note that the repeating structural unit is two carbon atoms, three hydrogen atoms and a chlorine atom as shown in Fig. 9.2.

Note also that the single lines in both figures indicate that there are single covalent bonds between the carbons and between the carbons and the hydrogens and the chlorine. You might recall that earlier it was stated that covalent bonds have a definite geometry. Thus, you might expect a PVC polymer to have a fixed, rigid shape. More careful study of the energy needed to bend a covalent bond indicates that there is sufficient thermal

```
       H    H    H    H    H    H
       |    |    |    |    |    |
 ...— C —  C —  C —  C —  C —  C —...
       |    |    |    |    |    |
       H    Cl   H    Cl   H    Cl
```

Fig. 9.1 The central portion of poly(vinyl chloride).

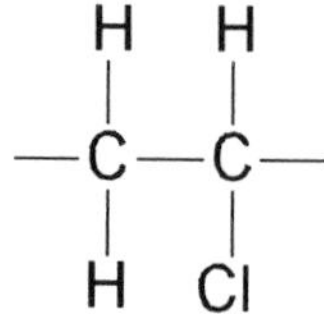

Fig. 9.2 The vinyl chloride monomer.

energy at room temperature for the bond angle between two atoms to vary from between 1 to 10 degrees. This is not a large effect for a small molecule. However, when one deals with a degree of polymerization of 10^2-10^5, such a small effect at each bond adds together to produce a very large effect. In fact, most polymers, when viewed on a large enough length scale, appear to be flexible. This flexibility allows us to go to another level of simplification in our structural models of polymers. We will now represent each monomer by a small circle or bead, and the polymer by a large number of circles as shown in Fig. 9.3. This is a model of a "polymer" with a degree of polymerization of 18. The effort necessary to draw a polymer with degree of polymerization 10^2 or 10^5 is clearly too great. For this reason, we go one step further in simplifying our model and represent the polymer by a line. Thus, we can represent a linear polymer (linear in the sense that the structure of the molecule can be modeled by a curving line) by a curve such as shown in Fig. 9.4.

Notice that at this level we are **not** worrying about the details of the polymer. This model does not allow us to distinguish between the Teflon in a non-stick fry pan and the polyethylene in a plastic bottle.

We may now summarize some of the unique properties of polymers:

(1) Polymers have a chain-like structure.
(2) Since each monomer is approximately 0.5 nm long and since a common degree of polymerization is 10^3 to 10^5, a typical polymer can be 500 to

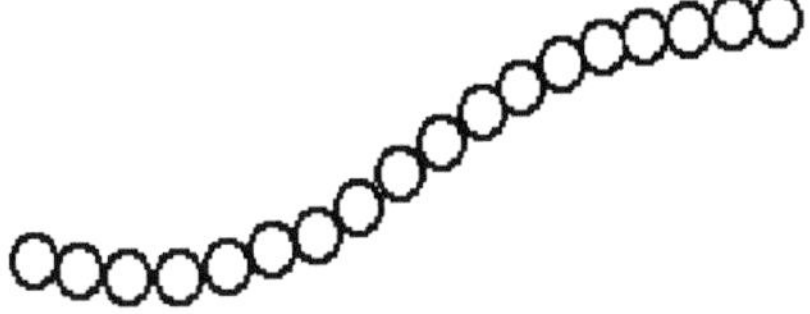

Fig. 9.3 The bead model of a polymer.

Fig. 9.4 Modeling a linear polymer as a line.

50,000 nm long.

(3) Because of this great length and the great number of bonds that it implies, small thermal variations in bond angles lead to polymers that are flexible.

Of course, nature is more complex than just described. A few complications will now be added to this brief sketch.

So far, the polymers which have been discussed contain only a single kind of monomer. These polymers are called *homopolymers*. Often, more desirable physical and chemical properties result when several different types of monomers are polymerized. These polymers are known as a *heteropolymers* or *copolymers*. Two biologically significant copolymers are proteins that have exactly 20 different types of monomers, and DNA, which has four distinct types of monomers.

Copolymers can be divided into two general types. The first is the *statistical or random copolymer*. Statistical copolymers could be described as the result of putting a typewriter with a key for every type of monomer in front of a monkey and letting the monkey type away. The sequence of monomers is completely random. For two types of monomers (A and B) a resulting polymer could be

ABABABBBBABBBBABABAAAABABA.

The second copolymer type is called a *block copolymer*. Block copolymers occur when there are groups or blocks of repeating monomers, as for example:

AAABBBBAAAAAAABBBBBAAAAAABB. . . .

More and less complex possibilities occur. For example, an A–B block copolymer could have the structure:

AAAAAAAAABBBBBBBBB.

An A–B–A block copolymer could have the structure:

AAAAAAAAAAAAAAAABBBBBBBBBBBBAAAAAAAAAAAAA.

Of course, other examples exist; some of these will be investigated in the exercises.

The next complication that frequently occurs is branched polymers. Branched polymers cannot be represented by simple linear curves, lines of letters or circles or the like. Figure 9.5 shows three examples.

In this figure the black dots represent covalent bonds or cross-links between different parts of the branched molecule. The branched molecule in (a) is called a *comb*. Indeed, it looks like a comb made of flexible backbone and flexible teeth. Example (b) is called a *star molecule*. Finally (c) is an example of a *randomly branched chain*.

Another common polymer structure is a *polymer network*. An example of a very strongly cross-linked polymer network is rubber or jelly. The network in such materials consists of a large number of entangled and covalently linked polymer chains. The size of such networks can be very large. A drawing illustrating a polymer network in two-dimensions is shown in Fig. 9.6.

Further complications are possible. For example, *charged polymers* are not uncommon, as are *ring shaped macromolecules*. These more complex polymers will not be discussed.

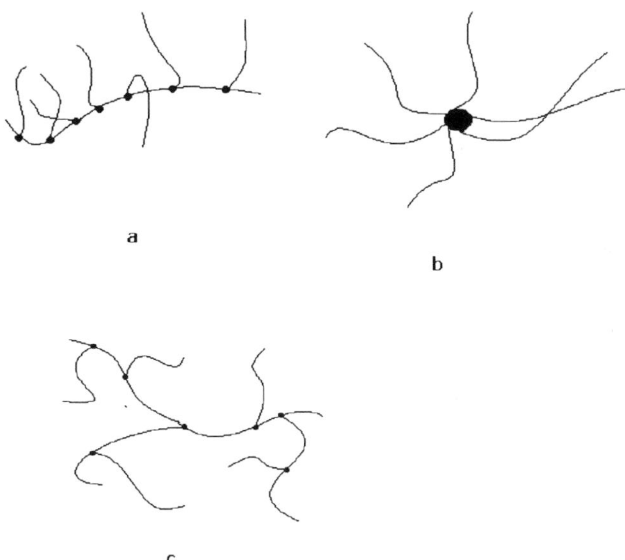

Fig. 9.5 Some examples of models of branched polymers.

9.5 Types of monomers

Polymers are ubiquitous in our daily lives. The purpose of this section is not to make you experts on the types of plastics, but to give you some awareness of the large number of different types of polymers and plastics.

Poly(vinyl chloride), PVC, is a very common polymer. The monomer is vinyl chloride which has the chemical formula $CH_2{=}CHCl$. The central portion of this polymer has the structure already shown in Fig. 9.1.

You are familiar with this polymer in two forms. The first is rigid PVC. It is used primarily for water pipe and pipe fittings that are used in home construction. It is occasionally used for electrical enclosures, too. In its plastic phase, PVC is corrosive to molds and molding machines. In its solid phase, PVC is non corrosive. The other form is flexible and is used in garden hoses, cheap raincoats and the like. This is the "vinyl" that everyone is familiar with. It is essentially rigid PVC with a *plasticizer* added to make it soft. A plasticizer is a polymer with a very low degree of polymerization. Once the polymer mixture is set, the plasticizer begins to work its way out of the compound. This "outgassing" is what gives new cars their "smell," as flexible PVC is used extensively in automobile interiors. It also explains

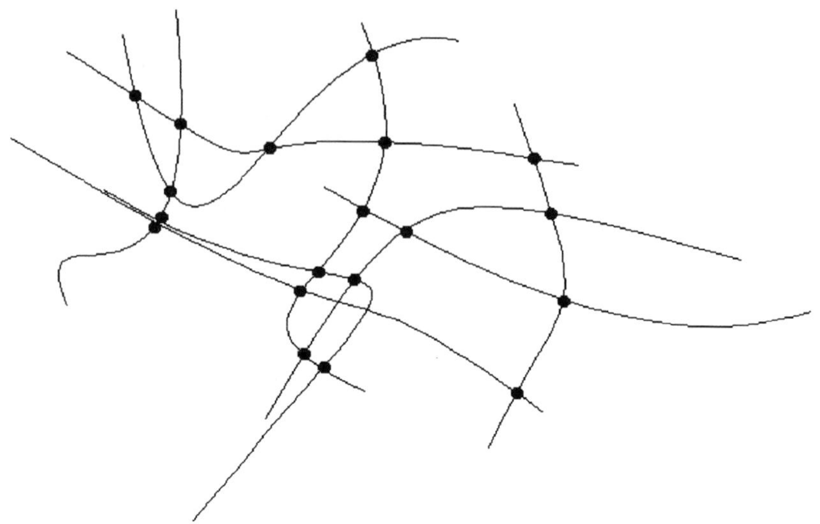

Fig. 9.6 A model of a two-dimensional polymer network.

why the flexible interior parts of a car tend to become less flexible with time.

Another common polymer is poly(ethylene terephthalate), PET. This polymer is made from two different monomers making it a heteropolymer. Its structure is illustrated in Fig. 9.7.

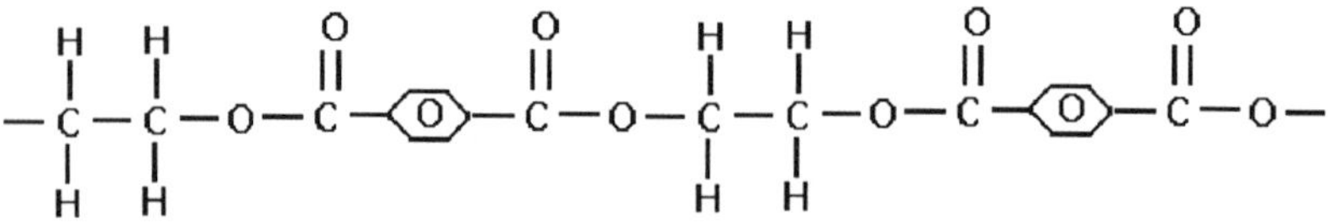

```
  H   H          O          O     H   H          O          O
  |   |          ||         ||    |   |          ||         ||
 -C - C - O - C -(O)- C - O - C - C - O - C -(O)- C - O -
  |   |                           |   |
  H   H                           H   H
```

Fig. 9.7 The chemical structure of poly(ethylene terephthalate), PET.

In this figure the hexagon with a circle in it represents C_6H_4, also called an *aromatic core*. PET is an example of a polyester. Polyester fibers have good wrinkle resistance, low moisture absorption, and good resistance to abrasion. Polyester films also have high resistance to failure on repeated flexing, fair tear strength and high impact strength. Some typical uses of polyester fibers are in garments[3] and rope. Poly(ethylene terephthalate), PET, is the best-known polyester, and is used as both a film and as a fiber. PET is blow molded to produce soft drink bottles. It is also used for a variety of manufactured objects including distributor caps, fender extensions, home appliances, plumbing components and sports equipment. Recycled PET is often used in carpeting. A biaxially oriented thermoplastic film made of PET is known as Mylar. It is employed in many applications in the electronics, magnetic media, imaging/graphics and packaging industries.

Teflon is the common name for polytetrafluoroethylene, PFTE. You are familiar with it as a coating to non-stick cookware. Its basic structure is shown in Fig. 9.8. This polymer has the same structure as polyethylene *except* that all of the hydrogen atoms have been replaced by fluorine atoms. PFTE has excellent electrical insulating properties, and is very corrosion resistant.

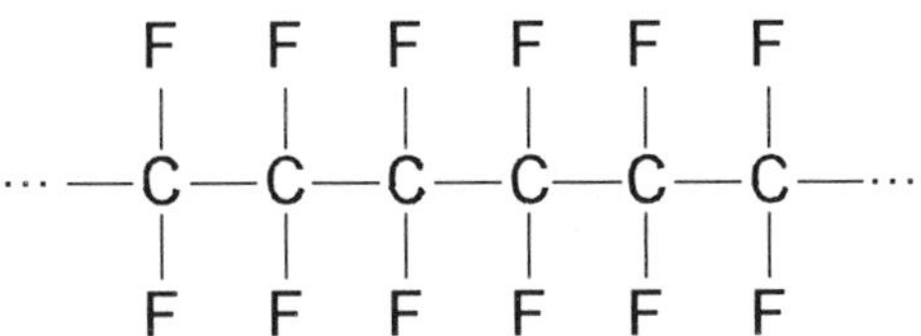

Fig. 9.8 The chemical structure of Teflon.

Polystyrene, an inexpensive, clear and easily colored polymer, is another plastic. The chemical structure of this polymer is illustrated in Fig. 9.9.

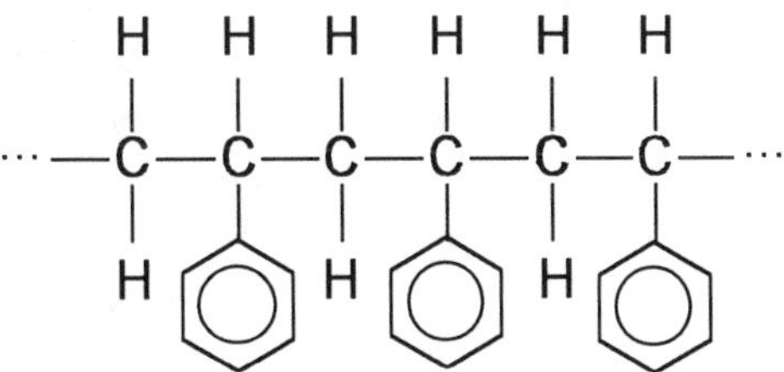

Fig. 9.9 The chemical structure of polystyrene.

[3]Both 1970s double-knit leisure suits and modern 'Microfiber' garments are made of polyester, only the texture of the modern fibers has been changed to protect our aesthetic sensibilities.

In this figure, the hexagon and circle represent C_6H_5[4].

Polystyrene is available in many forms. General-purpose polystyrene has no additives other than rubbers or copolymers. It is a brittle, transparent material with a smooth surface finish that can be printed on. It is used for injection molding containers for cosmetics, boxes and ball point pen barrels. Toughened or high-impact polystyrene is a blend of polystyrene and rubber particles and is used to produce disposable cups and casings for cameras, projectors, radios, television sets and vacuum cleaners. Polystyrene is attacked by many solvents.

The last polymer that we will discuss is acrylonitrile-butadiene-styrene, ABS. ABS is a grafted heteropolymer, in which polymer chains of acrylonitrile and styrene are grafted onto a butadiene chain. ABS will be further explored in the exercises. It is mentioned here because it is easy to process and is widely used for computer housings, small appliances, telephones and luggage.

9.6 Structures of polymeric phases

We now know the molecular configuration of some polymer molecules. Given these configurations, what structures and phases are possible? We will answer this question only for polymers, not for polymers in solution or solutions of polymers and another non-polymeric substance.

First, you might be tempted to suggest that there must clearly be solid, liquid, and gas phases just as with "regular" materials. Very well, but what do these phases look like? There are no polymeric gases. The reason is that at temperatures high enough to form a gas phase, the polymer dissociates into the monomers from which it was formed. Of course, you have seen solid polymers so that you know this phase exists. There are also liquid-like states of polymers. We now ask (and answer) what do these phases look like at the level of our standard cartoon? (Chosen because it shows sufficient detail without becoming too specific to any one polymer.)

The liquid phase may be thought of as being composed of highly mingled and entangled molecules that are able to easily move with respect to each other. A polymeric liquid is very *viscous*, viscosity being a quantitative

[4] It's now time to explain exactly what this symbol means. It is a chemical shorthand for six carbons and up to six covalent bonds, one at each carbon. Generally, the covalent bonds that are not connected to other moieties are connected to hydrogen. Thus, an isolated hexagon and circle represents C_6H_6, and whenever a covalent bond to another moiety is made, it replaces a hydrogen.

measure of a liquid's ability to flow or pour. Water has a low viscosity and pours easily. Pancake syrup has a higher viscosity and pours less easily. A polymeric liquid is generally *very* viscous which means it pours *very slowly.* A cartoon of a polymeric liquid in two-dimensions would appear similar to Fig. 9.10.

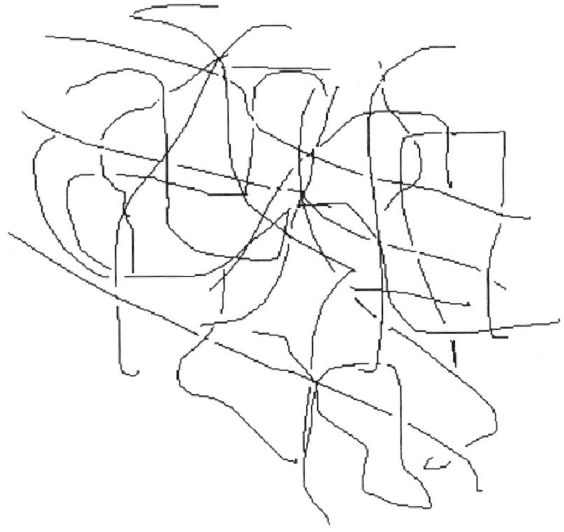

Fig. 9.10 A simple model of a polymeric liquid.

Fig. 9.11 A model of a crystalline polymer.

Going back to the solid phase, the situation is not as simple as the crystalline phase observed with smaller molecules. There are, in fact, three types of solid phases in non-cross-linked polymers: the crystalline phase, the semicrystalline phase, and the polymer glass or amorphous solid. The 'perfect' crystal is the lowest energy state. It would appear similar to the model shown in Fig. 9.11. Notice that, just as in crystals of smaller molecules, there is perfect periodicity. In practice, this structure is either impossible or very difficult to achieve, because when polymeric materials start to crystallize they often start to crystallize in several places at the same time. Thus, one does not obtain a single crystal. However, closer inspection indicates that this is also often the case with non-polymeric materials, which is why metals are generally polycrystalline—made of many randomly oriented crystallites. The situation in polymers is more complicated than metals because the molecules are long and flexible. Thus, it is possible for several different parts of the same polymer molecule to crystallize. This leads to a very common situation in which there are crystalline and non-crystalline sections in the same sample and within a single molecule. Such a material is called a partially crystalline or semicrystalline solid. For example, it is possible to obtain Teflon that is 90% crystalline.

A model of a semicrystalline polymer is shown in Fig. 9.12.

In this figure, observe there are three regions with four parallel lines, indicated by the arrows. The parallel lines indicate a crystalline portion of this material. The number of parallel lines, four, is arbitrary and was chosen simply for illustrative purposes.

We may now ask a fundamental question. If all polymers cannot be crystallized, can all polymers form semicrystalline solids? The answer is a resounding **NO!** There are two reasons for this.

The first reason involves a "kinetic bottleneck." Recall that while a crystalline solid is the lowest energy state, it may not be possible for the polymer to reach that state. For example, suppose a polymeric liquid is cooled rapidly. The equilibrium (lowest energy) state may be the crystalline phase. However, the thermal motion (kinetic energy) of the molecules may not be great enough to overcome the entanglements of the polymer molecules in the liquid state and thus a crystal will be unable to form. Simply put, the molecules do not have sufficient kinetic energy or time to move and form the crystalline or semicrystalline state. In fact, with rapid enough cooling, which may not be rapid in the usual sense of the word, the polymer can get trapped in a solid state that structurally appears to be liquid, but has very different physical properties. This is the *glassy* polymer or *amorphous*

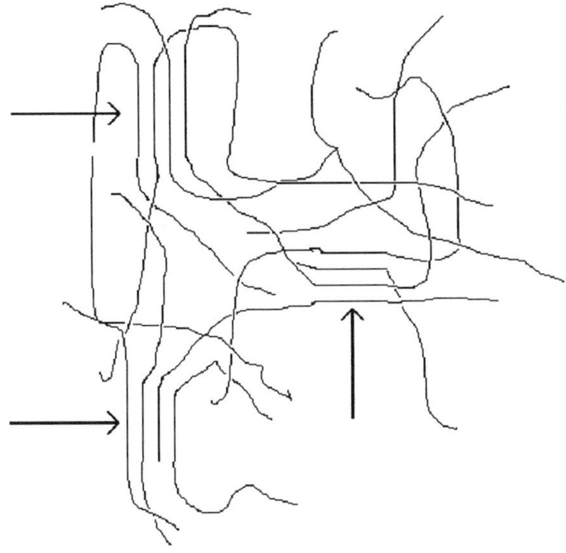

Fig. 9.12 A model of a semicrystalline polymer.

polymer state. The term "kinetic bottleneck" indicates that the system lacks sufficient time and/or energy to reach the equilibrium state.

The second reason polymers cannot form semicrystalline solids is even more fundamental. There are some polymers that can not be crystallized even in principle. How is this possible? Recall that a crystal has a periodic structure in *all* of its components; in this case, for *every* monomer. Thus, a more precise model of a crystalline polymer than Fig. 9.11 would be Fig. 9.13.

Fig. 9.13 Model of crystalline polymer showing some monomers.

In this figure, the monomers are indicated by circles enclosing the letter A. Consider a random copolymer. Each polymer chain is different from every other one because they have different arrangements of the constituent monomers. It is therefore impossible to form the periodic structure required to form a crystalline solid. Instead, the polymer will form an amorphous solid.

A similar situation exists when a sample consists of homopolymers with different three-dimensional structures. This is difficult to show without three dimensional drawings or space filling models. Figure 9.14 shows the central part of two different homopolymers of PVC. Both structures are PVC, but notice the difference in the placements of the Cl's and H's.

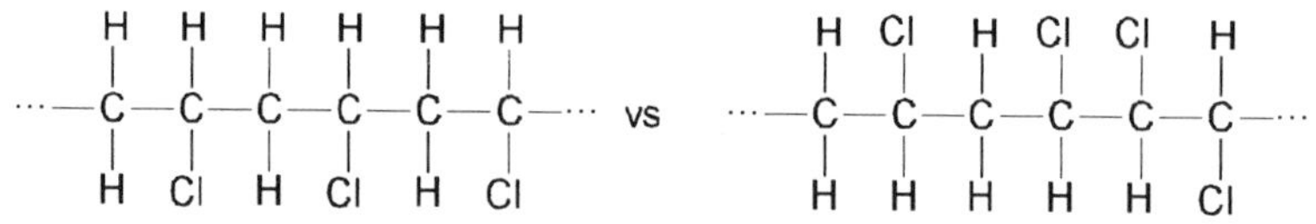

```
     H   H   H   H   H   H                H   Cl  H   Cl  Cl  H
     |   |   |   |   |   |                |   |   |   |   |   |
···—C—C—C—C—C—C—···   vs   ···—C—C—C—C—C—C—···
     |   |   |   |   |   |                |   |   |   |   |   |
     H   Cl  H   Cl  H   Cl               H   H   H   H   H   Cl
```

Fig. 9.14 Two different structures or configurations of PVC.

In a normal chemical synthesis of a polymer, the two configurations shown in Fig. 9.14, along with many more, would occur. In fact, many of the polymers you are familiar with are fundamentally uncrystallizable. Such polymers are said to be *atactic.*

What happens when one cools an atactic polymer? Recall that the kinetic energy of the polymer is proportional to the absolute temperature. Thus, as the sample is cooled, the mean kinetic energy of the molecules decrease. Since there is less kinetic energy there must be less motion (recall kinetic energy is energy of motion). As the polymer has less kinetic energy, smaller motions are available to attempt to overcome any energy barriers to its movement. Thus motions begin to get "frozen out." The large-scale motions become frozen out first, followed by the smaller scale motions. This "freezing" normally occurs over a relatively small temperature range (approximately 10-30°C). The midpoint of this temperature range is called the *glass transition temperature*, T_g. The glass transition is marked by a change in physical properties, near T_g. For example, the viscosity of a polymer decreases very rapidly with temperature near the glass transition. The glass transition is continuous, so for example, there are no discontinuities in the volume as a function of temperature at the glass transition. However, the ratio of the fractional change in volume for a given change in temperature is different above and below the glass transition.

There is yet a fourth type of solid that may form, the elastic solid or rubber. Recall that in rubber, the polymer chains are cross-linked and there can be no flow of the network. However, because the polymer chains are not stretched, one can temporarily change the shape of the network by deforming it. We say that such networks are *elastic*. Interestingly, there is no need for such a network to be formed by covalent cross-links. Topological entanglements, such as knots, and crystalline regions of a semicrystalline solid can serve the same purpose.

9.7 Plastics

The elastic polymer, polymeric liquid, amorphous polymer, and semicrystalline polymer states are all important and useful polymeric states. Following Grosberg and Khokhlov[5] we define plastics as those materials which are produced in either the elastic or viscous (liquid) polymer state and used in either the amorphous or semicrystalline state. There are various ways to transform a polymer from one of these states to another. A common and easy means to transform a polymer is to change its temperature. When the polymer is a viscous liquid at high temperatures and a solid at low temperatures it is called a thermosoftening plastic or *thermoplastic* polymer. This behavior is characteristic of linear polymers and occurs because the chains can easily move past one another at higher temperatures.

In contrast, a second common type of plastic, *thermosetting* polymers become hard and rigid upon heating. Significantly, this behavior is not lost upon cooling because a permanent network forms. Thermosets are materials that typically form by mixing two or more materials together. They form a network upon heating, or, in many cases, evolve their own heat when mixed. Examples include epoxies and polyurethanes.

These materials should be contrasted with elastomers that are essentially very highly cross-linked polymers. Common elastomers include rubber and rubber substitutes like neoprene and Viton.

When selecting a polymer to use for manufacturing a product, we are most frequently concerned about the polymer's physical properties, which should be commensurate with the product's intended use. The key difference between a semicrystalline polymer and a glassy or amorphous polymer is that the semicrystalline polymers are more deformable and elastic than

[5]See the references at the end of this chapter.

the amorphous polymers. They also tend to better maintain their shape under moderate deformation and are not as susceptible to fracture.

9.8 The elasticity of materials

The extent to which a given solid can be elasticly deformed[6] is characterized by various *elastic moduli*. One of the simpler moduli that shows great variation in different polymeric solids is the *Young's modulus*, E. To see how this modulus is defined, consider the following experiment. The sample to be deformed is a solid that has a uniform cross section area A, and length, l when no external forces are applied. Suppose that a pair of forces each of magnitude F is applied to each end of this solid. Under the action of these forces, the solid will elongate and have a length $l + \Delta l$, where Δl is the change in length of the solid. The zero force and applied force situation are shown in Fig. 9.15.

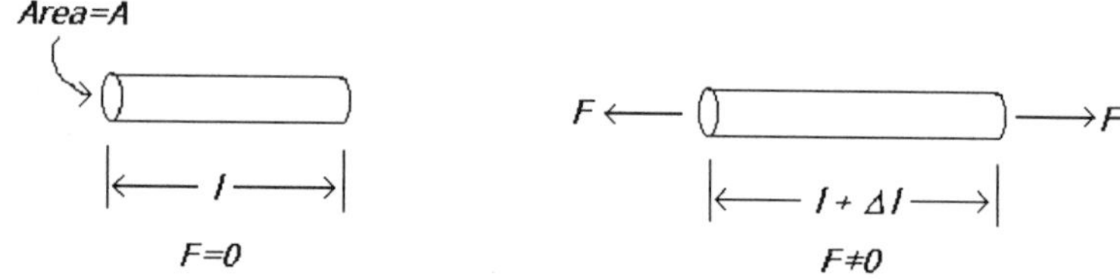

Fig. 9.15 Deformation of an elastic object.

For small enough changes in length, which depends on the material and its state, one finds that $F/A = E(\Delta l/l)$, or $\Delta l/l = F/AE$. This relation states that for a given force per area, the change in length divided by the original length is proportional to the reciprocal of E, the Young's modulus. Thus, for a large E, it is difficult to elongate a sample, while for small E, it is easy. Young's modulus is an intrinsic property of a material that is independent of the size and shape of the sample, but depends on composition, temperature and other properties of the material. It has the same dimensions as pressure, force/area. In SI units, it has dimensions of Pascals, Pa. Table 9.1 gives some representative Young's moduli. While it is not listed in the table, the maximum strain, $\Delta l/l$, before the material

[6] An elastic deformation is a deformation, generally rather small, in which the material returns to its initial undeformed state upon removal of the deforming force.

Material/State	Young's modulus, E (Pa)	Comments
Metals	10^{11} to 10^{12}	
Steel	$20 * 10^{11}$	
Window glass	10^{10} to 10^{11}	Typical is $6 * 10^{10}$ Pa
Polymeric crystal	10^{10} to 10^{11}	Deformation parallel to crystalline axis
Polymeric glass	10^{9} to 10^{10}	
Semicrystalline polymer	10^{8} to 10^{9}	Directional dependence too
Amorphous polymer	$\leq 10^{6}$	
Rubber	$< 10^{6}$	

fractures (fails), also depends on the state of the polymer.

How do we explain these numbers? In particular, why are polymers so much more elastic than metals? Moreover, why are semicrystalline polymers and amorphous polymers so much more elastic than polymeric glasses? We must first understand metals. Recall that in simple solids the molecules are very tightly bound and the potential energy of the molecules is much greater than their kinetic energy. There is so little motion of the molecules that the only way to stretch a solid is to stretch the chemical bonds between the molecules. This is very difficult to do and results in the large Young's modulus. The very low E of elastomers such as rubber and amorphous polymers is also easy to explain. In these two cases, there is sufficient thermal motion of the chains, that are generally not fully extended but entangled, to allow the chains to be stretched or to move past one another. The polymeric glass can be understood by recalling that, in this state, the chains have their largest motions frozen out, but there is still the possibility of some smaller chain movement. This chain movement allows the polymer to deform. The amount of movement is substantially larger than in window glass that also has a "frozen" structure, a consequence of the fact that T_g is much closer to room temperature for a polymeric glass than for window glass.

Lastly we must explain the semicrystalline polymer which is intermediate between the amorphous polymer and the polymeric glass. Recall that a semicrystalline polymer has regions of crystallized polymer and regions of amorphous polymer. It can be treated as a two-phase mixture that is a combination of these two types of polymer. This model suggests that one would expect that the Young's modulus would be between a crystalline polymer (E greater than or equal to that of a polymer glass) and an amorphous polymer. One would also expect that the Young's modulus would depend on the relative amount of each type of polymer, and the direction of the crystallites. These expectations are also verified by experiment.

9.9 Fibers

A major application of polymers is in fibers. This is clearly an important application since nearly all of our clothes and many 'natural' materials are fibers. For the present discussion there are two kinds of fibers: natural and synthetic.

A very common natural fiber is cellulose. The cell walls of plants contain cellulose, and we are very familiar with it as a major component of wood and wood products, especially paper. In fact, cellulose is also the 'fiber' in cotton and flax. Two examples of common animal based natural polymeric fibers are wool and silk. These two fibers consist of long protein chains that are, of course, polymers.

Synthetic fibers actually consist of two different classes of fibers. The first class is often called *artificial fibers.* These are natural materials that have been chemically modified so that the material has improved properties. Examples are acetate and triacetate fibers that can be produced from cotton cellulose. The second class encompasses truly synthetic polymers that have been synthesized from simpler chemical compounds. Examples of this class of compounds include nylon, terylene, orlon, and acrylon.

What is the structure of polymers in fibers? Since fibers must be reasonably tough, not deform too much during use, and be structurally robust, they cannot be viscous liquids or elastic polymers. By default, fibers must consist of amorphous solids or semicrystalline polymers. Yet even in these two phases, the polymers in fibers do not have the same structure as in the bulk structures discussed in the previous section.

Experimental studies of semicrystalline polymer fibers indicate that the crystalline regions of the fiber are not randomly oriented, as in a bulk sample, but are oriented parallel to the axis of the fiber. A similar situation occurs in amorphous polymer fibers where the polymer chains are also parallel to the fiber axis. These two configurations are shown in Fig. 9.16.

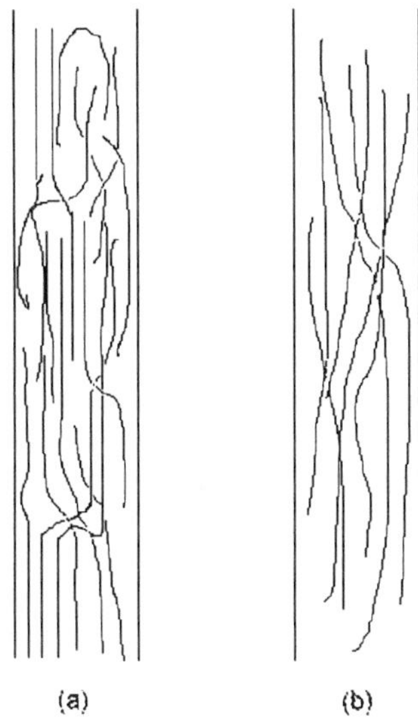

Fig. 9.16 Examples of models of polymer fibers. a) Semicrystalline b) Amorphous

The semicrystalline polymer fiber is on the left and the amorphous polymer fiber on the right. Notice that there are places in the semicrystalline fiber where the chains are straight and parallel indicating crystalline regions. On the other hand, the amorphous fiber has regions where polymer molecules are essentially parallel, but these regions are not crystalline. These regions are places where the polymer chains just happen to be parallel.

It is found that polymer fibers are always anisotropic with the polymer chains having a preferential orientation along the fiber axis. Furthermore, the greater the anisotropy, the greater the Young's modulus, the greater the stretching of the fiber, and the stronger the fiber.

How do we understand this unusual behavior? First, consider a polymer crystal with the chains aligned vertically. When one tries to stretch these chains, one is attempting to extend the distance between carbon-carbon covalent bonds. This is possible, but there is an upper limit. One material with a very high Young's modulus and covalent carbon-carbon bonds is crystalline carbon, or diamond. Thus, one might expect that a polymeric crystal would have a rather large Young's modulus parallel to the polymer axis. This is indeed the case as the previous listing of Young's moduli in Table 9.1 indicates. At the same time, the semicrystalline or amorphous polymer has a much smaller Young's modulus. Thus, when a fiber is stretched, any chains essentially parallel to the stretching direction absorb most of the strain while the other chains start to align in the stretching direction. Further stretching causes this process to continue. The result is that more and more chains become aligned parallel to the stretching direction and the Young's modulus increases. It is possible to increase the Young's modulus of a fiber by a factor of 10-100 by this technique. You are familiar with this if you have ever stretched a plastic "six pack holder" for soda cans. At first it is very easy to stretch. Then, as you stretch it further, it becomes more and more difficult to extend the 'fiber.'

The polymeric structure of fibers is all very interesting, but just how does one go about making fibers? Nature, of course, has solved the problem and the manufacturers of synthetic fibers solve this problem during production. Usually, a synthetic fiber is made by squeezing a melted polymer or a polymer dissolved in a solvent through small holes like toothpaste out of a tube. The flow of the material through the holes imparts some initial alignment to the polymer chains within the fiber. However, this is not usually sufficient. The fibers are further aligned by stretching them. This is performed at a high temperature to prevent the fibers from forming a glass.

9.10 Additives

The properties of polymers may be enhanced by the addition of various *additives*. This results in the formation of something like alloys in metals. We have already encountered one additive, *plasticizer*, that is added to soften a polymer.

Fillers are another common additive. Fillers serve several purposes. They can strengthen a polymer by reducing chain mobility. They are also used for "volume replacement," so less polymer is needed and costs are reduced. Fillers are also used to provide dimensional stability so the size of objects can be more tightly controlled. Fillers come in many varieties. Cellulose (natural filler), asbestos (inorganic filler), and carbon black (organic filler) are common examples. Roughly one third of a typical automobile tire is carbon black filler.

Reinforcements are also considered additives. These materials are used to improve the strength and stiffness of polymers. This is necessary for polymers to compete with metals in many applications. Such mixtures are called composites. A typical reinforcement is glass fibers. Typically, when additives are used, the material is still referred to as a polymer unless more than 50% by volume of the material is additive. Beyond this concentration, the properties may change so much that the word polymer is meaningless.

Stabilizers are a very important class of additives that are used to reduce polymer degradation. There are many ways a polymer might degrade. Examples include oxidation, combining with oxygen in the air, thermal degradation, heating may cause changes in the polymer such as outgassing of plasticizers, and light induced degradation especially due to ultraviolet light. Each degradation mechanism and application may require a different stabilizer, these represent a diverse and complex group of molecules.

Many polymers are composed of carbon and hydrogen. These are the same building blocks that occur in natural gas, gasoline and other highly flammable materials. Thus, it is often desirable to reduce the inherent flammability of **some** polymers by the addition of *flame-retardants.* It is important to realize that not all polymers are combustible. For example, PVC is not. It is believed that this is due to the chlorine atoms that are part of this polymer. The release of chlorine inhibits combustion.

Lastly, manufacturers do not want all polymers to be the same (bland) color. For this reason they add *colorants.* There are two types of colorants: pigments and dyes. A pigment is a colored material that is added in powdered form and mixes in with the polymer but does not dissolve in the polymer. A dye dissolves in the polymer and is usually an organic compound.[7]

9.11 Polymers and recycling

One unique feature of plastics is that many of them are essentially indestructible. The decomposition time for plastic foam is said to be forever. Because, by their nature, polymers tend to have long lifetimes, there have been significant efforts to recycle some polymers. You should recall that all plastics are polymers, but all polymers are not plastics. Examples of commonly recycled polymers (all of which are plastics), their recycling identification numbers and first and recycled uses are in Table 9.2. It should be noted that number uses of recycled polymers are limited in the U.S. by

[7] In a mixture the components retain their individual identity. When a substance dissolves into another substance the individual identity of the substances is lost.

the U.S. Food and Drug Administration's requirement that polymers that come in contact with food be sterile. Current technologies cannot sterilize most used plastics without destroying them.

Polymer Name, Abbreviation, and Recycling Number	**Typical First Use**	**Typical Recycled Use**
Polyethylene terephthalate, PET or PETE, 1	Bottles and jars	Bottles and jars, carpet, clothing
High Density Polyethylene, HDPE, 2	Milk jug	Pens, fencing, drainage pipes, park benches & tables
Polyvinyl Chloride, PVC, 3	Car seats, window frames, vinyl siding	Playground equipment, bubble wrap
Low Density Polyethylene, LDPE, 4	Bread bags, sandwich bags, grocery bags	Trash bags, trash cans
Polypropylene, PP, 5	Medicine bottles	Brooms, ice scrapers
Polystyrene, PS, 6	Foam cups, drinking cups, CD cases	Egg cartons, packing peanuts, insulation, concrete
Other, One or more of the above, or none of the above, 7	Containers for solvents automotive fuel additives	Bottles, plastic lumber

9.12 References

Polymers are a diverse group of matter, with a vast literature. Any list will surely offend an expert who feels that their favorite reference has not been included. With this caveat, the following books, which cover a wide range of levels, are suggested as places to gain a more complete understanding of polymers.

The Theory of Polymer Dynamics, M. Doi and S. F. Edwards, Oxford University Press, 1986.

Scaling Concepts in Polymer Physics, P. G. de Gennes, Cornell University Press, 1979.

Statistical Physics of Macromolecules, A. Yu. Grosberg and A. R. Khokhov, AIP Press, 1984.

These three books are for the mathematically inclined professional. However, they offer insights to novices who are willing to carefully read the text and realize that $\sum$ and $\int$ both resemble the letter "s" and mean skip.

Doi offers a mathematically simpler introduction in **Introduction to Polymer Physics**, M. Doi, Oxford University Press, 1996.

The following book is more of a textbook then a treatise: **The Physics of Polymers**, Gert Strobl, Springer, 1997.

Grosberg and Khokhlov have an introductory book on polymers that extends the ideas introduced here: **Giant Molecules**, A. Yu. Grosberg and A. R. Khokhov, Academic Press, 1997.

Another book that is intended for students is **Introduction to Polymers**, R. J. Young, Chapman and Hall, 1981.

A text that covers material at a range of levels is **Introduction to Physical Polymer Science**, L. H. Sperling, Wiley Interscience, 2001.

Many introductory materials science texts have chapters on polymers. The author used **Introduction to Materials Science for Engineers**, James F. Shackelford, Macmillin, 1988.

No list of references on polymers would be complete without **Principles of Polymer Chemistry**, Paul J. Flory, Cornell University Press, 1953.

9.13 Exercises

(1) Why are polymers organic molecules?

(2) Sketch a diagram similar to those in the text of a random copolymer

that has two different types of monomers. Let the two monomers be represented by the letters C and B. Note that there are many solutions to this problem.

(3) One of the polymer types that we did not discuss is the *alternating copolymer*. Alternating copolymers are fairly common and occur when the two different types of monomers alternate along the polymer chain, for example, ... BCBCBCBCBC .... Draw a cartoon of an alternating copolymer in which the two monomers are represented by a circle and a square. This problem requires you to draw the small-scale structure of the polymer, not just squiggly lines.

(4) Another polymer that was not discussed is the *graft copolymer*. Graft copolymers occur when a polymer of one kind is grafted to a **central portion** of a chain of a different polymer. Sketch such a polymer. You will need to represent two different types of polymers. This problem requires you to draw the small-scale structure of the polymer, not just squiggly lines.

(5) Sketch a diagram of the structure of a polymer mixture that consists of a 50-50 mixture of two different types of linear polymers. This problem requires you to draw at least two pictures showing both the small-scale structure of the polymers and the large-scale structure that is represented by squiggly lines.

(6) A semicrystalline polymer is mixed with small spheres of a pigment. Draw this structure. Do you expect any pigment in crystalline regions? Explain.

(7) Sketch a plasticized polymer. You will need two types of polymers, a very short one and a rather long one.

(8) Sketch a cartoon of the structure of a polymer in an automobile tire.

(9) List the items in your dorm room or house that are made of plastics. Try to identify the type of plastic. (This means make a serious effort. "I don't know," is not an acceptable answer. This question may require some research.)

(10) Extend problem 4 and research question 5 by drawing a chemical structure for the central portion on an ABS molecule.

(11) Do you expect a polymer with a high degree of polymerization to have a higher or lower viscosity than the same polymer when the degree of polymerization is smaller? Explain your reasoning. Is there a relationship between viscosity and degree of polymerization or molecular weight?

9.14 Research questions

(1) What are some of the properties and uses of neoprene? (This question will require a trip to the library and/or the Internet. You should include references).
(2) What are some of the properties and uses of Viton?
(3) Investigate and tabulate the production and use of several common polymers discussed in the text.
(4) What are some of the properties and uses of nylon?
(5) Investigate the structure, uses and properties of ABS.
(6) Investigate the production and properties of Kevlar.
(7) What properties of polymers make them particularly poor materials for use in high temperature applications? What is the highest temperature that a polymer may be used at continuously? What are the structures and properties of high temperature polymers?
(8) What are conducting polymers? Why is there interest in these materials?

Chapter 10

Liquid Crystals

10.1 Overview

Many of us are familiar with liquid crystal displays on watches or electronic devices. We also may have seen mood rings and indicating thermometers. In addition, who among us has not seen the sludge like residue at the bottom of a soap dish? These all contain or are liquid crystal materials. Yet, we rarely think about the materials that make up the display, the thermometer, the ring, or the sludge. This chapter will introduce liquid crystals. We will answer the following questions:

(1) What is a liquid crystal?
(2) What types of molecules form liquid crystals?
(3) What types of liquid crystals exist?
(4) What phases of liquid crystals exist?
(5) What are some properties of liquid crystals?
(6) What are polymeric liquid crystals?

10.2 Liquid crystals and ordering

We now consider another extension of the common phases of matter which we discussed earlier – solid, liquid, and gas. When certain organic materials (about one in 200) are heated, there is not a single phase transition from solid to liquid as is observed, for instance, in water. Instead, a cascade of phase transitions is observed involving phases with different mechanical and physical properties. Some of the physical properties of these new phases are intermediate between those of a crystalline solid and an isotropic liquid. For this reason, they are called *liquid crystalline phases.* A more accurate name is *intermediate* or *mesomorphic phases.*

To understand how these mesomorphic phases are both similar to and different from solids and liquids, it is useful to review some of the properties of solids and liquids that we discussed earlier. In solids, the molecules are very close together, $l \approx d$, where l is a typical separation between molecules and d is a typical molecular size. Furthermore, the molecules move large distances very infrequently. Generally, they just vibrate about their equilibrium positions. Most importantly, the molecules have a regular periodic arrangement, which, in principle, goes on forever. However, it should be noted that "forever" is a relative term. Even in a small crystal, 1 mm on a side, this periodicity repeats millions of times, effectively "forever." This periodicity means that **there is long-range positional order in the crystal. If the molecules making up the solid are not spherical, there is also orientational order in the crystal**. This order is contrasted with the molecules in a liquid that still have $l \approx d$, but can easily move about, and have a random arrangement. This means that in an isotropic liquid, a liquid whose properties are direction independent, **there is *no* long-range positional or orientational order.**

This characterization of the order present of a crystalline solid and an isotropic liquid allows us to define liquid crystals as phases of matter that have some liquid-like order, some degree of long range orientational order, and, in some cases, some positional order. We will become more specific later in this chapter.

With this definition of liquid crystals in mind, we can now begin to use our imagination to construct possible liquid crystalline phases. First, we will assume that we have orientational order along some direction in space. **Orientational order** is a necessary condition for a liquid crystal to exist. This necessarily means that the molecules that we are considering are anisotropic,[1] and cannot be represented in cartoons by spheres. We will come back and draw cartoons of these phases later. For now, we will simply use our mind's eye to picture them.

The simplest liquid crystal one could imagine is a phase in which the molecules are oriented along a common direction in space, and the positions of the molecules are totally random (subject, of course, to the constraints of molecular volume and forces). Stated another way, there is no long-range positional order between the molecules, but there is long-range orientational

[1]An isotropic object has the same form (size) in all directions – a sphere. An anisotropic object has different forms or properties in different directions – for instance, a cylinder or a box.

order. This phase is an oriented or anisotropic liquid. Such liquids are frequently observed in nature and are called *nematic liquid crystals.*

Suppose now that a phase forms in which the long-range orientational order is maintained and long range periodicity exists, not in three dimensions as in a crystal, but only in one dimension. This can be thought of as a stack of two-dimensional liquid planes. The spacing between these liquid planes extends over very long distances resulting in a phase with long-range orientational order and long-range positional order in one-dimension. This type of liquid crystal phase is also observed in nature and is called a *smectic liquid crystal.*

The next most complex phase would be a phase that has long-range orientational order and long range positional order in two dimensions. Such a phase can be described as a two-dimensional array of liquid tubes. Such phases are also observed and are known as *columnar phases.*

Lastly, at least for this discussion, we might expect a phase in which the molecules exhibit long-range orientational order and long range positional order in three-dimensions. Of course, this is a crystalline solid of anisotropic objects.

Nematic, smectic, and columnar phases are the only known liquid crystal phases. However, slight variations in these basic phases occur and lead to hyphenated names for some observed phases. The type of phase observed depends very strongly on the structure of the constituent molecules or, in some situations, the aggregates that constitute the phase. Nematics and smectics are often observed when elongated or rod-like molecules or objects order. Columnar and nematic phases are more often observed when disk-like molecules or objects order.

10.3 The building blocks of liquid crystals

There are many ways to obtain the anisotropic objects necessary to form liquid crystalline phases. In this section, we will very briefly explore three classes of materials that form liquid crystalline phases.

The simplest anisotropic object is an anisotropic molecule. Many simple and common organic molecules can be described as elongated. The classic example is *p*-azoxyanisole, also known as PAA. It has the chemical structure shown in Fig 10.1.[2]

[2]In earlier sections, one, two, or three lines have been used to indicate single, double, and triple covalent bonds. This molecule has an arrow. This indicates that the oxygen (O) is associated with the both nitrogen atoms (N). It is shared by both nitrogen atoms

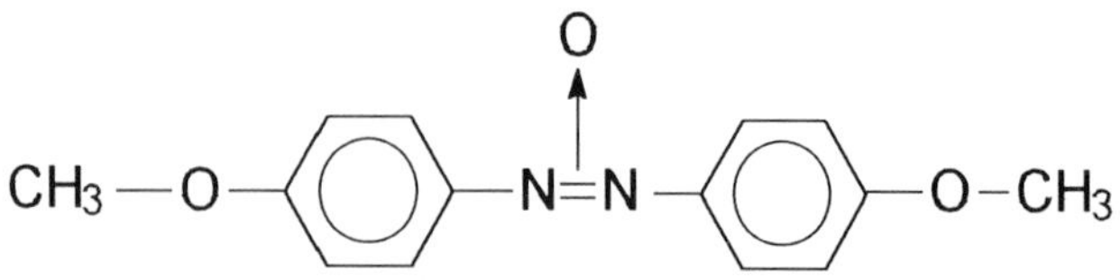

Fig. 10.1 The chemical structure of PAA, a classic liquid crystal forming molecule.

PAA can be thought of as having a rigid center or core and two short hydrocarbon tails. From a very rough geometric point of view, it is a rigid rod of length approximately 2 nm long and 0.5 nm wide. It could be represented in a cartoon by something like Fig 10.2.

Fig. 10.2 A simple cartoon for the PAA molecule.

Elongated molecules, such as PAA, are usually represented in cartoons or schematic diagrams by a line and not a rectangle.

The simplified, one-dimensional phase diagram for PAA is shown in Fig 10.3.

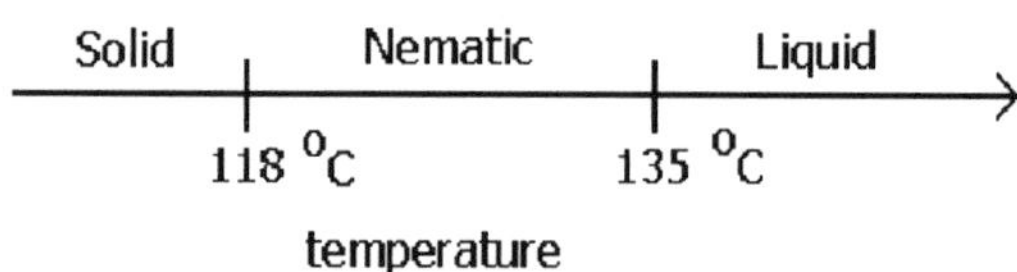

Fig. 10.3 The simplified, atmospheric pressure phase diagram of PAA.

Some general observations about the structure of the elongated organic molecules that form liquid crystalline phases have been made by scientists. First, the molecule must be elongated, and the elongation should be significant. Second, the molecule must have a rigid central core. Finally, the ends of the molecule should be somewhat flexible. A better, more accurate cartoon of this type of liquid crystal molecule would be a short unsharpened pencil with a piece of cooked spaghetti attached to each

and cannot be thought of as tightly bound to just one.

end. Simple arguments for why this structure works are not too difficult to make. The elongated shape of the molecule tends, from the point of view of both energy and entropy (disorder), to favor aligned phases. A rigid core encourages alignment, while a flexible core favors a phase more like an amorphous polymer. The flexible end chains work very much like the chains in a polymer. They allow the molecules to move about and align while not allowing microscopic holes to form in the material.

In this class of materials, the simplest way to change phases is to change the temperature. Therefore, these materials are also known as *thermotropic liquid crystals.*

The second common class of molecules that form liquid crystalline phases are small organic discoid molecules. These thermotropic liquid crystal molecules generally consist of a rigid planar core and typically six flexible hydrocarbon chains. A simplified steric model of these molecules would be a thin disk, like an unstruck penny. A more complex model would have a thin central disk with six pieces of cooked spaghetti attached about its circumference.

A third common class of materials that forms liquid crystalline phases is long rod-like objects in a liquid. Common examples are rigid polymers, DNA, and some types of viruses in water or an organic solvent. Changing the concentration of the rods rather than the temperature most easily induces phase transitions in these systems which are called *lyotropic liquid crystals.* We will focus most of our attention on thermotropic liquid crystals.

10.4 Liquid crystalline phases – a more detailed look

10.4.1 *The nematic phase*

The liquid crystalline phase with no long-range positional order but with long-range orientational order is the nematic phase. Schematic pictures or cartoons of this phase are shown in Fig. 10.4.

In this figure, a nematic liquid crystalline phase of elongated molecules is shown on the left and a nematic phase of disk-shaped molecules is shown on the right. The arrow indicates the average direction the long axes of the rods and the normals of the disks are aligned. The main characteristics of the nematic phase are:

(1) There is no long-range order in the positions of the molecules. Aside

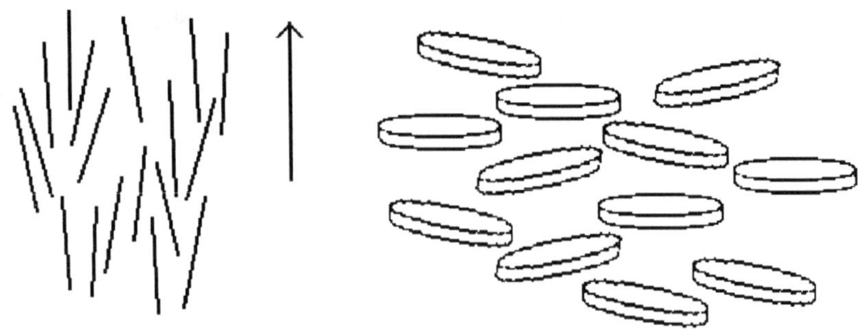

Fig. 10.4 A cartoon of the nematic phase for elongated objects (on the left) and disk-like objects (on the right).

from the anisotropy of these phases, due to preferred orientation of the molecules, they flow similarly to normal liquids. In fact, the viscosity of PAA and many other liquid crystals is approximately ten times that of water. This increase can be explained by examination of the molecular properties of these molecules.

(2) There is *some* orientational order in the direction of the molecules. The long axis of elongated molecules and a line perpendicular to the face of a discoid molecule (called the layer normal) tend to align parallel to a common direction. Note, however, that the orientational order is not complete; the molecules are not all aligned like uncooked spaghetti in a box. The common alignment axis is indicated by the arrow in Figure 10.4 and is called the *director*. The properties of the nematic phase are different parallel and perpendicular to the director.

(3) The director can point in any direction. In practice, the director's direction is determined by external forces, such as alignment layers on the walls of the container or external electric or magnetic fields.

(4) The direction of the director and the opposite direction are equivalent[3]. In the above figure, the director points towards the top of the page. This is equivalent to having drawn the director towards the bottom of the page.

[3]This means physically the same. A nematic whose director is pointed up and has a given degree of order has the same physical properties as a nematic with the same order and the director pointed down. This equivalency is reflected in mathematical models of nematics.

(5) Nematic phases only occur in non-chiral materials. Materials that form nematics must be either achiral or *racemic* mixtures. A racemic mixture consists of exactly 50% right-handed species and 50% left-handed species.

10.4.2 *The cholesteric or twisted nematic phase*

The situation changes when the molecule is intrinsically chiral or when a material called a chiral dopant is added to a nematic. In this case, **the nematic undergoes a helical distortion.** This helical structure corresponds to the minimum energy of a sufficiently large collection of such molecules in the temperature region of interest. This distortion was first observed in liquid crystals containing cholesterol, so, for historical reasons, this structure, which is a unique phase, is called the *cholesteric liquid crystalline phase.* There are many liquid crystals which exhibit a cholesteric phase yet have no connection to cholesterol. For this reason, a better name to use is *chiral nematic or twisted nematic.* **Locally**[4], a cholesteric looks like a nematic. It is only when the structure is studied over large enough distances that the cholesteric phase appears to have a structure different from the nematic phase.

A cartoon of the chiral nematic phase, encompassing only a few planes of the cholesteric, is shown in Fig 10.5.

This figure illustrates several features of the chiral nematic phase.

(1) Only four representative cholesteric planes have been drawn. There are in reality an infinite number of planes. At the same time, these planes are drawn only for our convenience. They have no physical significance.
(2) The director, indicated by the arrows, characterizes the average orientation of the molecules in each plane and **rotates continuously** from pointing to the right to pointing to the left in a distance $L = \pi / |q_o|$. The distance required for the director to point again to the right is twice this distance, and is called the *pitch*, p, of the cholesteric. Naturally, $p = 2L = 2\pi / |q_o|$.
(3) The structure of the cholesteric is periodic along the z-axis. This could also be the x or y-axis. Scientific convention places the periodicity along the z-axis.

[4]By locally is meant on length scales larger than the molecule, but shorter than the distance needed for the nematic structure to twist. This distance is the *pitch* and will be defined shortly.

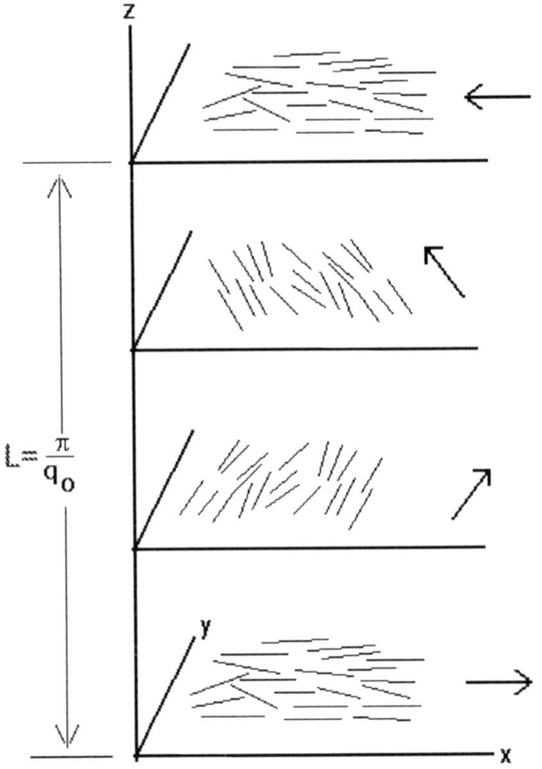

Fig. 10.5 A cartoon of the cholesteric or chiral nematic phase. The director is shown for only a few representative planes.

(4) The director pointing to the right is equivalent to the director pointing to the left. Again, equivalent means the two directions correspond to situations with identical physical properties. The period of this structure is the distance parallel to the periodicity for the physical properties of the phase to repeat. Thus, the period of the chiral nematic is the spatial repeat distance L which is one half the pitch p.

(5) Finally, the z-axis direction is arbitrary. This axis may point in any direction in space, and is usually experimentally determined by surfaces or external fields.

The terms period and pitch may be confusing to the uninitiated. The period is the distance for the physical properties to repeat, while the pitch is the distance needed for the director to repeat. Since, antiparallel directors are equivalent, the period is always one half the pitch in such systems. Since period and pitch both begin with "p" the letter L is used for the period and p for the pitch.

A typical value of L is 300 nm.[5] This is similar to optical wavelengths (400-700 nm), and much larger than a typical molecular dimension of approximately 3 nm. Since L is comparable to optical wavelengths, we might expect spectacular effects when light interacts with a cholesteric.[6] This is indeed the case.

You might have noticed that in our definition of the pitch there are two vertical lines around the quantity q_o. These lines are the symbol for absolute value. The absolute value of a number is always positive and is the magnitude of a number, colloquially, "how large the number is." For example, $|-5| = 5$. The magnitude and sign of q_o both have physical significance. The sign of q_o indicates whether the helix is right- or left-handed. A given material at a given temperature will always produce the same sign or handedness of helicity. The magnitude, or size, of q_o determines the repeat distance, L.

An interesting and useful property of many cholesteric liquid crystals is that q_o is a function of temperature. Thus, more properly, we should write $q_o(T)$. This temperature dependence is why, for instance, the display on a liquid crystal thermometer changes color with temperature. Lastly, we note that it is possible for the sign of $q_o(T)$ to continuously change from positive to negative or vice versa, with varying temperature. In summary, the following points should be noted about chiral nematics. First, the nematic and cholesteric phases are related. The cholesteric is essentially a slowly twisted nematic. At the same time, there are significant differences. A nematic phase, which we will designate by the letter N, can only be formed by achiral molecules or by a racemic mixture of chiral molecules. The cholesteric or twisted nematic phase, designated by Ch, or N* only occurs when the molecules are chiral, different from their mirror images, or when chiral dopants are added to a nematic. These two phases cannot occur in the same material except possibly at a unique temperature at which $q_o(T)$ changes sign.

[5] The length L is temperature and material dependent.

[6] These effects are due to interference of the reflected light and is discussed later in the chapter on liquid crystals and the arts.

10.4.3 *The smectic phases*

The third type of liquid crystalline phases that we shall discuss are the *smectic* phases. The word smectic was coined in the early 20th century because these phases have mechanical properties similar to those observed in soaps. In fact, smectic is derived from the Greek word for soap: $\sigma\mu\eta\gamma\mu\alpha$. All smectics are layered structures with a well-defined layer spacing. Thus, smectic liquid crystalline phases are characterized by long-range positional order in one direction, and some long-range orientational order. In the alphabet soup of liquid crystal phases, the main types of smectic liquid crystals are called smectic A (S_A), smectic C (S_C), twisted smectic C (S_{C*}), and hexatic smectics. We will discuss the first three in detail.

The Smectic A Phase

The smectic A phase is a layered structure and has long-range positional order in one-dimension as well as long-range orientational order. A cartoon of the structure of the smectic A phase is shown in Fig 10.6.

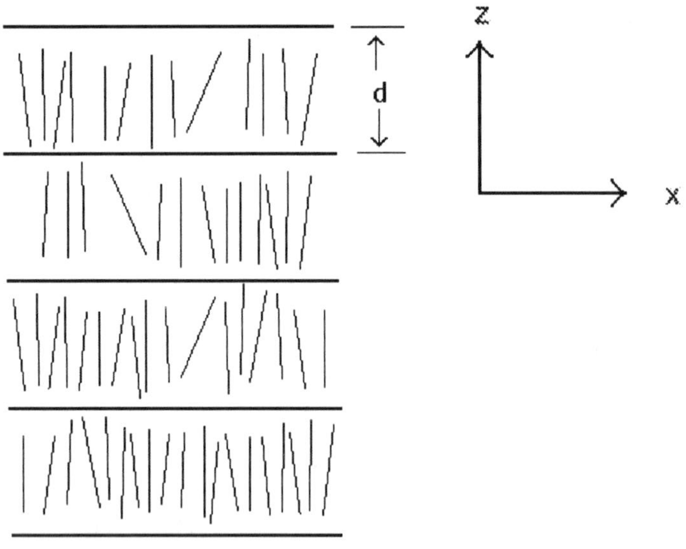

Fig. 10.6 A cartoon of the smectic A phase.

In this figure, the layer spacing is indicated by d. It is important to keep in mind that this figure is a schematic. The layers are not as well defined as in this figure, and the molecules may not be as oriented as shown.

The major characteristics of the smectic A phase are:

(1) The smectic A phase is a layered structure. There is long-range order (technically quasi-long-range, but we will not worry about this point) parallel to the layers. The layer spacing, d, depends on the molecules that form the phase. For thermotropic liquid crystals, d varies from about one to two times the molecular length.
(2) There is **no** long-range order within the layers. The molecules can freely move about within the layers and their positions within the layers are random. A layer may be thought of as a two-dimensional liquid. However, the molecules are not confined to a given layer; they move rather readily between layers.
(3) The molecules are, on the average, oriented perpendicular to the layers.
(4) The system has one unique direction and that is perpendicular to the layers. All directions parallel to or within the layers are equivalent. This means that the system exhibits different optical properties parallel and perpendicular to the layers. This will be the basis of any optical device using a smectic A.
(5) The z direction and the $-z$ direction are equivalent and physically indistinguishable. In practice, the alignment of the layers is determined by surface forces or external electric or magnetic fields.

Many chiral nematics form smectic A phases at lower temperatures. Technically, a smectic A phase formed from chiral molecules is different from a smectic A formed from achiral molecules, but this distinction will not be pursued at this stage.

The Smectic C Phase

The smectic C phase, like the smectic A phase, is a layered structure. It differs from the smectic A phase in that the molecules are, on the average, **not** oriented perpendicular to the layers, but rather at an angle that often depends on the temperature. An idealized cartoon of its structure is shown in Fig. 10.7.

The smectic C phase is characterized by the following properties:

(1) The smectic C phase is a layered structure.
(2) The molecules that form a smectic C phase are either non-chiral or form a racemic mixture.
(3) The molecules are randomly positioned within each layer. There is no long-range positional order within a layer. Each layer may be thought of as a two-dimensional liquid.
(4) The molecules within each layer show long-range orientational order.

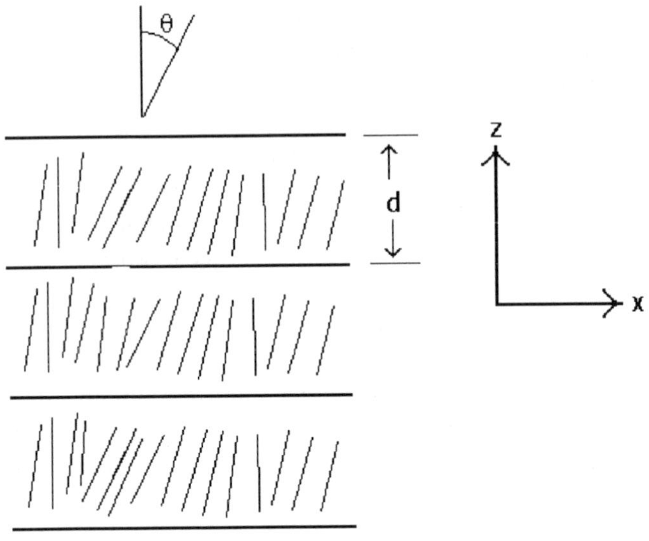

Fig. 10.7 An idealized representation of the smectic C phase.

On the average, the molecules make an angle theta, θ, with respect to a line perpendicular to the layers. (The layer normal.) This angle is called the *tilt angle* and is a function of temperature.

(5) An external field can orient the tilt angle. Surface forces may also determine the layer direction.

The smectic C phase has unique physical properties in three different directions. One of these directions is always parallel to the y-axis (pointing into the page in Fig. 10.7). The other two directions are perpendicular to each other and lie in the $x - z$ plane. These two directions in the $x - z$ plane may be different for different types of external forces, such as electric and magnetic fields. The phenomenon of direction dependent physical properties has practical applications.

The Smectic C* Phase

A very interesting and technologically important change in the smectic C phase occurs when the molecules that form the phase are chiral – a helical distortion is formed. The resulting phase has unique and interesting electric properties. Materials that form this phase have a permanent electric dipole (polarization) perpendicular to the long axis of the molecule. Phases that have a permanent polarization, even in the absence of an external electric field, are said to be *ferroelectric*. Very fast switching liquid crystal displays are made from thin surface stabilized ferroelectric liquid crystals. The structure of the smectic C* phase is illustrated very schematically in Fig 10.8.

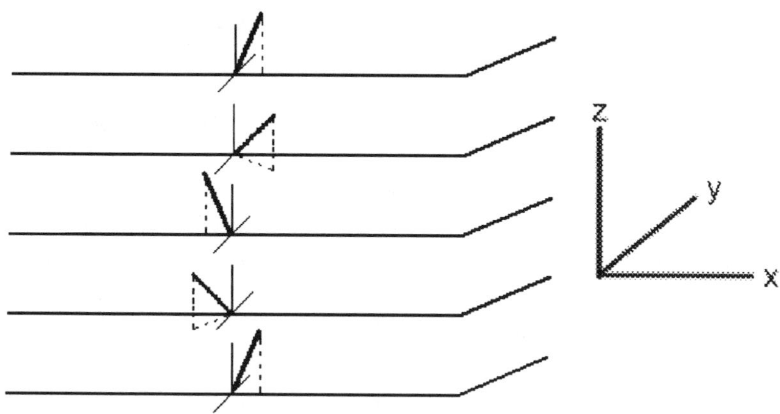

Fig. 10.8 A schematic cartoon of the smectic C* phase.

A smectic C* phase has a layered structure of layer thickness, d, and the molecules (indicated by the short heavy lines) always make an angle θ (theta) with respect to the layer normal. However, the direction of the molecules varies continuously from layer to layer. In the bottom layer, the molecules lie in the plane of the paper and point towards the right. In the second layer from the bottom, they point out of the plane of the paper and slightly to the left. In the middle layer the molecules again lie in the plane of the paper and point to the left. In the second layer from the top, the molecules point into the plane of the paper and to the right. Lastly, in the top layer, the molecules again are in the plane of the paper and point to the right as in the first layer. Thus, in this figure, the pitch, p, is five layer spacings, d ($p = 5d$). In real liquid crystals, the pitch is approximately 1000 layers ($p \approx 1000d$). Interestingly, the pitch need not be an exact multiple of the layer spacing. Thus $p = 999.3d$ may be a physically realizable pitch.

10.5 Polymeric liquid crystals

There is yet another class of materials that combines liquid crystals and polymers. These materials are known as *polymeric liquid crystals* and form when the polymer is synthesized from certain monomers that resemble molecules that form liquid crystals. Such monomers form phases between the isotropic liquid and the crystalline solid, which are liquid crystalline. You should recall that such phases are generically known as *mesophases* and the molecules that form them are called *mesogens.* Thus, you might quite correctly guess that monomers that are rigid and asymmetric, in the shape of rods or disks, are good candidates for polymeric liquid crystals. There are two major types of polymers that form polymeric liquid crystals: *main-chain* and *side-chain polymers.*

Main-chain polymers are composed of rigid mesogenic monomers attached to each other by flexible (polymer) spacers. Some typical structures are shown schematically in Fig. 10.9. In this figure, a rigid rod is indicated by a rectangle with rounded corners and a disk by the disk shaped object. The wiggly lines represent the flexible spacers between the mesogenic units.

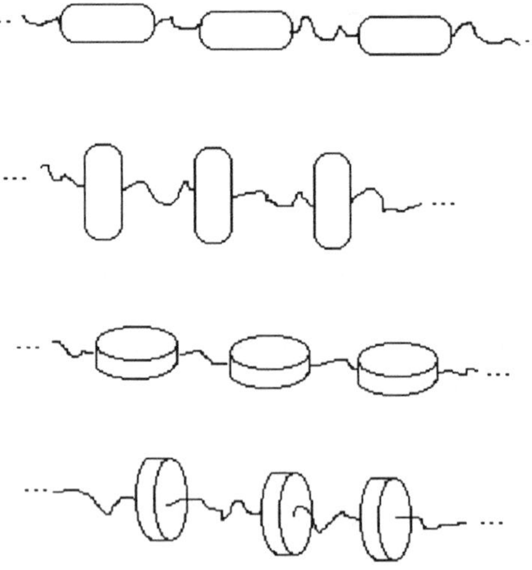

Fig. 10.9 Examples of main-chain polymeric liquid crystals. The rectangles are rod-like structures, the disks, disk like structures and the wiggly lines, flexible polymer spacers.

There are a very large number of possibilities for such polymers. The type of mesogenic units may be changed and the length and type of flexible spacers may be varied. Furthermore, the average degree of polymerization, the regularity or irregularity of the mesogenic units, and their orientation may also be varied. Thus, a wide range of materials may be synthesized and studied.

The second type of polymeric liquid crystal molecules that can form are called *side-chain polymers.* A side-chain polymer has the mesogenic units attached to the sides of the main chain. Once more, there are a wide variety of possible materials that can be synthesized. All of the variables that can be changed for main-chain polymers may also be varied in side chain polymers. A schematic figure of typical side-chain polymers, using the same convention as was used for main chain polymers, is shown in Fig. 10.10.

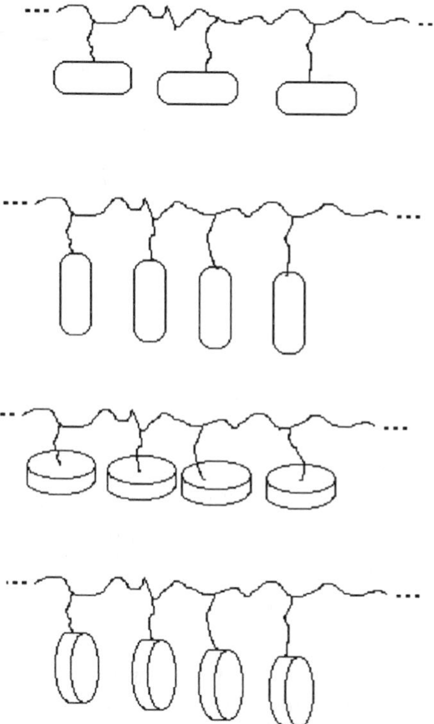

Fig. 10.10 Schematic examples of side-chain liquid crystal forming polymers.

Thermotropic mesophases exist for both types of polymeric liquid crystalline materials. These phases tend to be very stable, have very wide liquid crystal temperature ranges, and typically form polymeric glasses at temperatures below their lowest liquid crystalline phase. Nematic, chiral nematic and smectic behavior has been observed in such materials. In the nematic, phase the rigid mesogens tend to align parallel to a given direction in space. The flexible spacers are completely disordered and exhibit no orientational order. A sketch of the polymeric nematic phase is shown in Fig. 10.11.

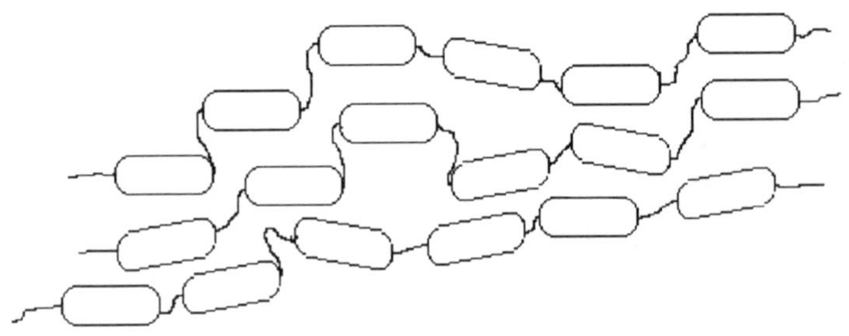

Fig. 10.11 An example of a polymeric nematic phase.

There are, of course, other polymeric nematics with side chains and disk shaped mesogens. These will be explored in the exercises. A smectic phase could have a structure similar to that illustrated in Fig. 10.12.

There are a number of uses for such materials. An important application you are familiar with is super strong fibers. Polymeric liquid crystals can achieve very high orientational order. By starting with an aligned nematic polymeric liquid crystal and using the inherent anisotropy of the solution, very highly oriented fibers may be produced. The order in such fibers is substantially higher than can be achieved by extrusion and orientational stretching. The Young's modulus of these fibers is similar to that of steel. A very common example of this process is Kevlar,^{TM} which is a fiber formed from the liquid crystalline phase of a polyamide. Kevlar has about the same density as nylon, but, pound for pound, is stronger than steel. It is used in bulletproof vests, automobile tires, and other "high demand" products. There are also electro-optical applications of these materials.

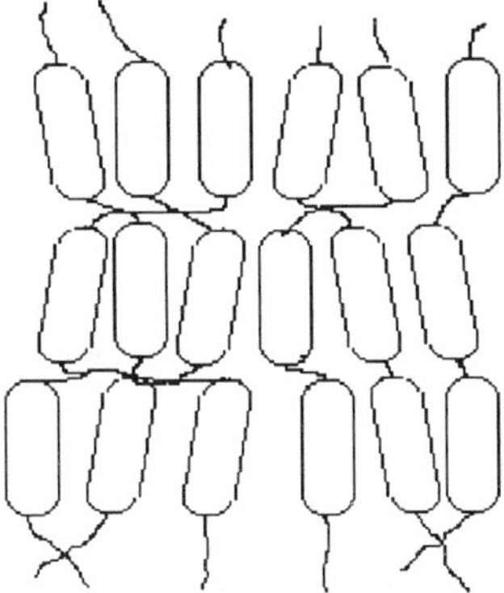

Fig. 10.12 An example of a smectic phase formed by polymeric molecules.

10.6 Lyotropic liquid crystals

There is a class of materials that only exhibit liquid crystalline behavior when in solution with a solvent. While the temperature is an important variable in these systems, the concentration of dissolved material is even more important. When such solutions form liquid crystalline phases the phases are said to be *lyotropic*.[7] Such phases are very common. For example, the goo in a soap dish and the membranes of cells in plants and animals. We discuss further biophysical examples later in the text.

For simplicity in this section, we will consider either binary (two component) systems consisting of dissolved molecules and water, or the slightly more complicated ternary (three component) systems consisting of dissolved molecules in water with an added salt. The solute molecules that form lyotropic phases in these situations are amphiphilic. This means that these are rather special molecules that have a polar (ionic or dipolar) head and one or more hydrocarbon tails. When a sufficient number of such molecules are added to water the following configuration occurs. The heads want to be in the aqueous environment because they will easily dissolve in water, a polar solvent. However, the hydrocarbon section(s) of the

[7]This is once more from the Greek. "Lyo" is from *lyein*-to loosen or dissolve, and tropic which is derived from *tropos*, changing in response to a stimulus. Thus liquid crystalline phases that depend on concentration of dissolved material.

molecules do not want to dissolve[8] in the water. They "prefer" to be segregated from the aqueous environment in a more oil-like environment. This dual behavior – *hydrophilic* (water loving) heads and *hydrophobic* (water fearing) tails on a single molecule causes the molecules to self-assemble into aggregates. The amphiphilic molecules in these aggregates are not chemically bonded to each other so the aggregates can change in size and shape with changes in temperature and other external physical constraints. These aggregates can then order and form lyotropic liquid crystalline phases.

Generally, amphiphilic molecules are classified into two types. The first, called Type I, consists of a head group and one hydrocarbon tail. Most soaps are examples of this kind of amphiphile. The second class, Type II, consists of a polar head and two hydrocarbon tails that generally lie side by side. These two types of amphiphiles are shown schematically in Fig. 10.13 where the head is indicated by a circle and the hydrocarbon tails(s) by zigzag lines.

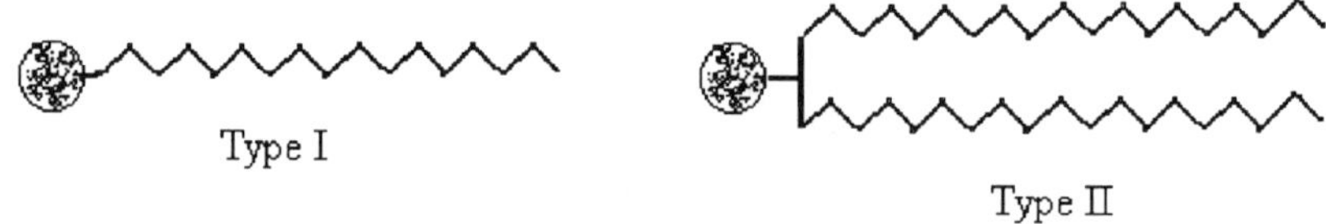

Fig. 10.13 Cartoons of Type I and Type II amphiphiles.

The structures and phases that form when these types of molecules are dissolved in aqueous solutions depend strongly on the concentration. At very low concentrations, the lowest free energy (equilibrium state) is obtained by forming a surface layer of these molecules. This reduces the surface tension of the air-liquid interface. For this reason, these molecules are also called *surfactants*. At a higher concentration, the solution's equilibrium state contains this surface layer, a very few molecules in solution and self-assembled aggregates. Type I amphiphiles generally form *micelles* in which the polar heads form a coating for a core of the hydrocarbon tails. Type II amphiphiles cannot geometrically form micelles, so they tend to form *vesicles*. Vesicles are small essentially spherical aggregates made of bilayers that have water on both the inside and outside. These three

[8]The situation is actually more complex than this. The energy for hydrocarbon and water to mix favors mixing. However, the distortion in the water structure is so great that the total free energy favors demixing. Thus, except for interfacial layers, the materials separate. See for instance, *http://www.princeton.edu/lehmann/BadChemistry.html*.

structures are shown schematically in Fig. 10.14.

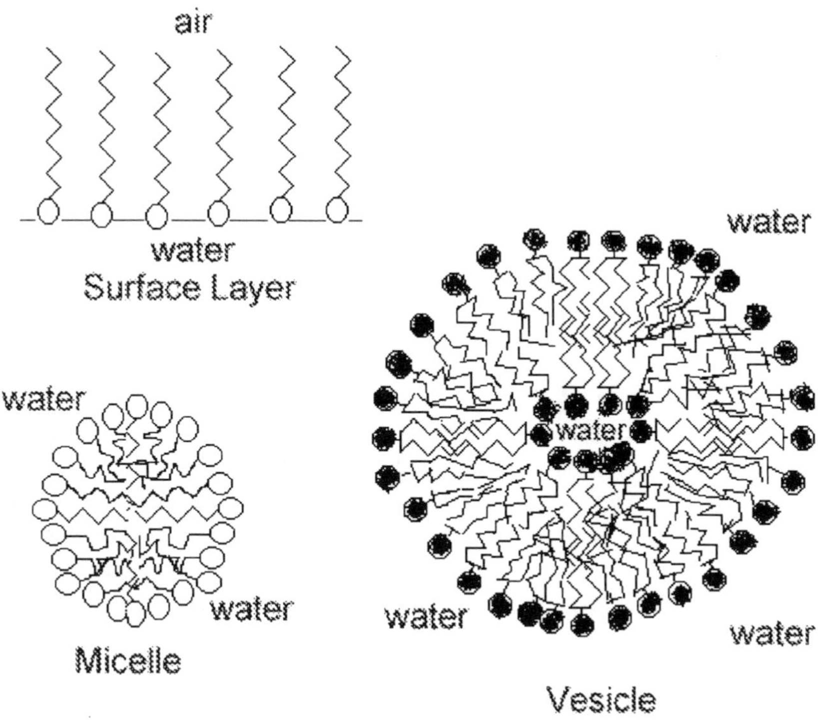

Fig. 10.14 Schematic structure of a surface layer, micelles, and vesicles formed from amphiphilic molecules.

With typical surfactants, and in the absence of additives, the number of molecules in a micelle is roughly 50. The number in a vesicle is even larger.

The phase behavior at higher amphiphile concentration can be rather complex. The key feature of the phase diagrams is that changing the concentration of amphiphile can lead to different liquid crystalline phases. Figure 10.15 shows a representative phase sequence.

This phase diagram represents an amphiphile-water system. Upon the addition of oil, this behavior can become somewhat more complex. The solid can be thought of as a crystalline phase of amphiphilic molecules. In the crystalline phase, the molecules will have long range positional and orientational order.

The lamellar structure in the phase diagram is also sometimes called

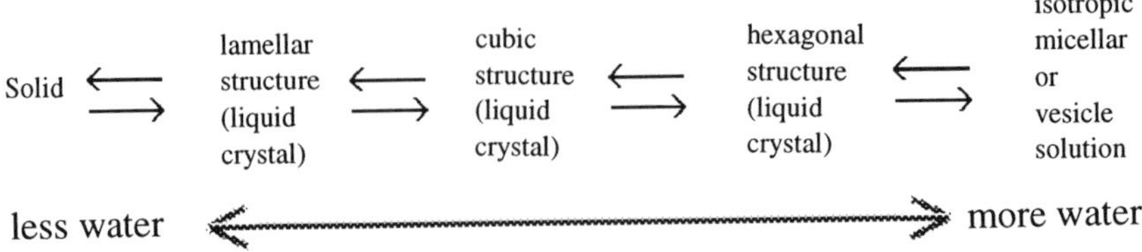

Fig. 10.15 A phase sequence as water is added or removed from a given amount of a generic amphiphile. All of these phases may not occur for a given amphiphile.

the neat soap phase or the lamellar phase, L_D. It is analogous to the smectic A structure found in thermotropic liquid crystals. The layers are very well defined by bilayers of amphiphilic molecules separated by layers of water. The thickness of the bilayers is generally less than twice the length of the amphiphilic molecules, and can depend on temperature and water concentration. The hydrocarbon tails are "melted" and behave in a manner analogous to a liquid hydrocarbon. A schematic figure of the lamellar phase is shown in Fig. 10.16.

At higher water concentrations, an optically isotropic cubic structure can form. In the cubic phase, the amphiphilic molecules tend to form spheres that then order within the aqueous environment to form a cubic pattern. A cross-sectional view of such a phase is shown in Fig. 10.17. The details of the amphiphilic spheres' structure depend on the solution. In an aqueous solution, the heads of the molecules point outward and the tails inward. This phase could be thought of as a cubic arrangement of micelles. In oil or oil-water solutions, inverted micelles with the head groups on the inside of the micelle, may form.

When more water is added to a system that has achieved the cubic structure, a hexagonal phase may form. In this structure, the molecules form cylinders with hydrocarbon tails on the inside and head groups in the aqueous environment. These cylinders then arrange themselves to form a two dimensional hexagonal structure which has a lower free energy than the cubic structure. In this hexagonal structure, the long axes of the cylinders are parallel. A cross-section of this phase looks similar to that observed in the cubic phase. The important difference is that this phase is formed of long parallel cylinders while the cubic phase is formed from spheres.

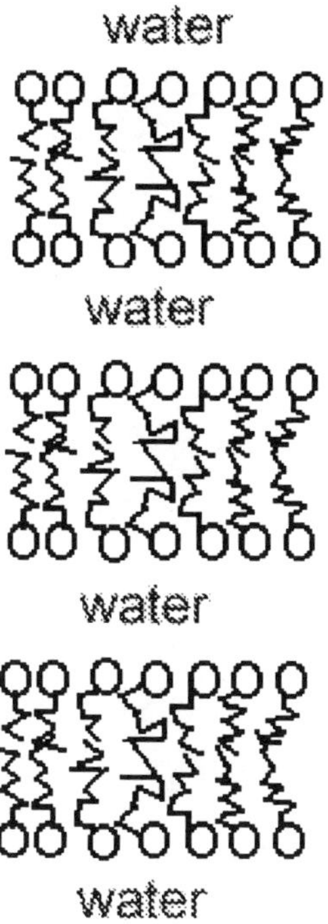

Fig. 10.16 A schematic figure of the lamellar phase.

10.7 Summary

Several new phases of matter[9] have been introduced in this chapter. The alphabet soup of letters labeling these phases can be confusing. In later discussions, we will focus on the nematic, smectic A, and chiral nematic (cholesteric) liquid crystalline phases.

The following figures will assist the reader in remembering the relationships between liquid crystalline phases. Keep in mind that all phases listed in each figure need **not** be present in a given material. As an example, consider a non-tilted thermotropic liquid crystal. Its particular a phase sequence on heating may be Crystal E to Smectic A to Isotropic. The other phases are not observed. Also, some of these phases have not been introduced, particularly the crystalline phases, as well as some of the more

[9]Recall that a phase of matter is a substance that is chemically and physically homogeneous. Thus, all of the examples discussed in this chapter are phases of matter.

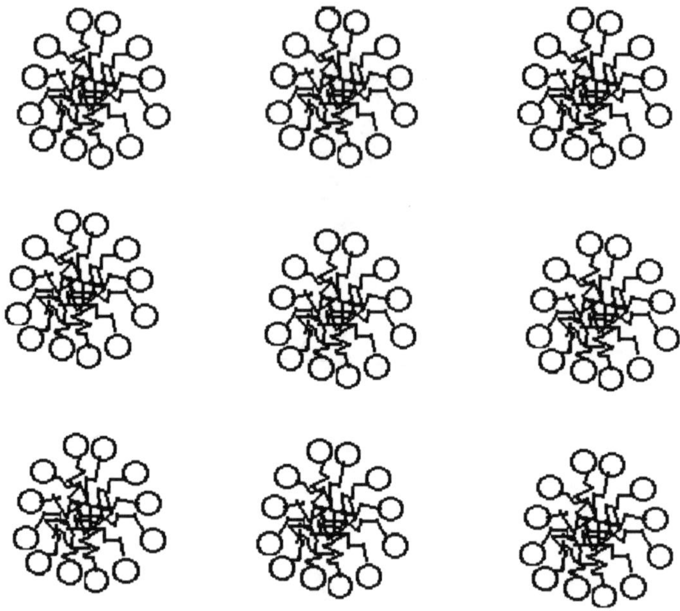

Fig. 10.17 The cubic phase of a lyotropic system.

complex smectic phases such as Hexatic B, Smectic F and Smectic I. The properties of some of these phases are left for the exercises.

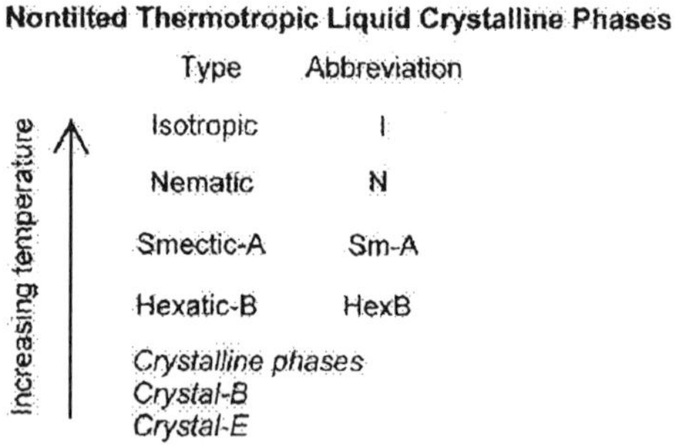

Nontilted Thermotropic Liquid Crystalline Phases

Type	Abbreviation
Isotropic	I
Nematic	N
Smectic-A	Sm-A
Hexatic-B	HexB
Crystalline phases	
Crystal-B	
Crystal-E	

Increasing temperature ↑

Fig. 10.18 Phases of non-tilted thermotropic liquid crystals.

Figure 10.19 summarizes the notation for the various tilted phases. Once more, all phases need not be present.

The notation for chiral molecules is summarized in Fig. 10.20.

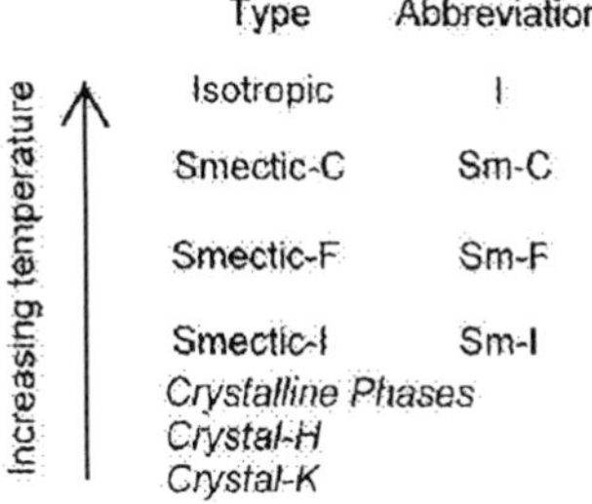

Fig. 10.19 Phases of tilted thermotropic liquid crystals.

Chiral Thermotropic Liquid Crystalline Phases

Increasing temperature ↑

Type	Abbreviation
Isotropic	I
Cholesteric or Chiral Nematic	Ch N*
Smectic-A*	Sm-A*
Smectic-C*	Sm-C*

*Other chiral states indicated by a **
Examples include Hex-I, Cr-J*, Cr-H**

Fig. 10.20 The most general phase sequence for chiral molecules that form liquid crystals.

The lyotropic phases are summarized in part in Fig. 20.21. There are several cubic phases and the notation can vary from source to source.

10.8 References

The literature on liquid crystals is diverse. Good introductions are **Liquid Crystal: Nature's Delicate Phase of Matter**, P. J. Collings, Princeton U. Press, 1990 and **Introduction to Liquid Crystals**, P. J. Collings and M. Hird, Taylor & Francis, Philadelphia, 1997. The canonical reference for researchers is **The Physics of Liquid Crystals**, P. G. de Gennes and J. Prost, Oxford, University Press, New York, 1993. The introductory chapter, ignoring equations, is quite readable. Case Western Reserve University has a web site with a tutorial on liquid crystals and polymers at

Lyotropic Phases formed by Aggregates

Type	Abbreviation
nematic of rods	N_C
nematic of disks	N_D
cubic phases	C( )
columnar-hexagonal phase	Hex
Lamellar phase	L_D or L_α

Fig. 10.21 Summary of lyotropic phases.

molecules in water is **The Hydrophobic Effect: Formation of Micelles and Biological Membranes**, C. Tanford, Wiley, New York, 1980. A particularly good series of books, published by Springer-Verlag, is on partially ordered systems. The volume **Micelles, Membranes, Microemulsions and Monolayers**, edited by W. M. Gelbart, A. Ben-Shaul and D. Roux, has several chapters that have phase diagrams and pictures. Lastly, the Liquid Crystal Institute at Kent State University has a site of liquid crystal educational items at http://olbers.kent.edu/alcomed/dhtml1.html.

10.9 Exercises

(1) Draw a structure for a nematic composed of side-chain polymers with rod-like mesogens.
(2) Draw a structure for a nematic composed of a main-chain polymer with disk-like mesogenic units.
(3) Repeat problems 1 and 2 for smectic phases.
(4) Draw the structure of an isotropic liquid composed of rod-like molecules.
(5) Draw the structure of a crystal composed of rod-like molecules.
(6) Do you expect the lamellar phase to be optically anisotropic? Explain.
(7) The cylinders in a hexagonal phase may be flexible. In what ways is this phase similar to a polymeric nematic?
(8) Sketch the hexagonal phase. Try to indicate that the aggregates are long cylinders.
(9) Sometimes the hydrocarbon chains in lyotropic systems become solid and ordered. Suppose a lamellar phase forms in water and all the hydrocarbon chains are straight and tilted at an angle so they are not

perpendicular to the layers. What thermotropic phase is this lyotropic phase analogous to? Explain.

(10) An interesting problem that combines the material of the previous chapter and this chapter is discussed in Sperling's book (see references in the previous chapter).

From your experience with oils, you know that oils tend to become significantly thinner (less viscous) at high temperatures. For example, butter is a hard solid at 35 °F, a soft solid at room temperature and a liquid at yet higher temperatures. You are also aware that an automobile engine uses oil to reduce friction. This problem explores one way engineers prevent this thinning effect from adversely affecting engine performance.

A typical motor oil carries a designation such as SAE 10W-40. The first number refers to the oil's viscosity at -18°C while the second number is the viscosity at 99°C. The - is a hyphen not a minus sign!

One way engineers reduce the decrease in viscosity with temperature is to add polymers to the oil. In particular, they can add star-like polymers that have many arms (7 to 15) that are block coploymers or homopolymers. These arms may be several different types of polymers. By design, the polymers aggregate at low temperatures and form structures similar to micelles. At higher temperatures, these structures dissolve and entangled free polymers in oil result.

(a) Which motor oil has less variation in viscosity with temperature: SAE 10W-40, or SAE 5W-30?
(b) In variable viscosity motor oils, is the viscosity greater at the higher temperature or the lower temperature?
(c) Draw a cartoon of a a star-like polymer with several different types of polymers for its arms. Include diblock coploymers and homopolymers.
(d) Draw a cartoon of an aggregate of several of these polymers. (This will look a bit like a furry micelle)
(e) Draw a cartoon of the free polymers. This system will be similar to a polymeric liquid except there will be much more room between the polymer strands. What is the space between strands filled with?
(f) Explain why the viscosity of the oil containing free polymers may be larger or at least similar to that of the aggregates in oil at a lower temperature. (Assume, that the viscosity of the pure oil decreases

very rapidly with temperature.)

10.10 Research questions

(1) Find several applications of liquid crystals. Be sure to list references.
(2) Find out about the structure, uses and properties of Kevlar. (Try the Dupont web site.)
(3) Discuss the properties of the hexatic-B phase or another of the alphabet soup of smectic phases.
(4) Study and write a report on why oil and water do not mix. If you find conflicting rationales, discuss this situation and summarize how you would go about finding the "real answer."
(5) Find phase diagrams of lyotropic or thermotropic liquid crystals and discuss them in light of your knowledge of these materials.
(6) Explain how detergents help remove dirt from clothes.

Chapter 11

Semiconductors

11.1 Overview

Almost all common electronic devices use semiconductors. Yet, many readers might not be able to answer the question, "What is a semiconductor?" Similarly, they might not know the most basic science and technology of this important class of materials. In the spirit of trying to build understanding from the basics, this chapter introduces the reader to semiconductors. The basic question, "What is a semiconductor?" will be addressed first. In practice, pure semiconductors are not as technologically useful as "doped semiconductors," semiconductors containing controlled amounts of impurities. Thus, we will ask:

(1) What is a semiconductor?
(2) What is a doped semiconductor?
(3) How are semiconductors doped?
(4) Why are semiconductors doped?
(5) How are semiconductors used to make electronic devices?
(6) What are some electronic devices?
(7) How do semiconductor devices work?

11.2 Introduction

Semiconductors are of great importance to our everyday life. Almost every appliance, electronic entertainment device, and automobile contains semiconductors, in the form of transistors, that are part of integrated circuits that are part of more complex systems. While the details of semiconductor technology are very complex, a basic understanding of what semiconductors are and how they are used is not too complex and is vitally important to our later understanding of the computer.

The macroscopic behavior of semiconductors may be distinguished from metals and insulators in two ways. First, as the name implies, the (electrical) conductivity, or ability to conduct electricity, at room temperature is between that of metals (much higher electrical conductivity) and insulators (much lower conductivity). It is important to realize that this range of conductivity is quite large. Copper, one of the best electrical conductors at room temperature, has a conductivity of nearly $6 * 10^7$ $\text{ohm}^{-1}\text{m}^{-1}$, while a typical insulator has a conductivity of 10^{-8} to 10^{-16} $\text{ohm}^{-1}\text{m}^{-1}$. Semiconductors have "intermediate" conductivities of about 10^{-2} to 10^1 $\text{ohm}^{-1}\text{m}^{-1}$. Secondly, the conductivity of pure semiconductors *increases* very strongly with increasing temperature. This behavior is in contrast to metals whose conductivity decreases with increasing temperature.

These observations may be understood using the band structure of solids that we discussed earlier. A metal (conductor) has partially filled bands, which allow electrons to move freely under the application of an external electric field; hence, metals have large conductivities. An insulator has a full band separated from the next band by a fairly large energy gap. Thus there are very few (ideally zero) electrons available in the conduction band to move in response to an external electric field; hence insulators have very low conductivity.

A semiconductor is mid-way between these two materials. It has a nearly filled valence band and a nearly empty conduction band. The gap between these two bands is small enough that electrons can be thermally excited from the valence band to the conduction band. An increase in temperature leads to a greater number of electrons becoming excited into the conduction band. Interestingly, the electrons and the "holes" that were left behind both contribute to the electric current. This explains why a semiconductor's conductivity is between that of conductors and insulators and that it increases with increasing temperature. These semiconductor properties change when even small amounts of impurities are added to a pure semiconductor.

11.3 Intrinsic semiconductors

The two elemental semiconductors of greatest importance to the electronic industry are Silicon, Si and Germanium, Ge. There are also mixed semiconductors such as Gallium-Arsenide, GaAs that are becoming more important. We will focus primarily on elemental semiconductors. Both Ge and Si are located in the fourth column, Group 14, of the periodic table and have four valence electrons. They form covalently bonded crystals with a tetrahedral structure in which each atom has four nearest neighbors. A two dimensional representation of this structure is shown in Fig. 11.1.

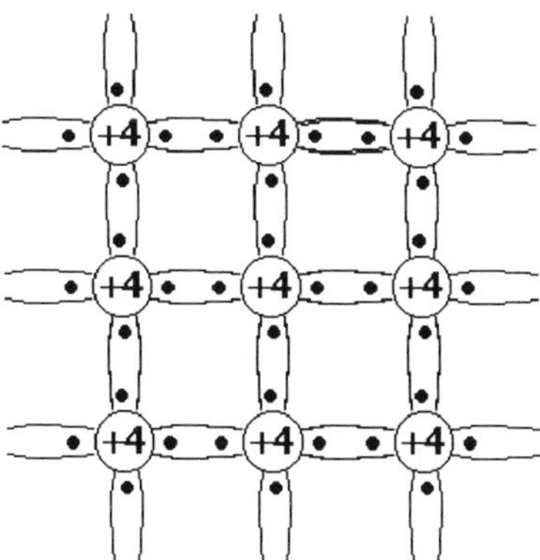

Fig. 11.1 A two dimensional representation of a tetrahedrally bonded semiconductor.

In this figure, the valence electrons are indicated by the solid black dots, the covalent bonds by the curved lines, and the central charged core by the open circle with the +4. The +4 indicates that the atomic core has a net charge of $4e$. This figure describes a semiconductor very close to absolute zero. There are no free carriers, because no thermally generated charge carriers exist and the semiconductor behaves like an insulator.

At finite temperatures the situation is somewhat different. The energy gap between the valence band and the conduction band is the energy required to break one covalent bond. This is the same as the energy required to "knock" an electron free. This energy corresponds to about 0.7 eV (electron volt) [1] for Silicon, for Germanium, it is about 1.1 eV, and for Gallium-Arsenide it is about 1.5 eV. At all temperatures above absolute zero, some covalent bonds will be broken. At room temperature, roughly

[1] An *electron volt* is a unit of energy. It corresponds to the energy gained by an electron moving through a voltage (potential difference) of 1 volt. 1 eV=$1.6 * 10^{-19}$ J. The electron volt is a very natural unit of energy when discussing atoms, molecules, and solids.

300 K, there is enough thermal energy that in silicon, a few atoms per 10^{12} have a broken bond. When a covalent bond is broken, an electron is free to move and a vacancy or *hole* is left behind, so that overall the semiconductor is still electrically neutral. This situation is shown in Fig. 11.2.

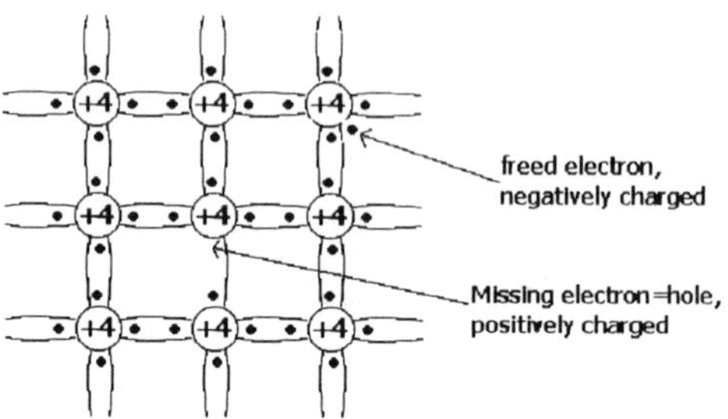

Fig. 11.2 A representation of a pure semiconductor at room temperature.

The region where the vacancy is created (a hole) has a net positive charge, while the region with the freed electron has a net negative charge. Both holes and electrons can move from atom to atom. Consequently, as stated earlier, both electrons and holes contribute to the electrical conduction. Holes and electrons move independently and have different *mobilities.* The conductivity, σ, of such an *intrinsic*[2] semiconductor can be expressed as $\sigma = n_e(\mu_h + \mu_e)$, where n_e[3] is the number of electron-hole pairs per unit volume, μ_h[4] is the mobility of the holes, and μ_e is the mobility of the electrons.

There are about $3.3 * 10^{12}$ silicon atoms per electron-hole pair at room temperature, and approximately $5 * 10^{22}$ silicon atoms per cm^3. Hence, at room temperature, there are approximately $1.5 * 10^{10}$ electron-hole pairs per cm^3 of silicon.[5] Thus, while there is only one electron-hole pair per $3.3 * 10^{12}$ silicon atoms, there are still, in an absolute sense, many charge carriers per cm^3. However, this number is rather small compared to copper, which has approximately $8.4 * 10^{22}$ free electrons per cm^3.

[2] An intrinsic semiconductor is a pure semiconductor. All charge carriers are intrinsic to the material; they are not added.

[3] n_e is sometimes replaced by the intrinsic number of electrons, n_i which equals the intrinsic number of holes, n_p. Mathematically, $n_i = n_p = n$.

[4] The mobility, μ (mu) of a charge carrier is the ratio of the carrier's drift velocity to the applied electric field. The drift velocity is the average velocity at which a carrier moves in response to an applied electric field. The mobility need not be the same for electrons and holes in the same material, and these mobilities are material dependent.

[5] This is n_i, the intrinsic number of electrons and holes per cm^3.

Generally, intrinsic semiconductors, such as just described, are not of great technological interest. One exception, the *thermistor*, is a temperature-sensitive device that is used in temperature control and measurement. It utilizes the fact that the number of electron-hole pairs, n, is a strong function of temperature. It increases about 5% per °C at room temperature. The most important applications employ semiconductors that are made extremely pure and then have their electrical properties very precisely modified by adding controlled, very small amounts of impurities. The process of adding controlled amounts of impurities is called *doping*. Transistors and, by extension, all integrated circuits use doped semiconductors.

11.4 Doped semiconductors

The normal dopants for silicon and germanium are pentavalent (5 valence electrons) and trivalent (3 valence electrons) elements which are in columns of the periodic table adjacent to silicon and germanium. The following table reproduces part of this area of the periodic table.

Column IIIB, Group 13	Column IVB, Group 14	Column VB, Group 15
3 valence electrons	4 valence electrons	5 valence electrons
Z=5, Boron, B	Z=6, Carbon, C	Z=7, Nitrogen, N
Z=13, Aluminum, Al	Z=14, Silicon, Si	Z=15, Phosphorus, P
Z=31, Gallium, Ga	Z=32, Germanium, Ge	Z=33, Arsenic, As

To illustrate how doping works, assume one adds a controlled amount of a pentavalent element to otherwise pure silicon. Only four of the five valence electrons of this element are needed for chemical bonding. At room temperature, the extra electron becomes free and goes into the conduction band and becomes a conduction electron. The dopant has donated an electron; therefore, this dopant is called a donor. Since the doped semiconductor has negatively charged charge carriers, it is called an *n-type* semiconductor. Its structure is shown schematically in Fig. 11.3. Once more, the material is electrically neutral overall.

Similarly, consider doping a semiconductor with a trivalent element. When the three valence electrons of this element are incorporated, there are too few electrons to form all the covalent bonds, resulting in an unfilled bond. This unfilled bond is a positively charged hole that can move through

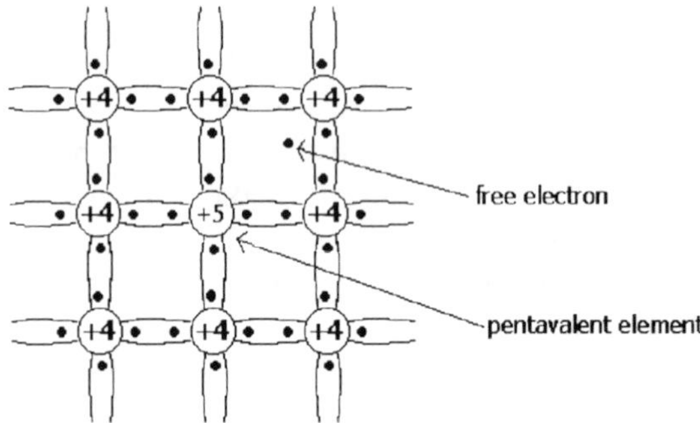

Fig. 11.3 Representation of doping with a pentavalent atom. A pentavalent atom donates an extra electron and is called a donor.

the crystal and conduct electricity. A trivalent dopant is called an *acceptor* because each dopant atom creates a deficiency that will accept one electron. Since the doped semiconductor has positive charge carriers, it is called a *p-type* semiconductor. A two-dimensional schematic of this structure is shown in Fig. 11.4.

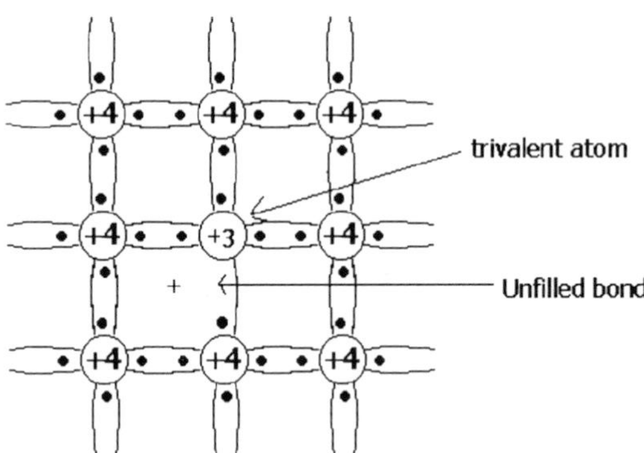

Fig. 11.4 Representation of doping with a trivalent atom.

It is important to realize that the electrons and holes, besides moving around within the semiconductor, can also recombine. When they do so, that particular pair is gone but a new pair is thermally generated. This generation occurs because the product of the number of electrons times the number of holes must remain constant. Thus, there is a thermally determined *equilibrium number* of electrons and holes.

Silicon is much more commonly used than germanium, and so the discussion will now focus on silicon. Suppose the silicon is n-doped creating an excess of electrons. Since there are many, many extra electrons, there is a high probability that the electrons will recombine with holes, automatically reducing the number of holes. Similarly, introducing extra holes reduces the number of electrons. Nevertheless, and maybe somewhat surprisingly, the addition of a dopant increases the conductivity of silicon. Interestingly, *n-type* and *p-type* dopants can cancel each other out leading to a doped semiconductor that has the same conductivity as the intrinsic semiconductor.

The relationship between the number of electrons and the number of holes is simple. It has been found experimentally that

$$np = n_i^2,$$

where n is the number of electrons per unit volume, p is the number of holes per unit volume and n_i is the number of electron-hole pairs in the pure semiconductor under the same conditions. This last quantity depends on temperature. For silicon at room temperature, $n_i \approx 1.5 * 10^{10}$ cm^{-3}. Furthermore, for *n-type* materials at room temperature, the number of electrons per unit volume, n_n[6] is given by

$$n_n \approx N_d.$$

Here, N_d is the number of donor atoms per unit volume. In *p-type* semiconductors the number of holes per unit volume, p_p is given by

$$p_p \approx N_a.$$

Here, N_a is the number of acceptors per unit volume.

Example 11.1

Just to give you some numbers, suppose one Si atom in 10^7 being replaced by In (indium). What type of semiconductor is formed? What are the number of electrons and the number of holes at room temperature? Since indium is trivalent, a *p-type* semiconductor is formed. The number of acceptors is the fraction of silicon atoms replaced by indium atoms times

[6]The notation may be getting confusing. The first n represents the number of electrons. The subscript n means that it is an *n-type* semiconductor.

the number of silicon atoms per cubic centimeter. If one atom in 10^7 is replaced, then the fraction of atoms replaced is $1/10^7 = 10^{-7}$. The number of silicon atoms per cm^3, $5 * 10^{22}$, was given earlier. Thus, the number of acceptors per unit volume is

$$p_p = 10^{-7} * 5 * 10^{22}\ \text{cm}^{-3} = 5 * 10^{15}\ \text{cm}^{-3}.$$

This is the number of holes per unit volume, and the subscript p reminds you that it is a *p-type* semiconductor.

The equilibrium number density of electron-hole pairs in pure silicon at room temperature is $n_i \approx 1.5 * 10^{10}\ \text{cm}^{-3}$. The number of electrons per unit volume is obtained from the equation, $np = n_i^2$, where in this case p is p_p. Thus,

$$n_p = \frac{n_i^2}{p_p} \approx \frac{2.25 * 10^{20}\ \text{cm}^{-6}}{5 * 10^{15}\ \text{cm}^{-3}} \approx 4.5 * 10^4 \text{cm}^{-3}.$$

Note that the electron concentration has been greatly reduced and the hole concentration greatly increased from that of pure silicon.

11.5 Technological applications of doped semiconductors

In technical applications, the two types of doped semiconductors are introduced in controlled amounts into adjacent parts of the same single crystal of substrate material. The details of the manufacturing process are well developed and described in other texts and will not be discussed here. We will however, explain the *pn junction* and briefly discuss two types of transistors.

To create a *pn junction*, you need to have a section of *n-type* material adjacent to a section of *p-type* material. The narrow region in which the semiconductor changes from *n-type* to *p-type* is the actual *pn junction*. At first, this might not seem like a very useful device, as the majority charge carriers in the p material are holes and in the n material they're electrons. However, the concentration differences tend to level out due to diffusion.[7] In a material containing *p-type* and *n-type* materials, the electrons and holes will **try** to diffuse and obtain equal concentrations. However, they

[7]Diffusion is a common physical mechanism for concentrations to become uniform. For example, suppose a sugar cube is very carefully placed, so as to not stir up the water into an unstired glass of water. The sugar will eventually dissolve and then move about or diffuse to fill the whole glass with sugar water.

cannot obtain completely equal concentrations because they are charged and like charges repel. Nevertheless the electrons and holes will diffuse and recombine near the junction creating a "depletion region" where there are not quite enough mobile electrons in the *n-type material*, and there is a slight deficit in the number of mobile holes on the *p-type* side. This equilibrium of separated charges results in electrostatic potential hills for the holes and the electrons. These potential hills are mirror images of one another and are illustrated in Fig. 11.5.

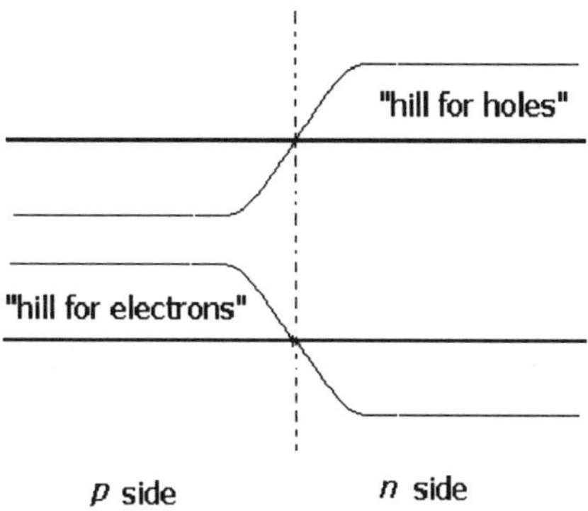

Fig. 11.5 The potential hills for n and *p-type* materials near a *pn junction.*

Suppose a voltage is applied to this junction. Experimentally, it is found that such a junction will allow large currents to pass only in one direction. This is called rectification, and the resulting electronic device is called a *diode.* To understand how this occurs, consider what happens when an electric battery is connected across a *pn junction.* Recall that a battery has two terminals, one marked positive and one marked negative. We will represent a battery by the symbol shown in Fig. 11.6.

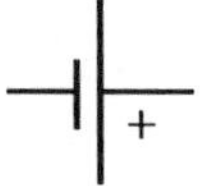

Fig. 11.6 The symbol for a battery.

The longer vertical line represents the positive voltage terminal, also marked with the plus, while the smaller vertical line represents the negative terminal. The horizontal lines connected to these vertical lines represent conducting wires. Since the battery has two terminals, one may connect it to the *pn junction* in two different ways.

First, consider the situation shown in Fig. 11.7, which is called *forward biasing.*

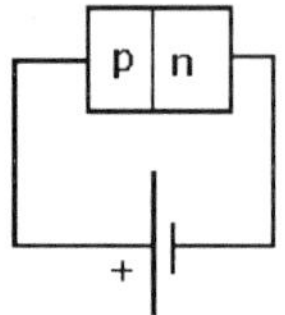

Fig. 11.7 A forward biased *pn junction.*

When the battery is applied as shown there is an electric force "pushing" the holes to the right,[8] and pulling the electrons in the opposite direction. This is the same as lowering the potential hills for both the holes and electrons. This makes it easier for current (charge carriers in the form of electrons and holes) to flow through the diode. This situation is illustrated in the Fig. 11.8.

Notice that the potential "hills" are lower. This means that it is easier for the carriers to cross from one side to the other. The larger the voltage the smaller the hill.

Now suppose that the battery is disconnected and then reconnected in the opposite manner, as shown in Fig. 11.9.

Both the holes and the electrons are now pulled away from the junction by electrostatic forces. In this "reverse biased" state, very little current flows through the diode. The current in the reversed bias region is limited by the relatively small number of charge carriers that can diffuse across the *pn junction.* The effect of the battery is to make the potential hills for electrons and holes even higher than they are when no voltage is applied.

The current voltage characteristics of a *pn junction diode* are illustrated in Fig. 11.10. Notice that there is a very small current for a reverse biased diode and large currents with modest voltage drops for a forward biased

[8]This follows because the positive terminal may be thought of as having positive charges, and positively charged holes are pushed away from positive charges, while electrons, which are negatively charged, are attracted to these positive charges.

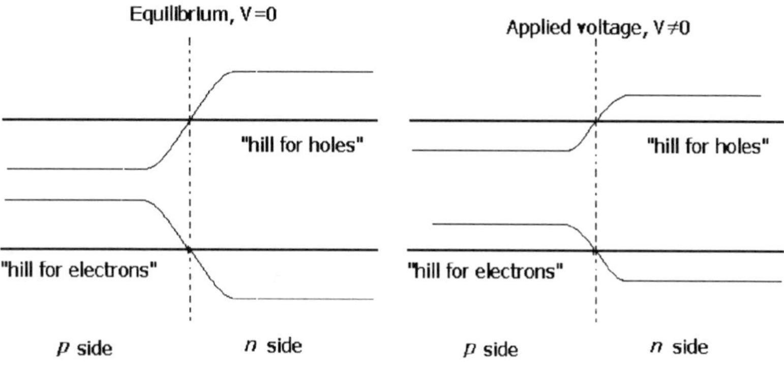

Fig. 11.8 Forward biasing lowers the potential hills and makes movement of electrons and holes easier.

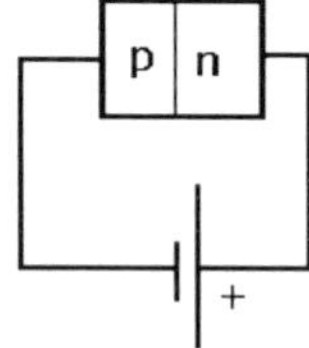

Fig. 11.9 A reversed biased *pn junction*. The potential hills are even higher than when no voltage was applied across the junction.

diode. To give you some idea of the numbers, a potential difference of 1 volt is less than the potential drop between the positive and negative terminal of a "D cell" flashlight battery, and 1 ampere is larger than the current that flows through a 100 watt incandescent light bulb that is connected to the standard 120 V outlet.

While semiconductor diodes are of great use, the real semiconductor "miracle" is the *transistor*. There are a number of types of transistors. Our discussion will be limited to *bipolar junction transistors* (BJTs) and *field effect transistors* (FETs). Even the bipolar junction transistor comes in two types. We will discuss the *NPN* transistor and skip the *PNP* transistor. Bipolar junction transistors are fabricated by making two *pn junctions* in

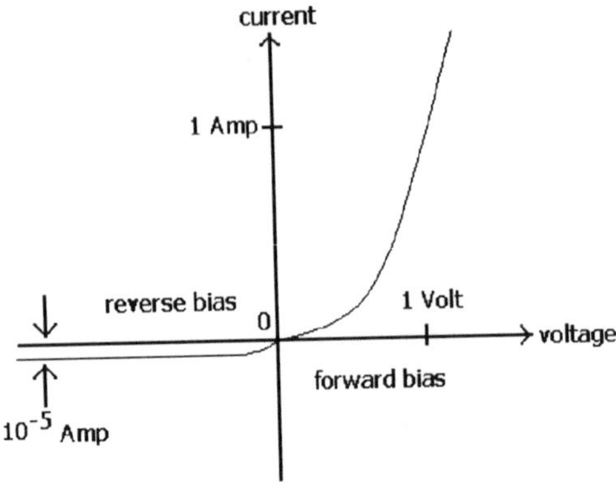

Fig. 11.10 Current-voltage characteristics for a *pn junction*. Observe that large currents (0.1 to 1.0 ampere) exist only for forward biasing.

close proximity. In an *NPN* transistor, a very thin section of *p-type* material separates two sections of *n-type* material, and all three sections are on the same crystal. A thin slice through such a device, including the "batteries" necessary to operate the device, is shown in Fig. 11.11.

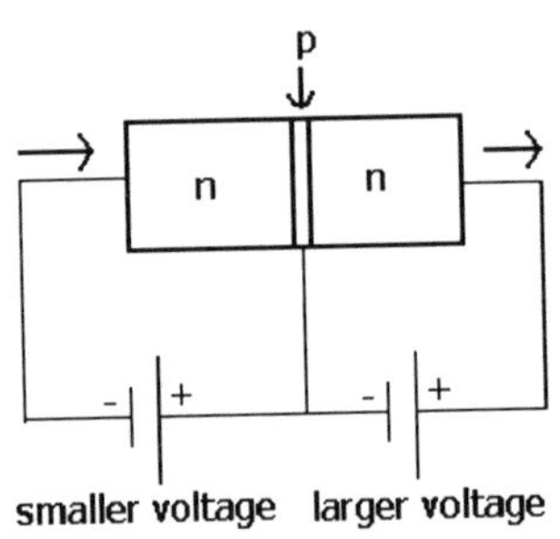

Fig. 11.11 An *NPN* transistor, showing the voltages, represented by batteries, needed for proper operation.

The **electron current** into and out of the two *n-type* materials is also shown by the arrows.[9] First, consider the left *pn junction*. This junction

[9]Conventional current, the current in everyday discussions, assumes that the moving charges are positive. Thus, conventional current flows in the opposite direction to

is forward biased. (Compare to the diode we just discussed.) The effect of the smaller voltage is to inject electrons into the p material where they are the minority carriers. Thus, it is easy for current to flow from the left *n-type* semiconductor into to the *p-type* semiconductor. The larger voltage from the battery on the right reverse biases the second *pn junction.* One might expect this to have the effect of leaving a small current. However, the smallness of the current in a normal reverse biased diode is due to the small number of electrons. In a *NPN* transistor the thinness of the p layer allows essentially all of the electrons injected into the p material to pass through the junction. Electrons that are in the right-hand *n-type* material are then pulled by the second battery through the semiconductor. The net result is that the output current on the right is essentially the same as the input current on the left.

Why go to all this trouble when the current into the *NPN* sandwich nearly equals the current out? The answer is found by looking very carefully at this circuit. (This will be left for an exercise.) Engineers have found ways to use this device for both current and voltage amplification. The basic idea is that the current into the p material, which was not shown above, can be used to control the much larger current flowing through the device as a whole. This allows current amplification and most importantly for the application to digital computers the performance of switching.

Another type of transistor is the field effect transistor or *FET*. These transistors are used extensively in digital electronics as switches. These devices operate somewhat more simply than bipolar junction transistors and are based on the ability of an external voltage to control the width of a conducting channel and hence the current flowing through the device. We will consider the simplest of a large family of field effect transistors, the *junction field-effect transistor*, or *JFET*. A crude drawing of a *JFET* is shown in Fig. 11.12.

This device is made of both *p-type* and *n-type* semiconductors on a single crystal and consists of three terminals called the **source**, the **drain** and the **gate**. Electrons flow through a channel of width w from the source to the drain. The quantity of electron flow from the source to the drain is controlled by the width of the conducting channel. The width of the conducting channel is determined by the voltage applied between the gate and the source. In fact, when the applied voltage is large enough, the width of the channel is zero and no current is conducted from the drain to the

electron current. Interestingly, this bizarre state of affairs is the result of the historical accident of assuming the charge carriers were positive.

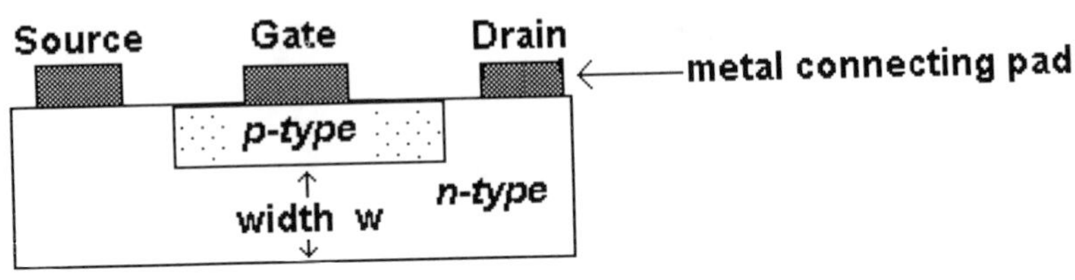

Fig. 11.12 A cross-sectional model of a *JFET*.

source.

Controlling the electron flow in the conducting channel is analogous to controlling the flow of water through a garden hose that is being constricted. The inlet to the hose is analogous to the source, and the outlet to the drain. The amount of constriction of the hose, determined by folding the hose and squeezing, varies the amount of water going through it. The amount of constricting force is analogous to the voltage applied between the gate and source.

The switching of voltages between two desired states, on and off or 1 and 0, is the fundamental process needed for operation of electronic digital computers. The circuits needed to perform switching between high and low voltages and for storing data use transistors, diodes and other simpler electronic components such as capacitors and resistors. (The Capacitor was discussed earlier in the chapter on forces and fields. A resistor is a device that can control the current that a given potential difference (voltage) can generate. The relationship between voltage, current and resistance is called Ohm's law and is given by $I = V/R$. It is not necessary that you understand the details of "electronics" to go to the next step of investigating digital logic and computers. However, now you know the basics of how some of the semiconductor components in electronic instruments function.

11.6 References

The information in this chapter is available in a number of books, and I have freely borrowed from several. The book, **The Nature of Solids**, Alan Holden, Dover, 1992, is written in the same spirit as the present book and has very good explanations of solids and, of course, semiconductors. The more mathematical part of this discussion parallels that in **Circuits,**

Devices, and Systems, R. J. Smith and R. C. Dorf, Wiley, 1992. Many solid state physics texts discuss semiconductors; one of the best known is **Introduction to Solid State Physics**, by C. Kittel, Wiley.

11.7 Exercises

(1) The conductivity of an intrinsic semiconductor decreases by 4% per degree Celsius. The conductivity of a particular piece of material is 1 ohm^{-1}m^{-1} at 20°C. Thus at 21°C its conductivity is $0.96 * 1 = 0.96$ ohm^{-1}m^{-1}, at 22°C, the resistance is (0.96 ohm^{-1}m^{-1})*0.96 $= 0.922$ ohm^{-1}m^{-1}. (The first 0.96 is the resistance at 21°C, and the second 0.96 represents the 4% that occurred between 21 and 22°C.) Make a table of conductivity and temperature for temperatures between 20 and 50 °C. Then graph your data. Describe the graph. Is this analogous to a situation in finance? If yes, what?

(2) A silicon crystal is doped with 10^{13} gallium atoms per cm^3. The chemical properties of the silicon are not significantly altered by by doping with gallium, but its electrical properties are greatly changed. Justify and explain this statement.

(3) 99% pure means that 99% of the material is the desired material and 1% is impurities. Very high purity for most materials is "5 9s" or 99.999% pure. This means 1 part in 10^5 is an impurity. Suppose I have 5 9s silicon, with aluminum as the impurity. Are the electrical properties of this silicon the same as "pure" silicon? Explain your answer. This exercise illustrates the level of purity that has been achieved in the production of silicon.

(4) Copper is not the only good metallic conductor at room temperature. Find the electrical conductivity of two other materials that are as good as or better than copper at room temperature. Do either of the materials you find have an advantage over the other when used as connecting wires between silicon wafers and the outside connectors? Explain.

(5) Consider three samples: a pure metal, a pure semiconductor, and a doped semiconductor, all at room temperature. The temperature is then increased by 10% on the absolute (Kelvin) scale. Which of the following most closely approximates the change in the conductivity of each sample?

 (a) The conductivity is about the same.
 (b) The conductivity increases by about 10%.

(c) The conductivity decreases by about 10%.
(d) The conductivity increases by more than 10%.
(e) The conductivity decreases by more than 10%.
(f) Explain your reasoning.

(6) A crystal of pure silicon is doped by adding 1 atom of aluminum for every 10^5 atoms of silicon.

(a) Is the doped semiconductor *n-type* or *p-type*? Explain.
(b) How many impurity atoms are there per cubic centimeter?
(c) How many minority carriers of opposite charge are there per cubic centimeter?

(7) A diode consists of a left-hand side that is doped with 10^{14} arsenic atoms/cm^3, and a right-hand side that is doped with $4 * 10^{14}$ boron atoms/cm^3. Sketch the diode, indicating the junction, the type of semiconductor in each part, and the number and type of all of the charge carriers in each part. Ignore the diffusion near the junction.

(8) In most junction transistors, the two regions that are of the same type are not doped at the same level. A transistor has one part doped with 10^{14} phosphorus atoms/cm^3. The other matching region is doped by replacing one silicon atom in 10^7 with antimony. What is the density of charge carriers in each of these parts of the transistor? Are these two regions *n-type* or *p-type*?

(9) This exercise explains how a resistor can be made by doping a pure semiconductor.
The resistance, R of a piece of material of uniform cross sectional area, A and length, l is given by the expression: $R = l/(\sigma A)$, where σ is the conductivity of the material. In the text we stated that $\sigma = e\,(n_h \mu_h + n_e \mu_e)$, where n_h is the number density of holes and n_e is the number density of electrons. Assume that there is a rectangular channel of doped semiconductor of length 100 μm and cross sectional area 0.1 μm^2. Also assume that there are so many holes that you can ignore the number of electrons.

(a) To what does the expression for the conductivity reduce?
(b) The mobility of holes in silicon is 0.048 m^2/(V$*$s). Following the example in the text, assume that there are $5 * 10^{21}$ holes per m^3 and calculate the conductivity.
(c) Use the size of the channel and the fact that 1 μm $= 10^{-6}$m to find the resistance of this resistor.

(d) Describe how and why this resistor works.

(10) This exercise further explores using *BJTs* for switches in a very simplified way. In practice, the BJT discussed earlier does not have equal doping in the two n regions. For this reason, one n region is called the emitter and the other the collector. The central p region is called the base. **In all useful configurations, the collector-base junction is reverse biased and the emitter-base junction is forward biased, as described in the text.**

(a) The statement in the text that the input current is essentially the same as the output current can be quantified by $I_C = \alpha I_E$, where I_C is the collector current, I_E is the emitter current, and α is a parameter describing the transistor and has a value less than, but approximately equal to, one; ($\alpha \leq 1$). The relationship between between the base current I_B and the collector current I_C is $I_C = \alpha I_B/(1-\alpha) = \beta I_B$. Calculate β for $\alpha = 0.98$. What happens if $\alpha = 1$?

(b) Consider the following circuit, where the symbol for an *NPN* transistor and a resistor have been introduced, and the currents are as shown. We will assume an idealization of a real transistor with $\alpha = 0.98$. The triangle like symbol represents ground and provides a fiducial point for measuring potential differences. In our simplified model, by Ohm's law and conservation of energy, $V_{out} = 5V - 1000 * I_C$.
Graph V_{out} vs I_C for $0 < I_C < 5 * 10^{-3}$ amperes.

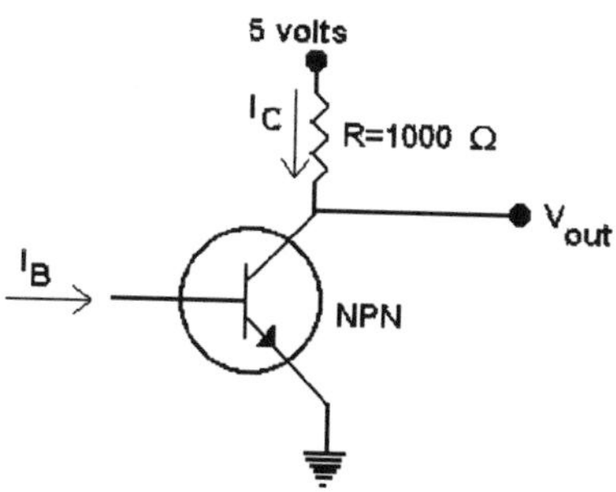

(c) For $I_B = 0$, calculate I_C and V_{out}.

(d) For $I_B = 10^{-4}$ amperes, calculate I_C and V_{out}.

(e) Explain how this "switch" works. In practice a voltage would be applied to a resistor that is connected to the base, and the voltage between the collector and emitter and other complications could not be ignored.

(11) The previous exercise investigated a *BJT* switch. In computers, field effect transistors are more frequently used as switches than *BJTs*. This exercise illustrates how an ideal *FET* switch operates. The metal-oxide-semiconductor *FET* or *MOSFET* is different than the *JFET* discussed earlier. The *n-channel enhancement-mode MOSFET* has the highest performance. At its simplest, this device is a voltage controlled switch. The current between the drain and source, I_{DS}, is controlled by the voltage between the gate and the source, (V_{GS}).

(a) In one such device the the "transfer characteristics" relating I_{DS} to V_{GS} is given by the expression

$$I_{DS} = K * (V_{GS} - V_T)$$

where K is a device parameter, V_T is the turn-on voltage, and $I_{DS} = 0$ for $V < V_T$. For a given device, $V_T = 2V$ and $I_{DS} = 3$ mA for $V_{GS} = 4$ Volts. Find K.

(b) The following circuit shows the symbol for a MOSFET that is configured as a switch. Find V_{out} for $V_{GS} = 2V$, $V_{out} = 5V - 1000 * I_{DS}$.

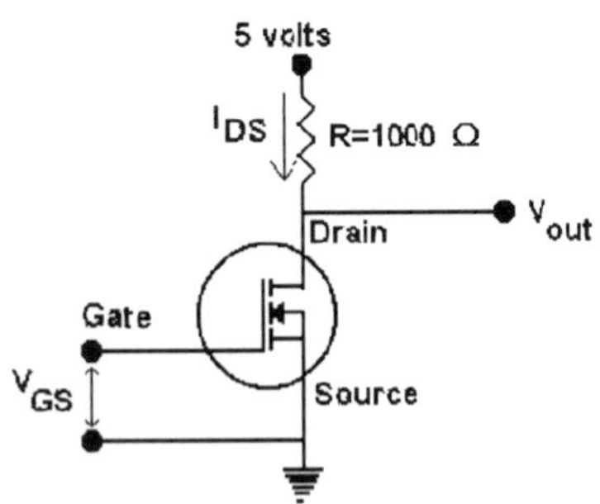

(c) Find V_{out} for $V_{GS} = 5V$.

(d) Find I_{DS} and V_{GS} for $V_{out} = 0$.

(e) Explain how this switch works.

11.8 Research questions

(1) What is diffusion? In what sense does it level concentrations out?
(2) Discuss how semiconductor integrated circuits are fabricated.

Chapter 12

The Molecules of Life

12.1 Overview

In keeping with our philosophy of explaining more complex phenomena in terms of simpler phenomena and the components that help explain the phenomena, we will now begin to explore the molecules that are essential to life. Understanding the molecules of life requires an understanding of chemical bonding, polymers, and ultimately, liquid crystals. Clearly, this is an active research area, and this chapter, the last chapter in this section on "materials," provides only the briefest of introductions. We will begin to address the following questions:

(1) Why are even the simplest living organisms so difficult to study?
(2) What molecules are especially important to life?
(3) What are the chemical structures of these molecules?
(4) What types of chemical bonds exist in these molecules that make them biologically active?
(5) Why is water so critical to life?
(6) How does the structure of these molecules influence their function?

12.2 Introduction

Biochemistry, the study of the chemical basis of life, and biophysics, the study of the physical basis of life, are both exciting, active fields of research. There are a number of reasons for this excitement. The first is probably that scientists understand the chemical basis of many of life's central processes. For example, the fundamentals of deoxyribonucleic acid (DNA) and how it relates to both the genetics and the physical properties of the organism are well known. A second reason is the commonality of all life on earth. All known organisms use the same building blocks to form the molecules of life and apparently use adenosine triphosphate (ATP) as part of their energy cycles.[1] Thus, the chemical processes at the core of an organism's metabolism (at least on earth) are universal. Furthermore, our growing knowledge of the relationship between disease states and their biochemical and biophysical underpinnings has allowed the development of diagnostic probes, new medicines, and disease treatments. These developments allow scientists to begin to study some of the most fundamental problems of biology, including how cell differentiation takes place, the causes of cancer, and the molecular basis of memory and mental illness, just to name a few. Finally, the emerging field of proteomics[2] is leading to an understanding, on the molecular level, of how genes "work", and how proteins interact and work inside cells. In part, this complex science is currently driving the next generation of supercomputers.

Like understanding liquid crystals and laptops, an understanding of life requires the bringing together many basic ideas from several different disciplines. The relationships we seek to understand and explain are more complex than those discussed so far. Thus, this chapter has the modest goals of providing a very basic introduction and setting the stage for later study of the relationship between liquid crystals and life. We will begin by reviewing the length, time, and energy scales associated with biological processes. This will naturally lead to discussion of the types of intermolecular forces that mediate the reversible interactions between biomolecules. Since water is ubiquitous in life as we know it, a brief discussion of its unique properties will be included. Lastly, the four major classes of biological molecules – proteins, nucleic acids, lipids, and carbohydrates – will be discussed.

12.3 The length, energy, and time scales of life

When we discuss most molecules or materials, there are relatively few characteristic lengths that are important. These would typically be a characteristic size of the molecule (we called this d), a typical distance between

[1] Apparently there are a few, very rare exceptions. However, they are so rare that this general statement is justified.

[2] This field consists of three activities: identifying all the proteins in an organism, determining how the proteins interact, and finding the 3-dimensional structure of the proteins.

molecules (l) and the size(s) of the sample (this has not yet been a consideration in our discussion). Even in our study of polymers, the most complex class of molecules that we have discussed, this was sufficient. Similarly, we have discussed only static properties with no time dependence. Even when time dependent phenomena are considered later in our discussion of liquid crystal displays, electronic memories, and the like, there will usually only be a few time scales that we must consider, and they will be similar in magnitude. There are typically three or four different energy scales that are of importance in our discussion of materials. These are characterized by the thermal energy, the intermolecular bond energy (covalent, metallic, or ionic), any external energy, and the intramolecular bond energy (covalent or ionic). In biological systems, the characteristic lengths, energies, and times are often not as well separated as those just noted and there is often no obvious cutoff to simplify the analysis.

Consider the various length scales that are characteristic of living organisms. First, because molecules are formed and destroyed in living organisms, interatomic distances, specifically, the distance between two covalently bonded atoms, are important. The next length scale is the small molecular level. These "small" molecules are roughly 5-10 times larger than a single covalent bond and include a large number of biomolecules that are about 1-2 nm in size, including glucose, amino acids (the building blocks of proteins), and lipids. Increasing the length scale by another factor of 5-10 brings us to the size of a typical protein. For example, hemoglobin, the oxygen carrying protein in our blood, has a diameter of approximately 6.5 nm. The next characteristic size includes assemblies of macromolecules. These assemblies are roughly 30-200 nm, or 5-20 times the size of protein molecules and include most viruses. Intracellular structures such as organelles and cells occupy the next characteristic size and are a factor of 10-20 times larger than these macromolecule assemblies. For example, a typical bacterium has a diameter of about 2 μm (2000 nm); a red blood cell has a diameter of approximately 7 μm. Multicellular organisms have a wide range of sizes. For example, 0.01 meter for a small bird (roughly 1000 times larger than a typical cell) to 10s of meters for large whales or trees.

These characteristic length scales vary over eleven orders of magnitude. Furthermore, there is no convenient cutoff size, well separated from other sizes at either the small or the large ends of the range. While the larger characteristic sizes clearly do not apply to a small organism such as a cell, there is still approximately a million-fold variation in important characteristic sizes for most organisms. This wide range of length scales complicates understanding biological systems, and is a first clue that these systems are significantly more complex than the other systems discussed in this work.

The range of time scales of importance in biological systems is also exceptionally wide. There is no single time scale that can be completely ignored while trying to understand biological phenomenon. One of the fastest biologically important events is the change in shape of a molecule involved in vision. This change occurs in a few picoseconds (10^{-12} seconds). The hinging motion of a protein takes about one thousand times longer, about 1 nanosecond (10^{-9} seconds). Go another thousand times slower and one arrives at the time needed for a DNA helix to unwind, about 1 microsecond (10^{-6} seconds). Yet another 1000 times slower is a typical enzyme-catalyzed reaction which occurs in about 1 millisecond (10^{-3} seconds). It takes about one thousand times longer for a protein to be synthesized – about one second. A generation of bacteria takes about 1000 seconds to grow. Ten days, about the time for a small cut to heal, corresponds to about one million (10^6) seconds. A human life of seventy years, is about two billion ($2 * 10^9$) seconds. Life began on earth about 10^{17} seconds ago. There is somewhat more flexibility with time scales than with the length scales one must understand and incorporate. Nevertheless, it is not unreasonable, even in a problem as "simple" as how cells differentiate, to consider times that vary by fifteen orders of magnitude.

The characteristic energies that are important to biological systems are similar to those encountered in non-biological systems. The lowest characteristic energy is thermal energy, $kT = 0.026$ eV. Roughly 5-10 times larger in magnitude (0.06 to 0.3 eV) are the non-covalent bonds that are essential to reactions and strongly influence the size and shape of biological molecules. A few times larger in magnitude is the energy of ATP (0.3 to 0.6 eV). Roughly 10 times larger than this energy is the characteristic energy of a covalent bond (2 to 6 eV).

From this comparison of characteristic energies, a first important conclusion may be reached: thermal energy can affect the non-covalent bonds in biomolecules but cannot affect the covalent bonds. Covalent bonds must be modified by other means. Thus, the non-covalent bonds that seemed so minor in our earlier discussions are now of paramount importance.

12.4 Non-covalent bonds in biomolecular systems

Life depends on reversible molecular interactions. The covalent bonds that bind the molecules of life together are generally too strong to participate in these reversible interactions. Not surprisingly, the weaker bonds that we discussed earlier, the hydrogen and the van der Waals bonds, are very important. Interestingly, the ionic or electrostatic bond is also important. These three bonds differ in strength, geometry, and specificity. All biomolecular processes depend on the interplay between all bond types and the structures that they form and stabilize.

The simplest bond is the ionic bond. In biological and biochemical literature, this bond has a number of other names including *salt bridge*, *electrostatic bond*, *salt linkage*, and *ion pair*. An ionic bond forms between two charges of opposite sign. Furthermore, the charges are usually on different chemical groups and entities. Recall that the force between two charges is along the line joining the two charges and has a magnitude F given by Coulomb's Law:

$$F = \frac{k}{\kappa} \frac{q_1 q_2}{r^2}.$$

In this equation, k is a constant, q_1 and q_2 are the two charges, r is the center-to-center distance between the two charges, and κ is the dielectric constant. The inclusion of the dielectric constant κ is significant because κ in water is about 80 compared to $\kappa = 1$ in vacuum. Thus, the magnitude of an ionic bond in water is significantly reduced compared to that obtained when the same two charges are in a vacuum or an ionic solid. This attraction will, of course, be resisted by intergroup or atomic repulsion. This competition between attraction and repulsion leads to a maximum in the attractive force at a separation of about 0.28 nm between the centers of the charges.

In many biologically relevant systems, the charges that interact are not simple ions such as Na^+, or Cl^- that we discussed earlier when introducing ionic bonds. Rather, in the biological case they are often groups of atoms. These groups of atoms are called *functional groups*, and occur so frequently in organic compounds that they are named. In solutions containing water, some atoms within these functional groups dissociate to form ions. This is process is called *ionization*. The chemical formula of some common functional groups encountered in biochemistry are shown in Fig. 12.1

tional groups encountered in biochemistry are shown in Fig. 12.1

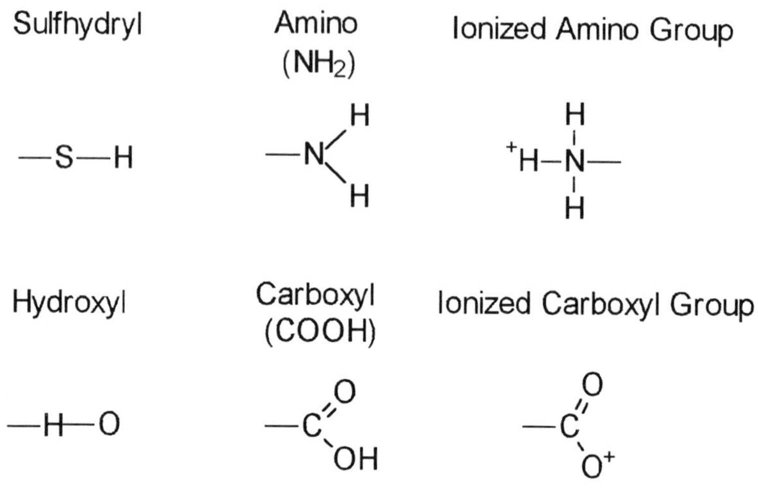

Sulfhydryl: —S—H

Amino (NH_2): —N(H)H

Ionized Amino Group: $^{+}$H—N(H)(H)—

Hydroxyl: —H—O

Carboxyl (COOH): —C(=O)OH

Ionized Carboxyl Group: —C(=O)O^{+}

Fig. 12.1 Some common functional groups.

Hydrogen bonds form when two atoms share a hydrogen atom. Hydrogen bonds are weak and can form between charged as well as uncharged bodies. In such bonds, a hydrogen atom is covalently bonded to one of the atoms, which is is called the hydrogen donor. The same hydrogen atom is *non-covalently* and more weakly bonded to the other atom which is called the hydrogen acceptor. In biological systems, both the donor and the acceptor are an oxygen or a nitrogen atom. Hydrogen bonds are also highly directional. The strongest hydrogen bonds occur when the donor atom, the hydrogen atom, and the acceptor atom are co-linear; that is, they are all in a straight line. When the donor atom, the hydrogen, and the acceptor atom are not co-linear, the bond strength is weakened. The energy of a hydrogen bond is between that of a typical covalent bond and a van der Waals bond. Additionally, hydrogen bond energies vary widely. The bonds in biological systems are sometimes called moderate or normal hydrogen bonds. Interestingly, hydrogen bond distances are not symmetric. Examples of hydrogen bonds and this asymmetry in distance are shown in the Fig. 12.2. In this figure, the three dots (...) indicate hydrogen bonds, while the solid lines indicate covalent bonds.

When the ionic bond is assumed to occur in water, the following relationships hold:

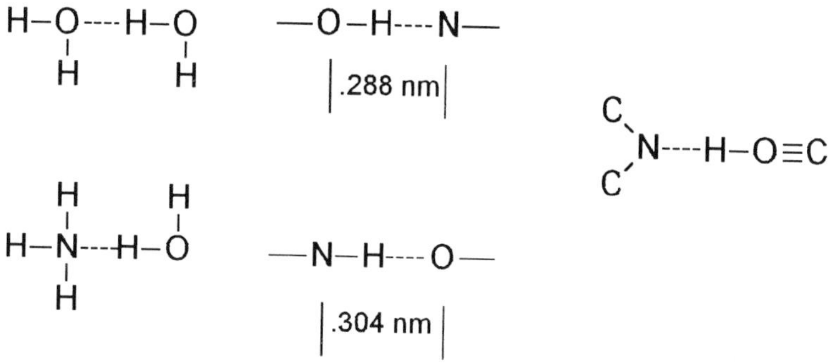

H–O----H–O —O–H----N—
H H |.288 nm|

C
N----H–O≡C
C

H H
H–N---H–O —N–H----O—
H |.304 nm|

Fig. 12.2 Common hydrogen bonds in biochemical systems. The hydrogen bonds are indicated by three dots (...).

$$U_{covalent} \approx 10 * U_{ionic\ in\ water} \approx 10\ to\ 20 * U_{hydrogen\ bond}$$
$$\approx 10\ to\ 100 * U_{van\ der\ Waals} \approx 200 * U_{thermal}$$

Thus, **ionic and hydrogen bonds are of comparable strength in water**.

The van der Waals bond is a non-specific attraction that occurs between all atoms. The van der Waals force is important when the atoms are very close together and falls off very rapidly as the molecular separation increases. Notice in the above energy comparison that a single van der Waals bond is only a little more energetic than the typical thermal energy of a molecule. Note further that a van der Waals bond is usually significantly weaker than a hydrogen bond or an ionic bond in water. Because of this small energy, there is no significant advantage to forming **a single** van der Waals bond. The energy advantage of this type of bonding, as well as the **biological** specificity that arises from it, is the result of the simultaneous formation of a large number of van der Waals bonds.

This advantage arises for the following reason: Because van der Waals bonds are weak and short-ranged, a large number of them are necessary to produce any appreciable effect. These conditions also mean that bonds form only when atoms are *very* close to one another. This effectively means that molecules can only interact with a specific conformational partner. Thus, the non-specific van der Waals attraction give rise to very specific molecular interactions.

12.5 The importance of water

Water is essential to life as we know it and influences all molecular interactions in biological systems. Water possesses two properties that are essential in this regard. The first property is that water is a polar molecule, meaning that the water molecule has a non-zero electric dipole moment, as was discussed earlier. The shape of the water molecule causes an asymmetric charge distribution and hence an electric dipole moment. The second property is that water has a very high affinity for itself. Ice has an ordered structure that maximizes the strength of the hydrogen bonds. While water is not as ordered as ice, it is still a highly organized liquid because of the large number of hydrogen bonds.

The polar nature of water and its hydrogen-bonding ability make water a strongly interacting molecule. As you are aware, table salt, an ionic solid, dissolves very easily in water. This property is more generally true; water is an excellent solvent for all ionic and polar molecules. This is because water has a high dielectric constant that greatly reduces the magnitude of electrostatic bonds between polar entities. Furthermore, water can very effectively compete for hydrogen bonds.

The existence of life on earth depends critically on the ability of water to dissolve polar molecules. A high concentration of these molecules can be sustained in water and it is easy for these molecules to move about within the liquid environment of the water. There is, of course, a down side to this behavior. Water also greatly reduces the interactions between polar molecules. Nature has solved this problem by creating water-free zones where the electrostatic screening behavior of water is eliminated. These special microenvironments exist in proteins.

The second essential property of water gives rise to the hydrophobic interaction which we discussed earlier. You are familiar with how oil and water separate. An analogous phenomenon occurs at the molecular level where non-polar molecules tend to cluster together in packets as isolated from water as possible. This is often said to be caused by a hydrophobic (water-fearing) interaction. This is not really so much an attraction of the non-polar molecules as it is the extreme affinity of water for itself. The addition of non-polar molecules to a water environment alters the (somewhat ordered) local structure of the water near the added molecules. This causes the water molecules near the added non-polar molecules to be more ordered than they would be in pure water. This reduction in disorder raises the free energy of the system. The fewer the number of isolated packets of non-polar material, the less disrupted water there is, and the less the increase in free energy. Thus, the system establishes an equilibrium in which the non-polar molecules have been segregated from the water to the greatest extent possible. This segregation is important in determining the structure of biological membranes and the conformation of proteins. It also leads to an effective attraction between two hydrophobic bodies.

12.6 Proteins

Jöns Berzelius coined the name protein in 1833. It is derived from the Greek word *proteios*, which means "of first rank," and is meant to stress the importance of this type of molecule. Along with nucleic acids, the molecules that control heredity and the synthesis of proteins, proteins are certainly of critical importance. Proteins are found in all living cells. They make up a large part of animal bodies, 49% of a human red blood cell, and plants as well, 70% of the chloroplasts in spinach lamella. Proteins are heteropolymers and usually consist of between 50 and 2000 polymeric units called *amino acid residues*. This section will briefly introduce the importance of these macromolecules, as well as a few details of their structure. While these molecules are complex, they can be understood using the basic guiding principles of organic and polymer chemistry that have already been discussed.

Proteins are essential to a broad range of biological processes. Following Stryer[3], we will briefly list some of these functions:

(1) Nearly all chemical reactions in biological systems are catalyzed by macromolecules called *enzymes*. The first question you might ask is "What does catalyzed mean?" A catalyzed reaction involves a *catalyst*, a substance that does not appear in either the reactants or the products of a chemical reaction, but greatly enhances the conversion of the reactants into the products. Some reactions are impossible without a catalyst, while the rate of other reactions are greatly increased. You have all heard of the catalytic converter in an automobile's exhaust system. This converter contains several metals that greatly enhance the conversion of hydrocarbons and nitrous oxides (NO_x) into water (H_2O), nitrogen (N_2), and carbon dioxide (CO_2). Biochemical catalysts, enzymes, usually increase a reaction rate by at least a factor of

[3]See references at the end of this chapter.

10^6. Several thousand enzymes have been isolated and almost all are proteins.

(2) Proteins play a central role in the transport and storage of small molecules and ions. For example, the protein *hemoglobin* transports oxygen in red blood cells, while *myoglobin*, a related protein, transports oxygen in muscles. Similarly, iron is transported in blood plasma by the protein *transferrin* and is stored in the liver as a complex with another protein, *ferritin.*

(3) The coordinated motion of muscles is accomplished by the sliding action of two kinds of protein fibers that make up the muscle. Coordinated motion also occurs at the cellular and intercellular level where proteins are also involved.

(4) The strength of skin and bones is due to *collagen*, a fibrous protein.

(5) Antibodies are highly specialized proteins that react with organisms that are considered foreign bodies by the host body.

(6) Proteins transmit nerve impulses.

(7) Hormones control many activities of cells, including energy and mass flow. Many hormones such as *insulin* are proteins.

The next reasonable question is "What is the structure of proteins?" Proteins are heteropolymers that are made of structural units called amino acids. An amino acid consists of a central carbon atom with a different functional group attached to each of the carbon's four tetrahedral bonding sites. There is always a hydrogen atom, an "amino group," NH_3^+, and a "carboxyl group," COO^-. (These functional groups are ionized because water is present.) The final site is bonded to a distinctive functional group, the "R group," (sometimes called a side chain) that determines the type of amino acid. There are twenty different functional (or side) groups. Since there are four carbon bonding sites and four different groups, amino acids are chiral. The two possible isomers of an amino acid are called the L-isomer and the D-isomer. Interestingly, *only L-amino acids are found to occur in proteins.*

The twenty different R groups vary in many aspects, including size, shape, charge, hydrogen bonding capacity, hydrophicity, and chemical reactivity. It is a remarkable fact that the proteins of all species, from bacteria to humans, contains the same twenty amino acids. Further, this "alphabet" of amino acids has existed for at least 2 billion years. Just as the twenty-six letters of the English alphabet allow us to make many words, these twenty amino acids allow a large number of proteins to form. The order in which the amino acids combine to form a protein has two effects. The amino acid sequence uniquely determines the protein, as well as its three-dimensional structure.

Proteins can be divided into two very broad classes: *fibrous proteins*, which tend to be insoluble in water, and *globular proteins* which are soluble in water and aqueous solutions of salts. This difference in solubility can be traced to the difference in the molecular shape of these macromolecules, which relates back to the amino acid sequence.

Fibrous proteins are typically the chief structural materials of animal tissues. Examples of this type of protein include *keratin* (a protein in skin, fingernails, and wool), some types of *collagen*, *myosin* (in muscle), and *fibroin* (silk). The molecules that form these types of proteins tend to be long and thread-like, lie side by side, and form fibers. These molecules can be cross-linked by covalent bonds or held together by hydrogen bonds. In either case, the intermolecular forces that need to be overcome to solubilize these molecules are very large.

Globular proteins consist of polymers that have coiled into a compact, often nearly spherical, shape. This coiling generally occurs so that the more polar portions of the protein are near an aqueous environment and the non-polar, hydrophobic portions of the protein are closer to each other and isolated from the aqueous environment. The coiling effectively creates a polar "coating" on these molecules and leads them to be rather soluble in water. Globular proteins are important to the regulation and maintenance of life. This is not surprising since they are soluble in water and can therefore be easily transported within the liquid environment. This group includes most enzymes, many hormones (for example, *insulin*), *albumin* (the white in eggs) and *hemoglobin*.

The irreversible precipitation and aggregation of proteins, most often by heat, is called *denaturation*. A common example of this is cooking of an egg, which denatures the protein *albumin* in the egg white. The relative ease with which proteins denature makes their study difficult. A further complication is that many proteins are "trimmed" after synthesis. Trimming means that some change occurs in the protein that changes its behavior. Examples of protein trimming include digestive enzymes that are stored as inactive precursors in the pancreas and activated in the intestine, and blood clotting, in which soluble *fibrinogen* is converted into insoluble *fibrin* by cleaving a peptide bond. Trimming is essential to the action of many proteins. Many proteins also contain non-peptide portions called *prosthetic* (helper) *groups*. These groups are related to the specific biological function of the protein. The most common example is the prosthetic group in *hemoglobin* – the *heme group*. The reversible heme-oxygen complex holds the oxygen that is transported by hemoglobin.

Protein structure is very rich and can be studied at several levels. The lowest level studies the *primary structure* of the protein. This is the study of the amino acid sequence and the covalent bonds of the protein. The next level, the *secondary structure*, refers to the spatial structure of nearby amino acid residues. It is found that when the side chains are rather small an arrangement called the β pleated sheet, in which chains lie side by side and are held together by hydrogen bonds, is observed. In regions where the side chains are larger a different arrangement called the α helix occurs. The α helix is similar to a spiral that rotates clockwise as one "moves up" the spiral. The next level, the *tertiary structure*, studies the spatial arrangement of amino acid residues that are far apart in the protein's linear structure. The dividing line between secondary and tertiary structure is, of course, somewhat ill defined. Lastly, when a protein is made of more than a single polypeptide chain, the *quaternary structure*, the spatial relationship between these "subunits," is studied. As interesting as all these levels of structure are, it is essential to remember that the amino acid sequence, the primary structure, ultimately specifies the three dimensional structure of the protein.

The secondary structure of amino acids and the combination of amino acids via a peptide bond are important and illustrate some of the ideas that we have been discussing. First, since a protein is a heteropolymer, we expect the various amino acids to be covalently bound. This is the case, and is shown schematically in Fig. 12.3

If one imagines this reaction occurring many times, a polymer molecule will result. However, as noted earlier, these are not the only interactions amongst the peptides. There are interactions between different functional groups in the same protein that produce structures that occur so frequently that they are called secondary structures. There are three very common interactions. The first is hydrogen bonding of the C=O on one amino acid with the N-H on another. The second is the side-chain to side-chain bonds between side groups. Finally, hydrogen bonding between a peptide bond in one amino acid and the side chain of a different amino acid can occur.

The α helix is one of the best known secondary structures. It is a rod-like structure, in which the carbons in the main-chain form a helix. This helix is stabilized by hydrogen bonding between all the main-chain CO and NH groups. In fact, the CO group of the *nth* residue is hydrogen bonded to

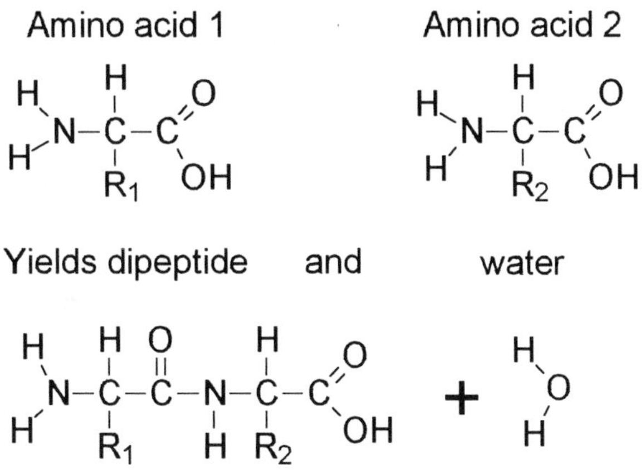

Amino acid 1 Amino acid 2

Yields dipeptide and water

Fig. 12.3 Two non-ionized amino acids and their formation of a dimer or dipeptide and water. R_1 and R_2 indicate different side groups.

the NH group of the $(n+4)th$ residue. This leads to a structure where each residue (or amino acid) is translated 0.15 nm along the helix and rotated by 100°. In principle, the α helix can either rotate clockwise or counter clockwise. The α helices found in proteins all rotate clockwise and are said to be "right-handed."[4] A right hand α helix is shown schematically in Fig. 12.4. In Fig. 12.4, only the main chain or backbone carbons and nitrogens are shown, and only the hydrogen bonds between the nth and the $(n + 4)th$ peptide bonds are shown.

Another common secondary structure that is stabilized by hydrogen bonds is the β pleated sheet. This is a planar like structure and occurs in two forms, parallel and antiparallel shown in Fig. 12.5. Once more, the side chains have been omitted for clarity.

12.7 Nucleic acids

DNA (deoxyribonucleic acid) is a very long, thin, thread-like macromolecule that is made up of a long chain or backbone that is the same (except for length and order) in all nucleic acid molecules. Attached to this backbone at regular intervals are one of four groups that are called *bases.* The sequence of these bases characterizes the specific nucleic acid. Alternately, we can

[4]This can be confusing. A good way to figure out handedness is to recall that a normal screw is right handed. This means that as you turn it clockwise it goes into the nut or the piece of wood. A right-handed screw has a right-handed helix.

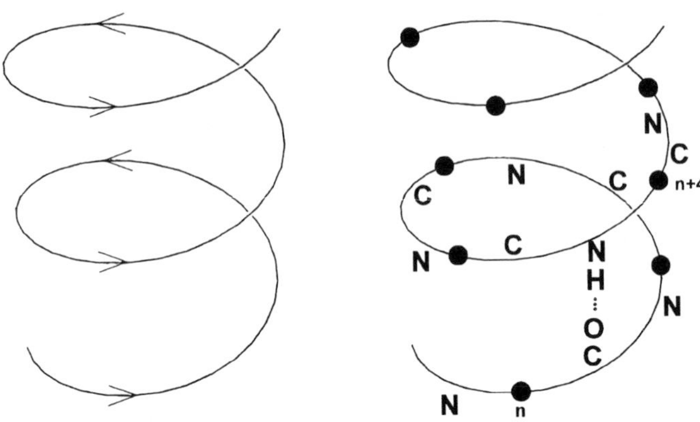

Fig. 12.4 The α helix structure. The left shows only the helical structure. For clarity, only one intra-chain hydrogen bond is shown and some moieties have been omitted on the right.

say that this heteropolymer is made of *deoxyribonucleotides* – a unit that consists of three functional groups: a base, a sugar, and a phosphate group (PO_4). The bases of the DNA molecule carry genetic information, while the sugar and phosphate are important as structural elements. This structure is shown schematically in Fig. 12.6.

The genes of all cells and some viruses are made of DNA. However, some viruses use RNA (ribonucleic acid) as their genetic material. These names follow from the sugars that are part of molecules. The sugar in RNA is D-ribose, while the sugar in DNA is 2-D-deoxyribose. The difference is simple: in RNA there is an OH^- group at a given position, while in DNA a hydrogen atom occupies the same position. The hydrogen in DNA leads to a more stable structure than the OH^- in RNA under physiologic conditions.

The variable part of either a DNA or a RNA molecule is not this backbone but the bases. In both DNA and RNA there are four different bases. DNA contains adenine (A), guanine (G), which are structurally related; and cytosine (C) and thymine (T), which are also structurally related. RNA contains adenine, guanine, cytosine, and a fifth base, uracil (U), that is structurally closely related to thymine and cytosine.

The primary differentiation between the structures of DNA or RNA consists of the order in which these bases occur. The secondary structure was discovered by Watson and Crick and is the famous double helix. The double helix model has the following features:

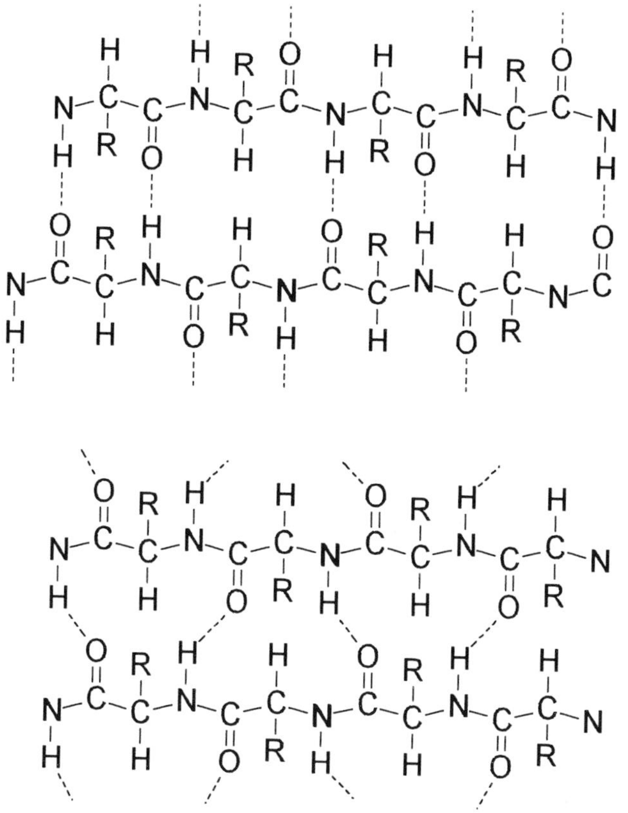

Fig. 12.5 The antiparallel (top) and parallel (bottom) β pleated structure. Both structures are stabilized by hydrogen bonds.

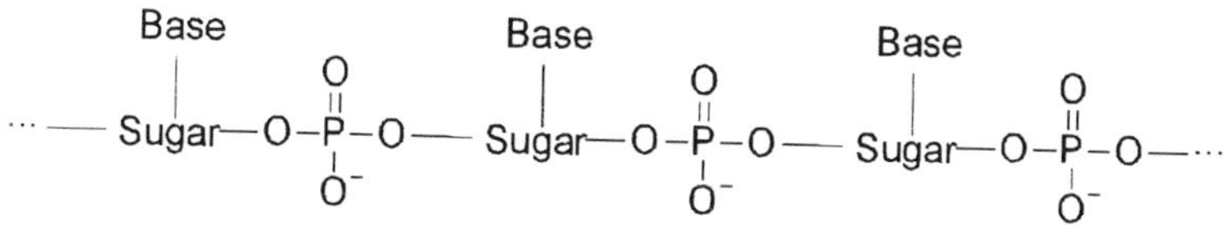

Fig. 12.6 The structure of DNA at the functional group level.

(1) DNA is made of two polynucleotide chains wound about a common axis. These chains are both right-handed and run in opposite directions.
(2) The double helix is 2 nm in diameter and has ten nucleotide elements per turn. The pitch of the helix is 3.4nm.
(3) The four bases are on the inside of the helix, while the phosphate and

deoxyribose functionalities are on the outside.

(4) The two chains are held together by hydrogen bonds between complementary pairs of bases. These are linear hydrogen bonds: two occur between adenine and thymine and three between guanine and cytosine. These may be indicated in diagrams, where the ··· indicates a hydrogen bond as in: A···T and G···C. The space constraints of the helix require these to be the only two pairings. Keep in mind that the two chains are not identical, but are complimentary, as illustrated in Fig. 12.7.

(5) The sequence of bases along the chain is not restricted in anyway and is what carries the genetic information.

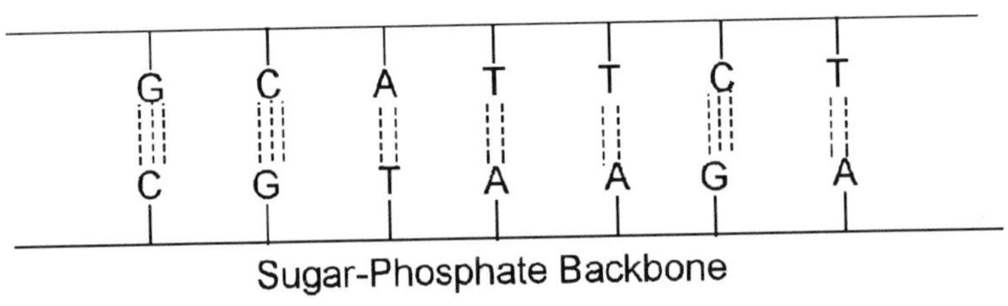

Fig. 12.7 Hydrogen bonding between complementary DNA base pairs.

Stryer states that "the most important aspect of the DNA double helix is the specificity of the pairing of bases." This structure allows DNA to play the dual role of director of protein synthesis and repository of genetic information.

The secondary structure of RNA helices is complicated. Though there is usually only a single stranded helix, the sizes of RNA molecules can vary greatly – some are very long like DNA; whereas others are rather short, only a few hundred bases long.

The tertiary and quaternary structures of nucleic acids are also of great interest, and justifiably so. To understand why, consider the following example. The DNA of the bacteria *E. Coli* is about 1.4 mm ($1.4 * 10^6$ nm) long (and still only 2 nm in diameter). This DNA must fit inside a cell that is only a few microns in size, and must be packed in such a manner that the cell can divide and perform the biochemical operations needed for it to maintain life. *E. Coli* is a rather modest organism with a rather short segment of DNA. In contrast, human DNA is roughly a meter long. Electron microscopy of intact DNA molecules has yielded some insights into the structure of DNA molecules. For example, the DNA of *E Coli* is continuous, rather like a rubber band. This is not always the case; linear DNA has also been observed. The axis of the double helix may further twist into a supertwisted structure called a *superhelix*. The superhelix appears to be biologically important for two reasons. First, a superhelix is more compact than a relaxed helix. Second, supercoiling affects the ability of the DNA to unwind and interact with other molecules.

How does DNA control the chemistry of the cell and heredity? Clearly, DNA is in charge at the molecular level. Furthermore, we know that the information is written in a language of four nucleotides represented by the four letters A, G, T, and C. A potential difficulty is that DNA has two functions. DNA must both preserve genetic information and use it. These two roles are possible because DNA molecules can duplicate themselves in a process called replication, and they can control the synthesis of the proteins that are characteristic of the organism.

The issue of replication is somewhat simpler since the sequence of one chain of the double helix pair necessarily determines the sequence of the other chain. The double helix "unzips" and the corresponding second chains forms so that two double helices result, each an exact copy of the original[5].

Protein synthesis is more complex. It begins with a process called transcription in which a double helix of DNA partially uncoils, and a chain of RNA forms about one of these chains. The DNA template determines the order of nucleotides in the RNA molecule. However, the sequence of nucleotides along the RNA chain are different than along the DNA chain since the base pairs are different. The difference is that opposite each adenine in the DNA, a uracil occurs in the RNA. The balance of the transcription is the same as in DNA. This type of RNA, called messenger RNA, carries information in a cell to the surface of ribosomes where protein synthesis takes place. Here the messenger RNA calls up a series of transport RNA molecules that carry the desired amino acids. While RNA is written in a code of 4 letters, there are 20 amino acids. Clearly, more than one letter is needed to determine which amino acid goes next in the sequence. It turns out that nature uses a three base code with redundancy, so that more than one three base sequence can generate the same amino acid. Not surprisingly, the difference of a single base in the DNA molecule or an error in reading the code can change the amino acid sequence in the resultant protein. Such errors lead to sickle cell anemia. Interestingly, producing incorrect proteins can have a positive side. Some antibiotics alter the cell

[5]Sometimes, (about 1 error per 10^6 base pairs) this replication is not exact. Such changes, called mutations are important to the evolution of species.

in such a way that the cell produces defective proteins that cause the death of the invading organism.

12.8 Lipids

In biophysics and biochemistry, it makes little sense to discuss lipids except within the context of their function in biological membranes. Biological membranes are organized, sheet-like, self-assembled units that consist mainly of proteins and lipids. Organic chemists define a lipid as a substance that is essentially insoluble in water and that can be extracted from cells by low-polarity organic solvents such as ether or chloroform. It is sufficient for our purposes to treat a lipid as a fat. This definition is far too broad to serve us for long but is sufficient to get us started. Membranes are important because they are highly selective permeable barriers that give cells their individuality. Membranes also contain receptors and signal generators for biological communication. Finally, energy conversion takes place within some types of biological membranes.

Stryer lists seven common features of all biological membranes. These are:

(1) Membranes are sheetlike structures that form closed boundaries between regions of different composition. Typically, they are only a few molecules thick – roughly 6 to 10 nm thick.
(2) Membranes consist mainly of proteins and lipids. The composition varies from 80% lipid to 80% protein in most membranes. There are also carbohydrates attached to the lipids and proteins.
(3) Membrane lipids are relatively small and are amphiphilic. Because they have both hydrophobic and hydrophilic portions, they spontaneously form closed bilayer structures in water, with polar surfaces and hydrophobic cores.
(4) Specific proteins mediate various functions of the membrane by acting as gates, pumps, transducers, and other similar functions.
(5) Membranes are non-covalently bonded self-assembled units.
(6) Membranes are asymmetric – the inside and outside are not the same.
(7) Membranes are fluid structures that can be regarded as two-dimensional liquids composed of proteins and lipids.

Although lipids have a variety of biological functions including their use as fuel molecules, signal molecules, in this chapter, we will focus on their use as constituents of membranes. There are three major kinds of membrane lipids: *phospholipids, glycolipids*, and *cholesterol.*

Phospholipids are common in all biological membranes. They are derived from glycerol, a simple alcohol[6] or sphingosine, a more complex alcohol. Phospholipids derived from glycerol are fairly common and have the general form shown in Fig. 12.8.

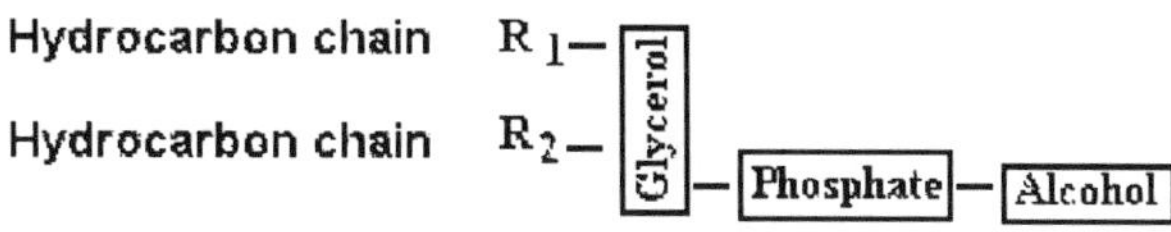

Fig. 12.8 The general form of a lipid molecule.

In Fig. 12.8, R_1 and R_2 represent two, possibly different, hydrocarbon chains of the form $(CH_2)_nCH_3$, where n is an integer typically between 12 and 18. A variety of alcohols are possible. Glycolipids are sugar-containing lipids that are similar in structure to phospholipids. However, there is no phosphate group and the alcohol is replaced by a sugar. The third major membrane lipid is cholesterol, a complex molecule whose structure we shall not discuss. It serves a number of purposes in the body, including that it helps maintain the flexibility of bilayers. Interestingly red blood cells are approximately 25 % cholesterol.

The key feature of phospholipids and glycolipids is that they readily form bilayers. This occurs because the molecules are amphiphilic, containing in a single molecule both hydrophobic and hydrophilic moieties. You might have guessed this because you know oil (the hydrocarbon tails) and water don't mix and alcohol and water do mix. More precisely, the hydrocarbon chains in these molecules are highly hydrophobic, while the alcohol and phosphate group are hydrophilic and form what is called a "polar head group." Furthermore, the geometric structure of these molecules separates these two distinct moieties to different parts of the molecule. A very schematic structure for these molecules is shown in Fig. 12.9.

When a suitable number of these amphiphilic molecules are in an aqueous environment, they self-assemble to form a biomolecular sheet. The reason these molecules form sheets is that the hydrocarbon chains are too bulky for other shapes of self-assembled objects to form. This self-assembly

[6]An alcohol is an organic compound of the general form R-OH, where R is a hydrocarbon group.

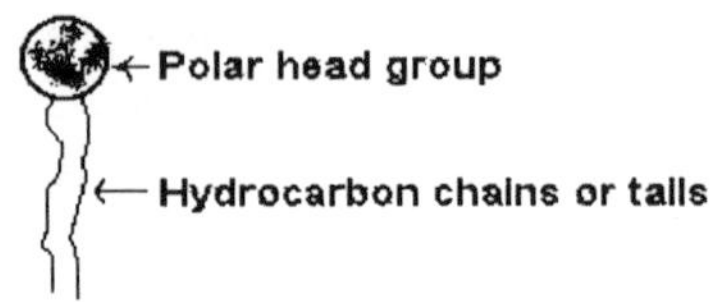

Fig. 12.9 A simple model of a lipid indicating its hydrophobic tails and polar head.

is driven by all of the non-covalent forces that we discussed earlier. However, the hydrophobic interaction between the hydrocarbon tails and water is the primary driving force. A schematic figure of part of a bilayer is shown in Fig. 12.10.

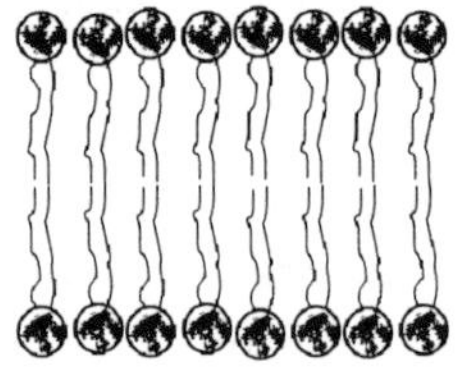

Fig. 12.10 A bilayer of lipid molecules.

It is import to realize that these bilayers have several unique features and properties. First, they are **not** covalently bonded. The forces that create these bilayers are strongly cooperative – van der Waals forces and the clustering tendency of water. These membranes also tend to be closed structures and are self-healing, because of the tremendous energy cost of a hole or open end of a bilayer. Of equal importance is the fact that the inner and outer surfaces of all known membranes differ in both composition and function. Furthermore, it is extremely difficult for most ions and polar molecules to pass through these lipid bilayers. An exception is water, which is found on both sides of most bilayers. This difficulty causes a great deal of variation in the "permeability" of ions. Water will cross a bilayer between 10 million and 10 billion times faster than small ions such as Na^+, K^+, or Cl^-. This means some other mechanism is needed for ion transport to occur. It turns out that proteins carry out most of the dynamic processes that occur in biological membranes. A simple model that works well is one where the membrane lipids form a closed structure that establishes a permeability barrier. The proteins then control the balance of the membrane functions. This model has been verified by studies that indicate that membranes that perform different functions contain different proteins.

The proteins that are associated with the membrane are classified as being either *integral* or *peripheral*. Nearly all integral proteins span the lipid bilayer, while peripheral proteins tend to be bound to integral proteins via hydrogen or ionic bonds. This structure is illustrated in Fig. 12.11.

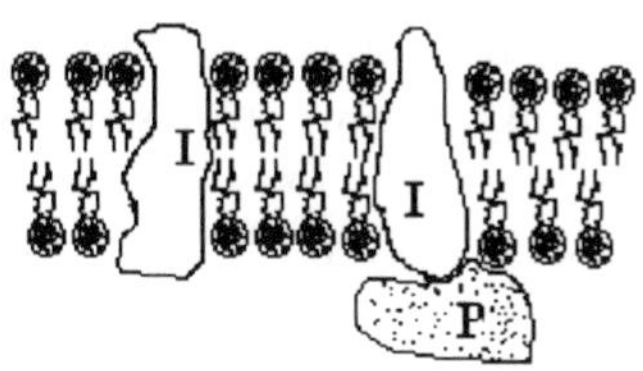

Fig. 12.11 Schematic figure of a bilayer showing lipids and both peripheral (P) and integral (I) proteins.

In this figure the **I** indicates an integral protein, while the **P** indicates a peripheral protein. Peripheral proteins can be on both sides of the bilayer even though they are only shown on one side.

The proteins and the lipids in a membrane can move about. Their movement is not in a straight line, but is analogous to a person strolling aimlessly from one point to another. Such behavior is called a random walk or diffusion and is characterized by a diffusion coefficient, D. Since this motion is confined to the membrane it is called transverse diffusion. The diffusion coefficient of lipids in membranes is about 10^{-8} cm^2/sec. This is a rather small number compared to the mass diffusion coefficient of water, roughly 10^{-5} cm^2/sec. Nevertheless, this is still sufficiently large that a lipid molecule can diffuse from one end of a cell to the other in about 1 second. The diffusion coefficients of proteins are more variable. A very mobile protein has a diffusion coefficient similar to lipids, while a low mobility protein might have a diffusion coefficient 1000 times slower. Typically, proteins do not flip from one side of the membrane to the other. This is not the case for lipids. However, the time for lipid flipping is long compared to lateral diffusion over a comparable distance.

12.9 Carbohydrates

The final major category of biological molecules that we will discuss is carbohydrates. Carbohydrates are the ultimate source (either directly or indirectly) of most of our food and makes up most of the organic matter on earth. Carbohydrates are important because they serve several roles to all forms of life. These include serving as an energy store and an intermediate in metabolism. They also are important as the sugars in DNA and RNA, where it is apparent that their size and flexibility is critical to the expression of genetic information. A class of carbohydrates, *polysaccharides*, is a structural element in cell walls of bacteria and plants and in the exoskeleton of insects and crustacea. Finally, many carbohydrates are covalently linked to proteins and lipids.

The ubiquitous nature of carbohydrates in life will be discussed in the next few paragraphs. We will then discuss the chemistry of these materials and some of their biological functions.

D-glucose is a carbohydrate formed in the leaves of plants from carbon dioxide, CO_2, and water, H_2O, via the process of photosynthesis that utilizes energy from light and the catalyst chlorophyll. D-glucose is used for a variety of purposes. For example, thousands of D-glucose molecules may be combined into larger molecules called cellulose, which provides support for the plant. D-glucose can be combined in a different way to form large molecules of starch that is stored in seeds to serve as food for sprouting plants. Starches are polymers that may be either branched or unbranched. D-glucose can also combine with other molecules to form ATP, adenosine triphosphate, the core of the energy transport cycle that is central to all life processes.

Of course, plants are grown for many reasons – beauty, food for people and animals, and building materials. If eaten for food, animals take starch and, in some cases, cellulose, and break them down into the original D-glucose molecules. These can be used by the animal as a source of energy (through ATP) or carried, in higher animals, to the liver where it is used to form glycogen, a very large branched polymer composed of glucose residues. In cells, D-glucose is converted to fats and amino acids which ultimately become proteins and the body of the animal.

These four materials, D-glucose, cellulose, starch, and glycogen, are all carbohydrates. Our life is dependent on these materials in many ways. For example, we eat plant seeds to obtain starches. Similarly, the animals we eat at some stage of their food chain ate starches or cellulose to obtain energy and grow. Most natural fibers and fabrics are composed of cellulose. The most common examples are cotton and linen. Our houses are largely built of wood – yet another form of cellulose. In recent decades, the emergence of synthetic polymers has decreased our use of these materials except for food, paper products, and the like. However, they are still central to our lives.

Carbohydrates are a complex group of molecules that can be characterized using the following scheme. A *monosaccharide* is a carbohydrate that cannot be broken into simpler compounds upon reaction with water. A *disaccharide* can be broken down into two monosaccharide molecules. A polysaccharide can be broken into many monosaccharide molecules. Further classifications are possible and useful in more advanced studies of these molecules. An important observation is that most naturally-occurring monosaccharides contain five or six carbon atoms. The common monosaccharide that we just discussed, D-glucose, is a six carbon sugar and has the chemical formula $C_6H_{12}O_6$. Its structure is somewhat complex as Fig. 12.12 shows.

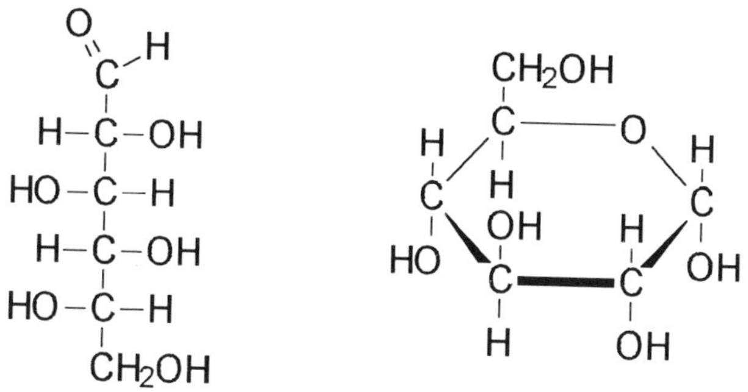

Fig. 12.12 On the left, the structure of a linear glucose molecule. The right side illustrates one of the two cyclic structures of D-glucose. The other structure interchanges the OH and H on the carbon on the far right.

The "D" in the designation indicates the absolute configuration of the carbon that is double bonded to the oxygen, but is of no great importance in the present context. The predominant forms of most five and six carbon monosaccharides in solution is a closed ring structure not the chain shown in Figure 12.12. This is a rather complex issue and will not be discussed further.

You are familiar with many disaccharides, two sugars joined through a special bond called an "O-glycosidic bond." Sucrose is common table sugar and is a combination of two different types of monosaccharides, glucose and fructose. Lactose is the disaccharide in milk and consists of glucose and yet another sugar, galactose. The detailed names are not important; that you understand that common sugars are disaccharides is important.

Cellulose is a polysaccharide that serves a structural role in plants and some animals. Cellulose is an unbranched polymer of ring shaped glucose residues that are joined in a different configuration than that obtained in starches. The result is very long straight chains that give structural integrity to plants. The configuration of this polymer leads to high tensile strength fibers. The corresponding polymer in insects and crustacea contain chitin, a similar molecule except that a different chemical moiety is substituted at one position in each glucose residue of the structure.

Finally, sugars in the form of oligosaccharides are often covalently bonded to the extracellular surface of integral membrane proteins. Oligosaccharides are also attached to many antibodies and clotting factors. The glycoproteins in blood appear to provide some time keeping functions that indicate when the proteins carrying them should be removed. Other carbohydrates appear to have a role in inter-cell interactions. All the roles of carbohydrates are yet to be discovered.

12.10 Further reading and references

This chapter has been far too brief for more than an introduction to the terminology and a few examples. This material is covered in many texts. I am particularly indebted to **Biochemistry**, by Lubert Stryer, W. H. Freeman and Company, NY, 1988. A 1999 edition is available. The first chapter of **Cellular Biophysics**, by Thomas Fischer Weiss, MIT Press 1996, is a brief yet outstanding higher level discussion of the material presented here. Most organic chemistry books also discuss this material. Because of a historical accident, I am most familiar with **Organic Chemistry** by Morrison and Boyd, but other texts are similar. **Biochemistry and Molecular Biology**, by W. H. Elliot and D. C. Elliot, Oxford University Press, 1997 is very readable, and was a constant companion during writing. The figures in this book are very well done. G. A. Jeffrey's book, **An Introduction to Hydrogen Bonding**, Oxford University Press, 1997 is very detailed and not for the novice. Nevertheless, the author found parts very useful while writing this chapter. Readers who are particularly interested in this area will find it very useful.

12.11 Exercises

(1) A protein has a molecular weight of 70,000 Dalton. A Dalton is the same as a unified atomic mass unit. The average mass of an amino acid residue is 110 Dalton. This protein has two strands of α helix chains that are the same length.

(a) What is the mass, in Dalton, of one of the chains?
(b) How many amino acid residues does this chain contain?
(c) Each residue is 0.15 nm long. How long are the chains?

(2) A mutation occurs to a single amino acid residue. This causes a protein to lose its ability to catalyze a reaction. Clearly, the primary sequence has been changed. Do you think that the secondary structure might have changed also? Explain.

(3) Glycine is the smallest amino acid residue. It is found that the position of glycine residues in a given protein does not change during evolution. Given that glycine is not charged and that all other residues are larger or charged, why might this be the case? Hint: think steric reasons and secondary structure.

(4) A scientist studies double helix DNA and finds that on one of the helices, 10% of the bases are A, 30% of the bases are C, 25% of the bases are T and the balance are G.

(a) What percentage of the bases are G?
(b) What are the percentages of A, G, T, and G on the complementary chain?

(5) The DNA of a virus has a length of 65,000 nm. The distance between base pairs is 0.34 nm. How many base pairs make up this DNA?

(6) Part of a strand of DNA has the following base pair sequence: AACGT-GTACTGGTAGC.

(a) What is the sequence of base pairs on the complementary part of the DNA chain?
(b) What is the sequence of base pairs on an RNA chain synthesized from this fragment of DNA?

(7) The average distance, s, in cm, that an object transversely diffuses in t seconds is given by $s = \sqrt{4Dt}$, where D is the diffusion coefficient and has units of cm^2/sec.

(a) What is the average distance a lipid molecule diffuses in 1 second if D $= 10^{-8}$ cm^2/sec.?
(b) A cell is typically $2 * 10^{-4}$ cm across. Could the lipid travel from one end of the cell to the other in one second?
(c) How long does it take for a protein with D $= 4 * 10^{-12}$ cm^2/sec to diffuse an average distance of 10^{-4} cm?
(d) How long does it take a lipid molecule to diffuse $5 * 10^{-7}$ cm? The time for a lipid molecule to flip from one side of a membrane to the other is 10^9 times longer than the time needed for it to diffuse $5 * 10^{-7}$ cm. What is the flip-flop time of a lipid molecule?

(8) A typical lipid occupies an area of 0.1 nm^2. The area of a cell is roughly $3 * 10^6$ nm^2. Assume that 90% of the surface is lipid. About how many lipid molecules are there on this cell membrane? Hint: Think bilayer!
(9) Cartoons of most carbohydrate molecules will look somewhat like polymers. Sketch a generalization of the cell membrane discussed in this chapter that includes short carbohydrates and glycoproteins.
(10) The interaction energy between two hydrophobic species in water is given by the following equation:

$$E_{hydrophobic} = -B\frac{R_1 R_2}{R_1 + R_2} e^{-d/d_0} \text{ kJ/mole}, \qquad (12.1)$$

where R_1 and R_2 are the radii, in nm, of the assumed spherical particles, that are separated by a distance *d* nm, B=84 kJ/mole-nm, and d_0= 1 nm. (From **The Forces Between Molecules**, M. Rigby, et al. Oxford University Press, 1986, pg. 207.)

(a) Is this energy attractive or repulsive? Explain. (Hint: What does the minus sign tell you?)
(b) Assume that R_1=R_2 and d=1 nm. Calculate $E_{hydrophobic}$.
(c) For R_1=R_2 graph $E_{hydrophobic}$ as a function of d. Estimate the range of this interaction.

(11) Convert kJ/mole into J/molecule.

12.12 Research questions

(1) What are the chemical structure of the nucleotides that form DNA and RNA?

(2) Discuss the structure of a super helix.
(3) Discuss the structure of cholesterol.
(4) Why are the β sheets called parallel and antiparallel?
(5) "β turns" connect the ends of two adjacent segments of an antiparallel β sheet. What is its structure? How is it stabilized by hydrogen bonding?
(6) Why are there only L-amino acids?
(7) It may seem amazing that all proteins can be formed from 20 amino acids. A corresponding situation is the following. There are 26 letters in the English alphabet. How many words are there in the English language? Discuss your answer.

Chapter 13

Digital Devices and Microcomputer Basics

13.1 Overview

This is the first of two chapters on applied technologies. It briefly discusses the basics of binary numbers[1] and their mathematics. Since computers are digital machines[2] Binary numbers are "the language computers speak." While at first blush this subject may appear difficult, most of this chapter consists of definitions. Many of the logic gates may be familiar to the reader from other contexts, and so we will merely formalize or restate what is already known. By combining these logic gates, the operation of the basic building blocks of a computer will become clear. We will address the following questions:

(1) What is the binary number system? Why is it used by computers?
(2) How does one perform binary mathematics?
(3) What is a logic gate, and what are the basic types of logic gates?
(4) What is computer memory?
(5) What is a computer bus?
(6) What are the minimum number of parts needed to make a computer? What task do these parts perform?

[1]Each digit of a binary number may have only two possible values, usually taken as 0 and 1.

[2]This means the numbers that are processed by such instruments are discrete as opposed to continuous. The simplest digital signals are binary signals–signals that are off or on, 0 or 1.

13.2 Introduction

The fuel gauge on many cars has a needle that continuously moves from E to F. The speedometer needle on most cars tells us how fast we are going by continuously changing where it is pointing. Both gauges are examples of *analog* indicators. An analog signal, like an analog indicator, can take on a continuous range of values. This should be contrasted with the signal stored on a compact disk, which is stored as a sequence of spots that either reflect light or do not. These two reflective states mean that the signal from each bit (**b**inary dig**it**) of a CD is *discrete*. Discrete signals are also known as *digital* signals. Digital signals have the virtue of having only a fixed and finite number of allowed values, determined by the sampling algorithm and the design of the electronics. Most often, this number is limited to just two values. Computers use digital numbers because they are particularly easy to process. A major result of the evolution of microelectronics has been the development of electronic circuits that process digital signals with high speed.

Most of our everyday experience is with analog signals or with digital signals that have been rendered nearly analog through *digital-to-analog* conversion. In order to understand and manipulate digital information, a special numbering system, the binary number system, and a "new" algebra called Boolean algebra are required. Both are used to design and explain the various logic operations that are performed by circuits called *logic gates*, the basic components of logic systems. In practice, computers and other digital circuits consist of *logic gates* and *memory elements*. Such digital circuits temporarily store data and/or instructions in special devices called *registers*. Other circuits may use *counters* to perform calculations or keep track of the progress of a process or a program. Finally, data and instructions are stored in *memory*. The purpose of the first part of this chapter is to provide a very brief introduction to this topic so that you can better understand how a computer functions.

13.3 Binary numbers

The numbering system utilized by digital equipment is the binary number system, where there are only two states or numbers, 1 and 0. Sometimes these are also called high and low, or on and off. It might appear at first that it impossible to represent any number as a binary number. As we shall see later, some numbers can only be approximated. Nevertheless, any number that you can write in normal decimal notation can be written in binary notation to very good accuracy.

To see how this works, we will first look carefully at numbers in the decimal system, where the ten digits 0 through 9 are allowed. In the decimal system each digit of number represents a value (0 through 9) and a position that can be related to the power of ten. For example, consider the number 314.29. This can be written as

$$300 + 10 + 4 + (2/10) + (9/100),$$

or

$$3 * 100 + 1 * 10 + 4 + 2 * (1/10) + 9 * (1/100),$$

or, using the powers of ten notion,

$$3 * 10^2 + 1 * 10^1 + 4 * 10^0 + 2 * 10^{-1} + 9 * 10^{-2}.$$

It is useful to recall that any non-zero number raised to the zeroth power is equal to one, that is, $z^0 = 1$ for all non-zero z. Since 10 is the base, each succeeding position to the left of the decimal point corresponds to one more positive power of ten. Likewise, each succeeding position to the right of the decimal point (also called the *mantissa*) corresponds to one more negative power of ten. This same procedure or representation can be used for any decimal number. We can write any number in the form $a_n * 10^n + a_{n-1} * 10^{n-1} + \dots$, where all the a_n's tell us how many of a given power are in the number and, by definition, each a_n is a whole number between 0 and 9.

Now, consider applying this scheme to binary numbers. By analogy with the situation just described, you would expect that any binary number may be written as $b_n * 2^n + b_{n-1} * 2^{n-1} + \dots$, where now each b_n is either 0 or 1.[3] For example, consider the binary number 10111. This may be written as:

$$1 * 2^4 + 0 * 2^3 + 1 * 2^2 + 1 * 2^1 + 1 * 2^0.$$

Now, since $2^4 = 16$, $2^3 = 8$, $2^2 = 4$, $2^1 = 2$, and $2^0 = 1$, the binary number 10111 is the same as the decimal number

[3]This is a general technique for writing a number in any base. Base ten was picked because it is the basis of the decimal system and everyday mathematics. Base 2 was picked because the binary system is used in computers.

$$1 * 16 + 0 * 8 + 1 * 4 + 1 * 2 + 1 * 1 = 16 + 0 + 4 + 2 + 1 = 23.$$

Similarly, the binary number 111.11 can be written as

$$\begin{aligned} &1 * 2^2 + 1 * 2^1 + 1 * 2^0 + 1 * 2^{-1} + 1 * 2^{-2} \\ &= 1 * 4 + 1 * 2 + 1 * 1 + 1 * (1/2) + 1 * (1/4) = 7.75. \end{aligned}$$

Based on these two examples, you should correctly conclude that conversion from binary to decimal number is straightforward. The conversion rule may be stated as follows. "To convert a binary number to its decimal equivalent, add the decimal equivalent of 2^n for each position occupied by a 1." The conversion of non-integer decimal numbers to binary is more complicated and will not be discussed. We will, however, discuss one important aspect of decimal-to-binary conversion, approximation. For example, 7.76 is only 0.01 larger than 7.75, but it cannot be expressed **exactly** as a binary number since:

$$0.01 = 1/100 \approx 1/128 + 1/512 + 1/8192 + 1/16384 + 1/32768 = 0.00997925.$$

Including more digits in the mantissa of the binary number will yield an even better approximation, but not exactly 0.01. By expressing 0.01 in powers of 2^{-n} for $n \leq 22$, one can approximate 0.01 to within 10^{-8}.

13.4 Binary arithmetic

Binary arithmetic is similar to decimal arithmetic. The key point that one must keep in mind is that in binary math $1 + 1 = 10$ is pronounced "one zero" or, more generally, "0 carry the 1." One performs addition column by column, starting at the right, just as in decimal math Subtraction is also performed column by column, again starting at the right, borrowing if necessary. Multiplication is straightforward once one knows the rules for partial products:

$$0 * 0 = 0,\ 0 * 1 = 0,\ 1 * 0 = 0, \text{ and } 1 * 1 = 1.$$

Division is more complicated and will not be discussed.

The basics are illustrated in the following examples of both decimal mathematics and the corresponding binary operations.

(1) decimal: $15 + 10 = 25$
binary: $1111 + 1010 = 11001$

check: $11001 = 16 + 8 + 1 = 25$

(2) decimal: $14 - 11 = 3$

binary: $1110 - 1011 = 0011$

check: $0011 = 2 + 1 = 3$

(3) decimal: $8.5 * 2.5 = 21.25$;

binary: $1000.1 * 0010.1 = 10101.01$

check: $10101.01 = 16 + 4 + 1 + (1/4) = 21.25$.

Other examples will be explored in the Exercises.

There are other number schemes used by computers. The two most common are *octal* or base 8 and *hexadecimal* or base sixteen. These are used extensively in microprocessors. There are also more complicated bases which will be explored in the research questions at the end of this chapter.

13.5 Basic logic gates

There are three elementary logic gates described by the names **AND**, **OR**, and **NOT**. These gates have standard symbols and are defined in terms of *truth tables* that list the gate output for each of the allowed combinations of inputs. These elementary gates also correspond to standard *Boolean expressions.* More complicated logic gates can be formed from combinations of these fundamental logic gates.

13.5.1 *The AND gate*

The first logic element we will consider is the **AND** gate. The standard symbol for a two input **AND** gate and its truth table are shown in Fig. 13.1

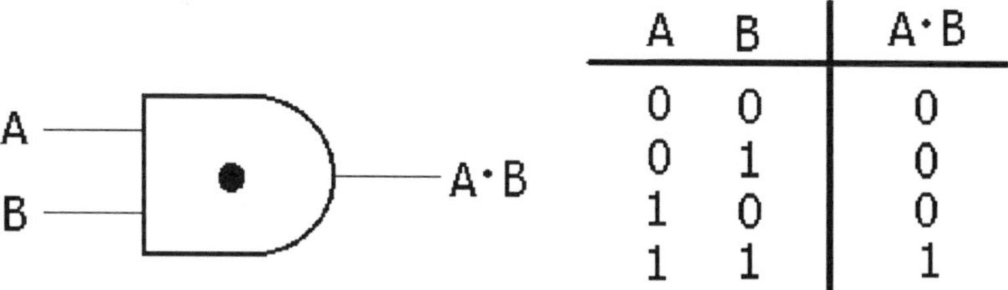

A	B	A·B
0	0	0
0	1	0
1	0	0
1	1	1

Fig. 13.1 The symbol and truth table for the two input AND gate.

The expression A• B in the symbol and the truth table is read "A and B." In this figure, the standard convention of having the inputs on the left and output on the right has been followed. Thus A and B are inputs and A• B is the output. Notice that the output is 1 (high, on, or true) only when **both** input A and input B are high at the same time. If either or both inputs are 0, (low, off, or false) the output is 0. This type of gate can be expanded to more than two inputs. In such a case, the output is high only when **all** of the inputs are simultaneously high. Finally, you can tell this is an **AND** gate by its symbol, which contains the "•" and has a characteristic shape.

13.5.2 *The OR gate*

The symbol and truth table for a two input **OR** gate is shown in the following figure.

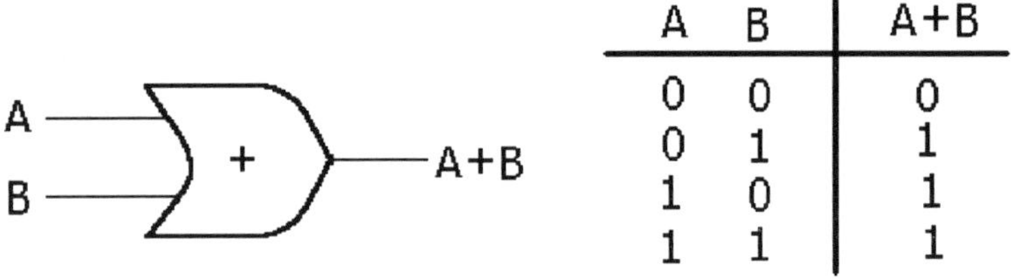

A	B	A+B
0	0	0
0	1	1
1	0	1
1	1	1

Fig. 13.2 The symbol and truth table for a two input **OR** gate.

Notice that this gate uses the word "or" in a slightly different manner than is used in normal language. The **OR** gate gives a high output when either input A or input B is high or when **both** input A and input B are high. In common language, this would be the same as saying, "I will go the football game or take a nap or both." Usually, we use "or" to mean "either-or" (one or the other), but not both. This common use of language can be represented by a different logic gate and operation called the **EXCLUSIVE OR**. The **EXCLUSIVE OR** gate can be constructed from the three elementary gates. You will explore this in an exercise.

13.5.3 *The NOT gate*

The third basic gate is the **NOT** gate. To explain this gate we must introduce the symbol $\overline{\text{A}}$, an A with a bar over it. This symbol is read "not A" and means that if A = 1, then $\overline{\text{A}} = 0$. Similarly if A = 0, then $\overline{\text{A}} = 1$. The symbol and truth table for the **NOT** gate are shown in Fig. 13.3.

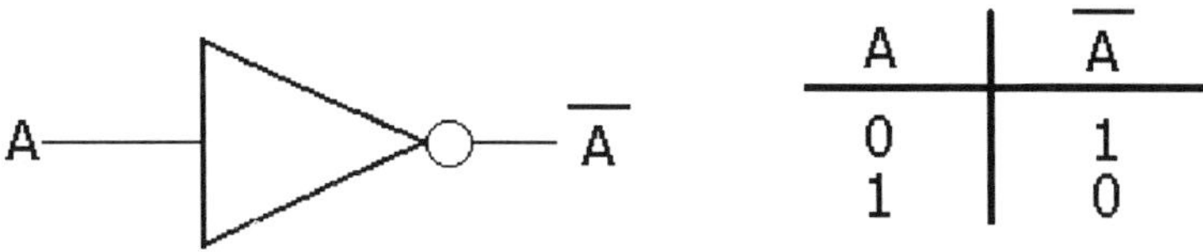

A	$\overline{\text{A}}$
0	1
1	0

Fig. 13.3 The NOT gate.

Clearly, the purpose of a **NOT** gate is to give an output which is the opposite of the input.

13.5.4 *The NOR and NAND gates*

Two compound gates occur often enough that they should be discussed. The first is the **NOR** gate, a combination of an **OR** gate followed by a **NOT** gate. The output is NOT **OR** or **NOR**. The combination of these two gates and the symbol for a two input **NOR** are shown in Fig. 13.4, along with the **NOR** truth table.

The symbol for a **NOR** is the similar to that for the **OR**, but there is a circle before the output. This circle is the key. It tells you to take the **NOT** of the input to it. It can be thought of as shorthand for a **NOT** gate. Once more, the bar over the A + B means "perform the **NOT** operation." Notice that if the A input is connected to the B input, the output is **NOT** the input as is illustrated in Fig. 13.5.

The second compound gate is the **NAND** gate, a combination of an **AND** gate followed by a **NOT** gate. The output is that of a **NOT AND**. Once more, a circle at the output of the **AND** logic gate is a key that this is a **NAND** gate. The symbol and truth table for a **NAND** gate is shown in Fig. 13.6.

Interestingly, a suitable combination of **NAND** logic gates may be used

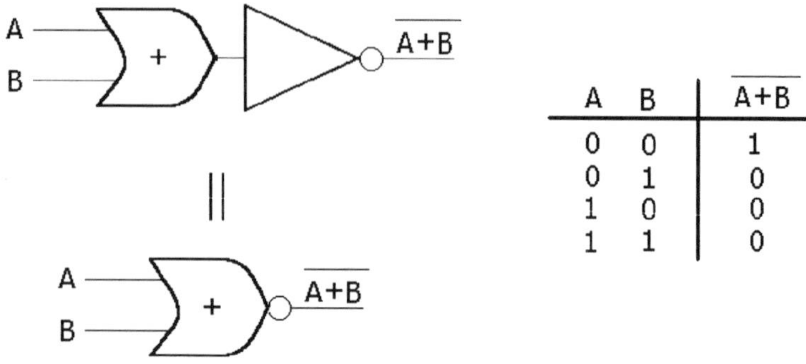

Fig. 13.4 The symbol and truth table for the two input NOR gate.

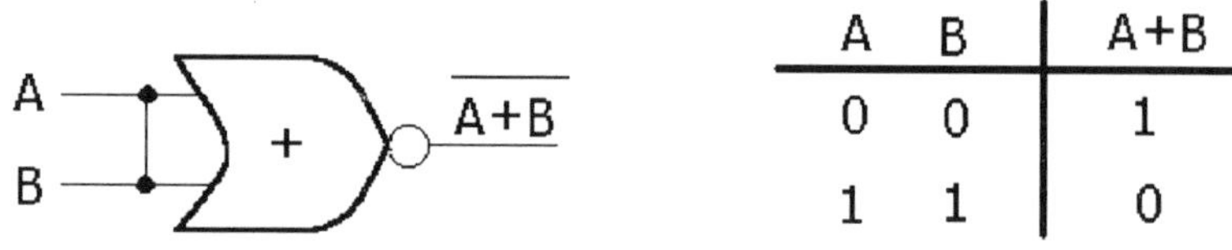

Fig. 13.5 A **NOR** gate wired as as a **NOT** gate and its truth table.

to replace the four other gates considered earlier. Thus, one may not need to make all sorts of different gates when building a digital device.

13.6 Memory elements

Various circuit elements may be constructed by combining logic gates including memory elements. A memory element must have two stable states: 1 and 0. It must retain these values until it is told to change to a different value. Ideally, it should change values very quickly and reliably. A memory element must often incorporate both past and present values of variables. A simple device that performs all of these tasks and is the basis of additional devices is the *flip-flop*. There are many types of flip-flops that serve as the building blocks of registers and counters, along with addressing

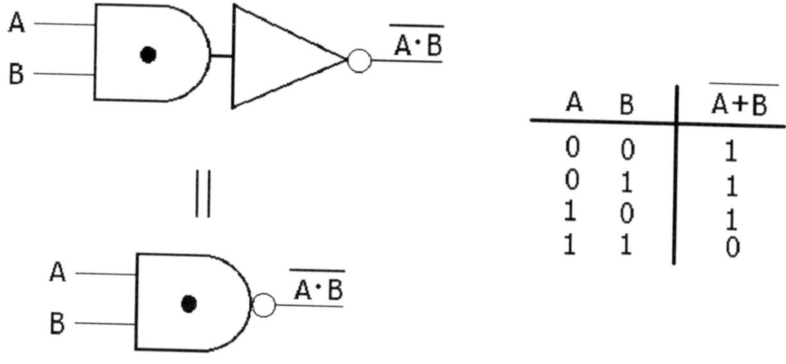

A	B	$\overline{A+B}$
0	0	1
0	1	1
1	0	1
1	1	0

Fig. 13.6 The symbol and truth table for the NAND gate.

hardware and memory elements. We will not dwell on the details of these more complicated logic elements. One of the simplest flip-flops is discussed in the exercises.

13.7 Boolean algebra

Boolean algebra is the algebra that allows us to simplify and manipulate binary variables using **AND**, **OR**, and **NOT** operations. From these operations one may construct adders, which add binary numbers, and subtracters, that may be thought of as adding a negative binary number to a positive binary number. Since multiplication is repeated addition and division is repeated subtraction, adders and subtracters allow the basic arithmetic operations to be performed. An understanding and detailed use of Boolean algebra allows one to synthesize and simplify various expressions. The basic postulates of Boolean algebra are the truth tables for the **AND**, **OR**, and **NOT** gates. And just as "standard" algebra extends arithmetic by allowing variables, Boolean algebra extends the simple gates by also allowing variables, which of course can only have two values: 1 or 0. A few Boolean theorems will be presented to illustrate the point.

In the following example, we will let A be a binary variable. This means A is not specified but can only take on values of 1 or 0. Then by constructing truth tables one can prove $A + 0 = A, A + 1 = 1, A + A = A, A \bullet A = A$, and other **one variable** relationships. The multivariable relationships are somewhat more complex and will not be explored here.

13.8 Memory

In a digital computer, instructions (programs) and binary numbers are stored in *memory.* Memory is an organized arrangement of individual memory cells in which each cell can store one binary digit or bit of information. There are two basic kinds of memory, *read-and-write* or *random-accessmemory* (RAM) and *read-only memory* (ROM). In random-access-memory (RAM), data may be stored (written) at a selected memory location, and later retrieved (read). In read-only memory, (ROM) the data or instructions are permanently stored, and the computer or instrument can read the data, but cannot change it.

A diagram of a very general integrated circuit memory is shown in Fig. 13.7.

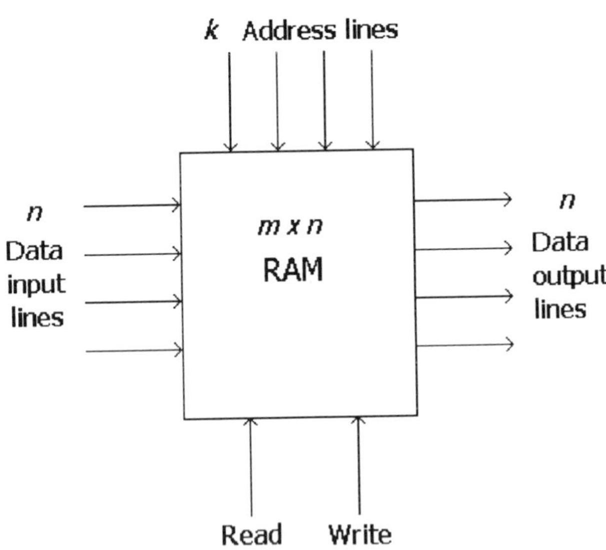

Fig. 13.7 A drawing of the essential inputs and outputs of a memory circuit.

A number of features of this memory should be noted. First, there are

n data lines for both the input and the output. For 16 bit data one needs 16 data lines, while for 32 bit data one needs 32 data lines. The address lines tell the computer where to look in memory for data. Note that for 1 line, one can specify two addresses, 0 or 1. Similarly for 2 address lines, 4 addresses (00, 01, 10, 11) are possible. In general, k address lines can designate $m = 2^k$ addresses, each address containing m bits of information. A memory circuit also has two control lines, the *read* and *write* lines. These lines are used to tell the chip whether it should read the data from address m or write data to address m. Thus, a memory operation requires the computer to specify whether to read or write, and the location or address that is to be written to or read from. It is not uncommon for other control lines to be included in memories. Finally, the data can be accessed in any order, hence the name random access memory.

The addressing of particular logical locations will also be of interest when liquid crystal displays are discussed. For this reason we will investigate memory addressing in more detail. Suppose a memory chip has 1k (actually $1024 = 2^{10}$) memory addresses. One can envision many ways to arrange these addresses. For example, one could lay the memory out so that the addresses are in a single row. The addresses would look something like that shown in Fig. 13.8.

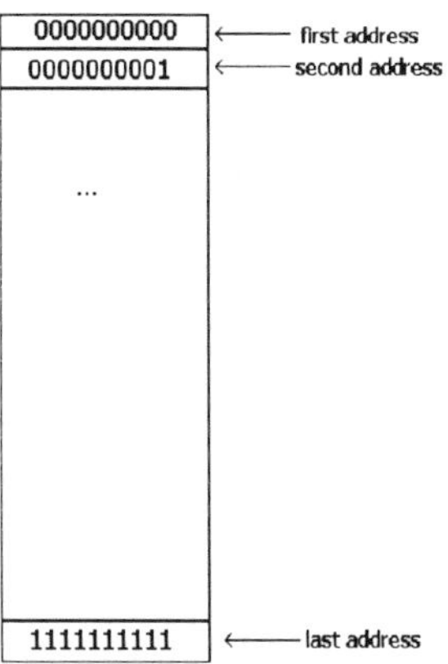

Fig. 13.8 A linear addressing scheme.

One can also use *coincident* or *matrix addressing*. In this addressing scheme, the memory is laid out on a grid and a memory location, or address, is specified by giving its X address, which selects its row, and its Y address, which selects its column. Using this scheme,a 1k memory chip one could be laid out in grids of 2x512, 4x256, 8x128, 16x64, or 32x32 memory cells. A similar matrix addressing scheme is used with liquid crystal displays. A simplified picture of a section of matrix addressed memory is shown in Fig. 13.9.

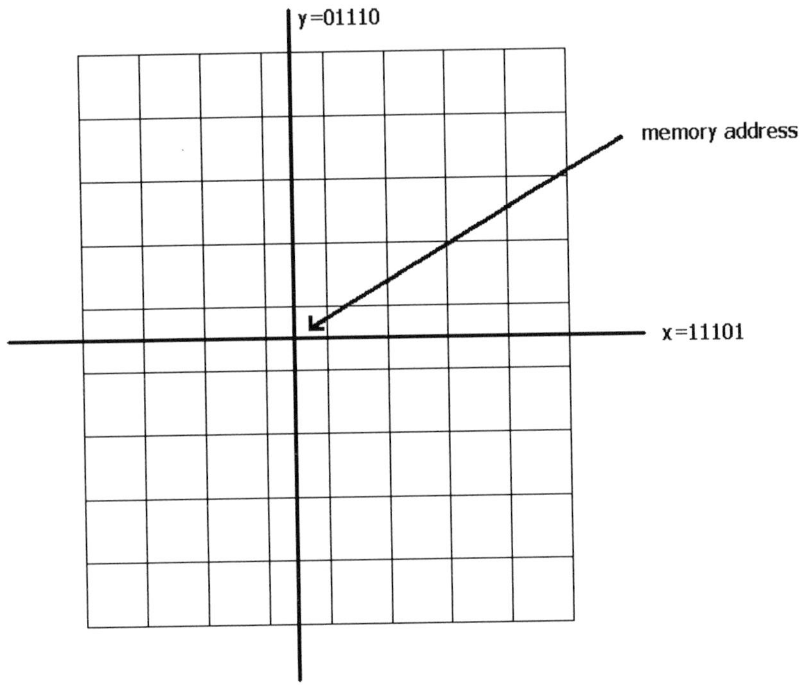

Fig. 13.9 An example of matrix addressing.

The addressing of read-only memory (ROM) is similar to that of RAM. Again, the important difference between these types of memory is that in ROM the data is permanently stored in the memory locations and can only be read.

13.9 Routing and transfer of data

In a typical digital system there are a large number of sources and destinations for the information. It is very inefficient and difficult to provide each possible source-destination pair with its own interconnects. The approach that engineers have taken to solving this problem is to transfer information along a common path or *bus*, and then, using digital switches called *multiplexers* and *demultiplexers* to connect the correct source and destination to the bus. This approach is described as *multiplexing*. To understand this more clearly, consider the multiplexing bus data transfer system shown in Fig. 13.10.

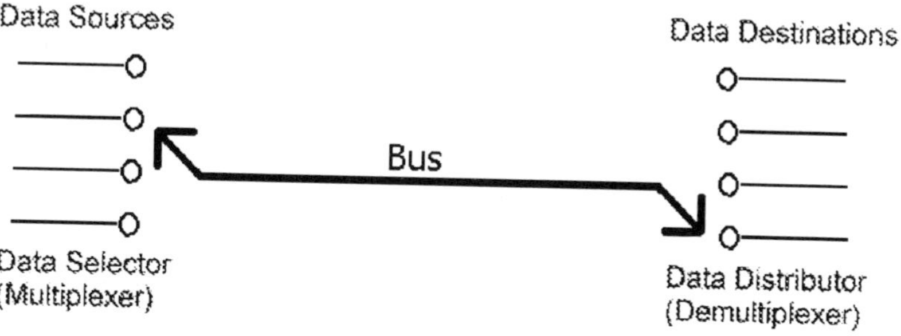

Fig. 13.10 An example of a data bus.

In this example, data can come from any one of the four sources. The source whose data is carried by the bus is determined by the position of the data selector or multiplexer on the left side. On the right side, the data can be distributed to any one of four different destinations depending on the position of the data distributor or demultiplexer.

13.10 The basic computer

The digital computer is a very large number of simple logic gates and memory devices that perform complex calculations at high speed. A diagram of the functional components of a general purpose computer is shown in Fig. 13.11.

The input component represents all sorts of input data and instructions. These can come from ROM, RAM, the keyboard, floppy disks, and CD-ROMs. All data that enters the computer for later use as either instructions or data comes through the input.

The memory component is where each instruction or datum is stored until it is retrieved for use. In contemporary Pentium class computers,

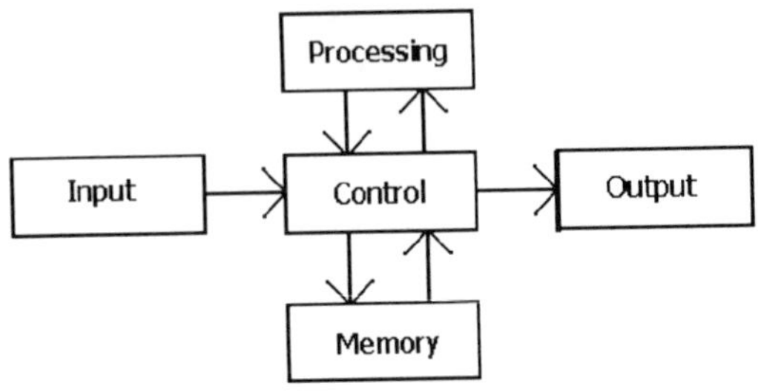

Fig. 13.11 The functional components and their arrangement in a general purpose computer.

memory can consist of several million addresses.

The control component is the heart and soul of the computer. The flow of instructions, data and output is managed by the control unit. It also controls the processing unit. The controller ensures that data and instructions are at the right place at the right time for the processor to act upon them. Furthermore, the controller sends intermediate results to memory for later use and "results" to the output.

The processing component is also called the arithmetic unit. It consists of several parts that perform arithmetic and logic operations on data in a special register called the *accumulator*. It is common for the processing unit to have a data bus, multiple inputs and outputs and several special registers.

The last functional component is the output. Once the computer has obtained a result or performed a task, the output is displayed. Common output devices include liquid crystal displays, cathode-ray tubes, printers, and in special applications, switches that control external equipment.

13.11 A more detailed look at a computer

We will now discuss one component of the computer in greater detail. First, consider Fig. 13.12 in which the importance of bus lines has been emphasized. This figure also emphasizes the details of the control and processing units shown in the previous figure.

The CPU or central processing unit is the heart of the computer. It has two main subsystems. The first is used for control of the timing. This was called simply "control" in the functional component diagram in the previous

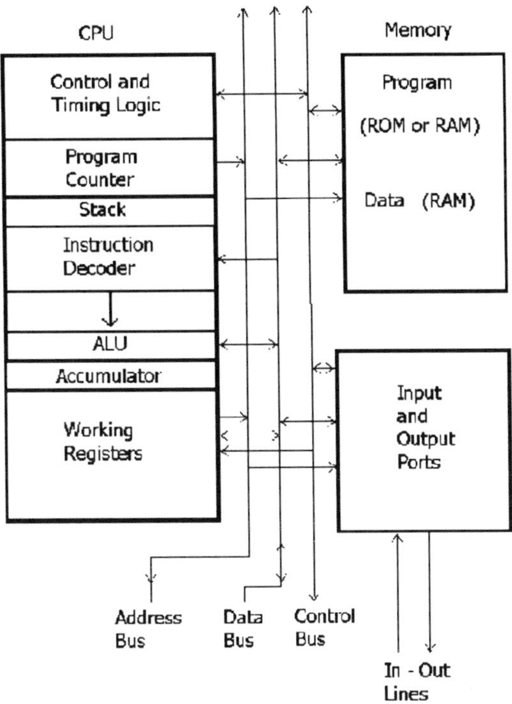

Fig. 13.12 Details of the CPU subsystems and bus lines of a computer.

figure. Notice that it actually has several parts. The *program counter* gives the memory address of the next instruction that will be executed. The *stack* is useful when separate sub-programs are run. The second portion of the CPU performs the arithmetic and logic operations and consists of the *arithmetic logic unit* (*ALU*), the *accumulator*, and several special *working registers*. The control subsystems tell the ALU and registers what to do, and is indicated by the arrow pointing from the working registers to the ALU. Also observe that some of the data lines have one-headed arrows while others have two-headed arrows. The arrows indicate the direction that data can flow.

13.12 References

There is a wealth of information available on digital electronics and computers. I particularly like **Circuits, Devices and Systems**, R. J. Smith and R. C. Dorf, Wiley, 1992 in large part because I taught from it and know it well. A tremendous book covering all aspects of electronics is **The Art of Electronics**, P. Horowitz and W. Hill, Cambridge University Press, New York, 1990. At a simpler, yet very clear level, is **How Computers Work**, R. White, Ziff-Davis Press, Emeryville, CA., 1997.

13.13 Exercises

(1) Find the decimal equivalent to the following binary numbers:
a) 11001, b) 101.01, c) 1.111,
d) 1111, e) 100000, f) 10101.

(2) Perform the indicated binary math operations. Verify your results by converting the numbers to decimal and performing the decimal math.
a) $1111 - 1010$, b) $1010 + 1010$, c) $111 - 011.1$,
d) $1101.1 + 0011.11$, e) $11.1 * 111$, f) $101 * 101$

(3) By expressing 0.01 in powers of 2^{-n}, where $n \leq 22$, show that one can approximate 0.01 to within 10^{-8}.

(4) A two input **AND** gate has the time varying signals on inputs A and B as shown in Fig. 13.13. Sketch the output as a function of time.

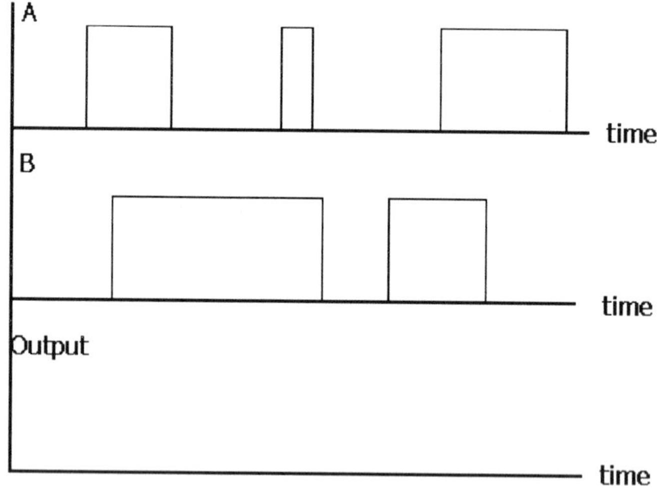

Fig. 13.13 Problems 3 and 4.

(5) A two input **OR** gate has the same time varying input signals shown in Fig. 13.13 as in the previous exercise. Sketch the output as a function of time.

(6) Find the truth table for the **NAND** gate circuit shown in Fig. 13.14. What other gate has the identical output? (The solid dots on the lines indicate that the wires are connected.)

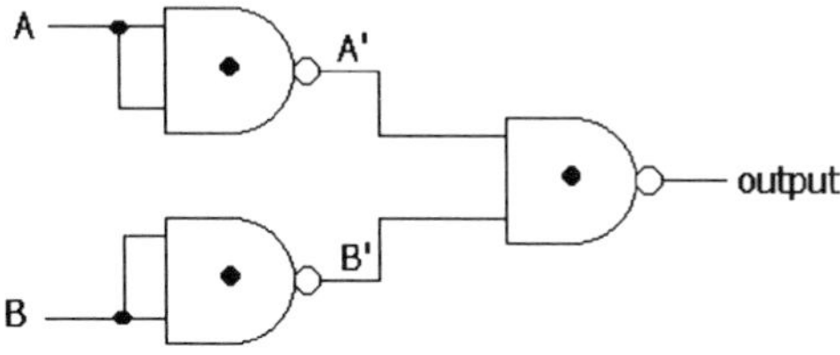

Fig. 13.14 Problem 5.

(7) The **EXCLUSIVE OR** gate discussed in this chapter is described by the logic function $(\mathrm{A}+\mathrm{B})\bullet\overline{\mathrm{AB}}$ which can also be written as $\overline{\mathrm{A}}\mathrm{B}+\mathrm{A}\overline{\mathrm{B}}$. This corresponds to "A **OR** B **AND NOT** A **AND** B." Construct the truth table for the **EXCLUSIVE OR** gate. This gate is sometimes called the "inequality comparator." Explain.

(8) In Fig. 13.15 below, which circuit operates as an **EXCLUSIVE OR**? Explain the function of the other combination of logic gates. (Hint: Find all the intermediate outputs.)

(9) From your experience with computers give several examples of RAM and ROM.

(10) Give an example of an analog signal and a digital signal that is not given in the chapter.

(11) Draw a three input **AND** gate and construct its truth table.

(12) Use truth tables to prove the following Boolean theorems. In all cases A is a binary variable.
a) $\mathrm{A}+0=\mathrm{A}$, b) $\mathrm{A}+1=1$, c) $\mathrm{A}+\mathrm{A}=\mathrm{A}$,
d) $\mathrm{A}\bullet\mathrm{A}=\mathrm{A}$, e) $\mathrm{A}\bullet 1=\mathrm{A}$.

(13) In the simple, generic memory chips discussed in the chapter, separate read and write lines were drawn. Are separate lines necessary? Explain your reasoning.

(14) A memory chip has 12 address lines. How many memory cells can be

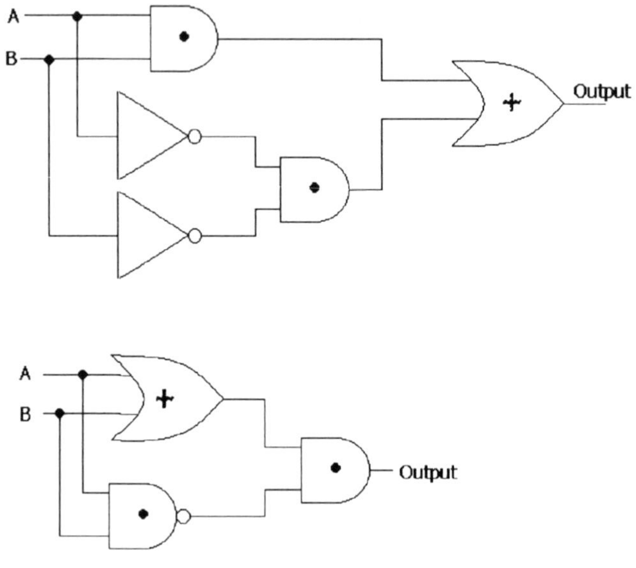

Fig. 13.15 Problem 8.

addressed?

(15) A simple two channel multiplexer is shown in the figure on the next page. Explain by use of truth tables and words how it works.

(16) A small memory chip is seven cells wide and twenty cells high. How many data lines are necessary for linear addressing? How many data lines are needed for matrix addressing? Which would be easier to implement? Explain.

(17) Dynamic read and write memory (DRAM) is a simple type of memory that is commonly used. In an implementation of this each memory cell is a MOSFET switch and a capacitor. The information is stored as a charge on the capacitor. Because of charge leakage from the capacitor, the information must be re-inputted or *refreshed* thousands of times per second. The electrical schematic diagram of two cells of such memory is shown in Fig. 13.17.

(a) To store a one in the top capacitor, the select line and the data lines must both be high. If the data line is held at seven volts and the *MOSFET* is ideal, what is the voltage on the capacitor?

(b) By repeating this circuit, draw a circuit with eight memory cells that uses single row addressing.

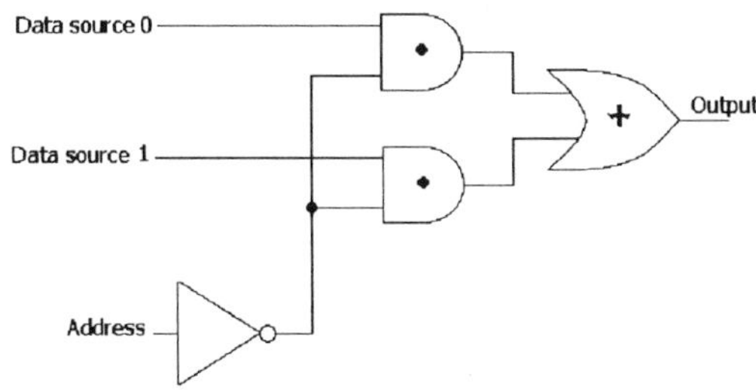

Fig. 13.16 Problem 15.

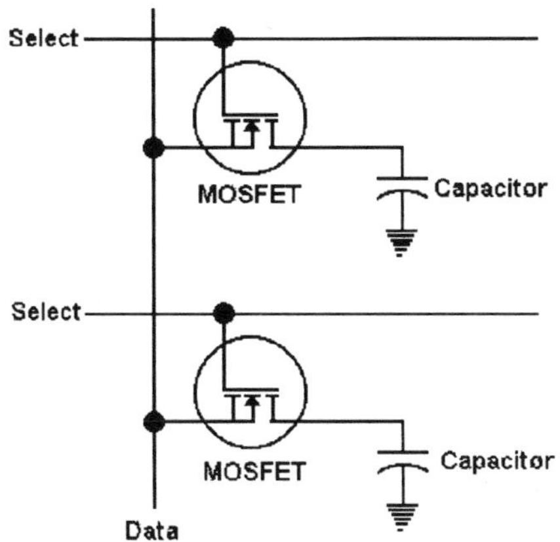

Fig. 13.17 Problem 17.

(c) By repeating this circuit, draw a circuit of eight cells that uses coincidence addressing.
(d) Discuss the difference, if any, in the number of lines needed in these two schemes.

13.14 Research questions

(1) Discuss how to convert decimal numbers to binary numbers. Include several examples.
(2) Discuss division of binary numbers.
(3) Identify the following: DRAM, EPROM, PROM.
(4) What are the octal and hexadecimal number systems? How are they used in computers?
(5) Another type of memory uses a flip-flop to store information. A very simple example of such a circuit is the *RS flip-flop* (RS stands for set-reset). This circuit can be made from two NOR gates as shown below.

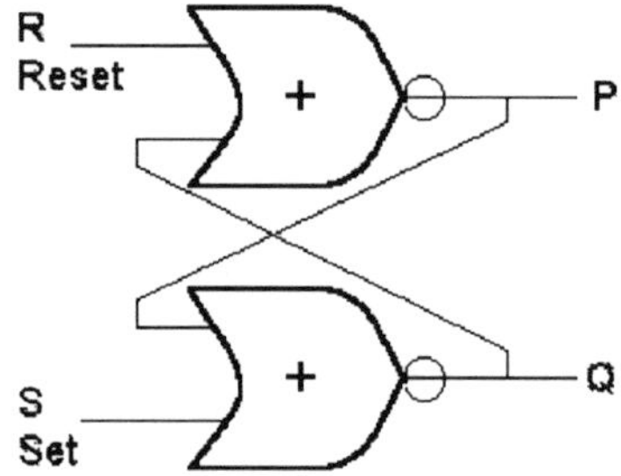

This circuit has *feedback*, which means that some of the output is returned or "fed back" to the input. Analysis of this device is more complex than some of the earlier devices because of this feedback. In practice, due to small differences in manufacturing, all **NORS** are not equally fast. This fact must be included in the present analysis.

(a) Begin by assuming that R = S = P = Q = 0. Furthermore, assume that the lower **NOR** operates first. Since P = 0 and S = 0, we have for the output of the lower **NOR**: $0 + 0 = 0$, $\overline{(0)} = 1$, so Q = 1. Then for the top **NOR**: $1 + 0 = 1$, $\overline{(1)} = 0$, so P = 0. Thus, the output has changed from P = 0, Q = 0 to P = 0, Q = 1. Now repeat the process on the bottom **NOR**: $0+0 = 0$, $\overline{(0)} = 1$, so Q = 1; and then for the top **NOR**: $1+0 = 1$, $\overline{(1)} = 0 =$ P. Thus, a constant state has been achieved.
Now assume the same starting condition, but the top **NOR** op-

erates first. Then $0 + 0 = 0$, $\overline{(0)} = 1 = P$. The lower **NOR** will yield Q = 0. Thus the outputs are now P = 1, Q = 0 (R and S still are both 0). This is a constant state. Which state will be achieved in a real device? Either one! We don't know in advance since this is an ambiguous state.

(b) Assume that S = 0, R = 1. There are 4 possible starting output states. We will assume that the bottom **NOR** is faster and use a table for this analysis. The previous P and Q are labeled P_p and Q_p. The analysis is just like in part a) and is shown in the following table.

Step	**R**	**S**	$\mathbf{P}_p$	$\mathbf{Q}_p$	**New P**	**New Q**	**Comment**
Start 1	0	1	0	1	—	—	
1	0	1	0	1	1	0	
2	0	1	1	0	1	0	Stable
Start 2	0	1	1	0	—	—	
1	0	1	1	0	1	0	Stable
Start 3	0	1	0	0	—	—	
1	0	1	0	0	1	0	
2	0	1	1	0	1	0	
Start 4	0	1	1	1	—	—	
1	0	1	1	1	1	0	
2	0	1	1	0	1	0	Stable

Thus, for S = 1, R = 0 the output when the lower **NOR** acts first always lead to P = 1, Q = 0.

(c) Verify that the same result occurs when the top **NOR** is faster (operates first).

(d) What are the stable outputs for S = 0, R = 1?

(e) Is there a stable output for S = 1, R = 1?

(f) From parts c) and d), note that once a stable state has been reached it doesn't change unless either R or S change. If S = 1 and R = 0, then P = 1. P can only be reset to 0 by S = 0 and R = 1. This behavior makes this circuit a memory element.

(g) The state R = S in the RS flip-flop is ambiguous. This ambiguity may be resolved by a somewhat more complicated circuit. Find out how. (Hint: What are J-K and D flip-flops?)

Chapter 14

Liquid Crystal Displays

14.1 Overview

The electronics that perform calculations inside a computer are not all that a computer needs. Calculating an answer and keeping it hidden in memory serves no useful purpose. Thus, all computers have an output device. Since one of the goals of this book is to understand a laptop computer, the output device we'll focus on is a display. The most common display in laptops is a liquid crystal display (LCD). This chapter builds on many earlier chapters and the reader will see how much of the earlier basic science and materials discussion is important to the present discussion. This chapter will address the following questions.

(1) What are liquid crystal displays (LCDs)?
(2) What types of LCDs exist?
(3) How do LCDs work?
(4) How does a computer address an LCD?
(5) Where is LCD technology going?

14.2 Introduction

One of the most common, everyday uses of liquid crystals is in displays. The simplest of these LCDs are temperature indicators and the digital displays in calculators and wristwatches. More sophisticated types of LCDs are used in laptop computers, projection TVs, overhead projector modulators, and similar devices. New types of LCDs are continuously being developed.

The commercial development of liquid crystal devices began in the 1960s. The first devices were not displays but chiral nematic temperature indicators and nematic electro-optic devices. LCDs first became viable in the 1970s. Their growth paralleled and depended, in part, on the great strides of the semiconductor industry. The first major commercial market was watches and calculators. This chapter will discuss some liquid crystal devices and how they work. It will bring together and extend the topics of the earlier chapters on forces and fields (Chapter 4), light and waves (Chapter 5), polymers (Chapter 9), liquid crystals (Chapter 10), and some ideas from the section on digital devices (Chapter 13). The truly interdisciplinary nature of liquid crystal devices and displays will be evident from the various sciences and technologies that are brought together in the simplest devices.

14.3 Cholesteric temperature indicators

As discussed earlier, the cholesteric or chiral nematic phase is characterized by order that is locally very similar to that of a nematic. However, these materials exhibit a long-range helical variation in the average direction of the molecules (also called the director). Recall that a helical structure is similar to the threads on a bolt. It can be either left- or right-handed, and it has a periodic variation in a physical property. In the bolt, the periodic variation is in the height of the metal. In a cholesteric liquid crystal there is a periodic variation in the index of refraction of the material. The variation in index of refraction is actually periodic in one-half of the pitch because two directors pointing in opposite directions are equivalent. The pitch length is often comparable to optical wavelengths.

Suppose white light is incident on a cholesteric sample. Most of the light will be transmitted through the sample. However, light with a wavelength *in the liquid crystal* equal to the pitch will be strongly reflected.[1] This is called *selective reflection.* The dispersion of the refractive indices of the cholesteric allows a small band of wavelengths about the central wavelength to be reflected. This selective reflection is the physical basis of cholesteric temperature indicators, mood rings, and "stress indicators."

This phenomenon is actually subtler than it appears. Recall that natural light may be modeled as a combination of equal parts of right- and left-handed circularly polarized light. Similarly, the helix of the cholesteric liquid crystal also has a handedness. One finds that when right circularly

[1] This behavior will be discussed in greater detail in chapter 16 on liquid crystals and the arts.

polarized light (the electric field rotates counterclockwise as it approaches you) is incident onto a cholesteric with right-handed pitch, light of the proper wavelength is reflected. When left-hand circularly polarized light is incident on the same chiral nematic it is transmitted without reflection. This polarization selectivity is illustrated in Fig. 14.1.

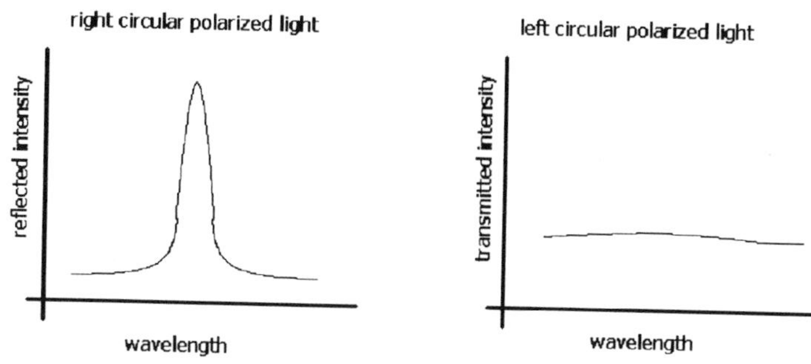

Fig. 14.1 Transmitted light vs. wavelength plots representing selective reflection.

Selective reflection alone is not sufficient to produce a temperature indicator. A good indicator requires a physical property that is a strong function of temperature. While the index of refraction of a cholesteric depends on the temperature, this effect is not large enough to produce a satisfactory temperature indicator. The cholesteric property that varies very strongly with temperature is the pitch length. The pitch length generally decreases with increasing temperature. The color change from violet to red can occur in a temperature range as small as 0.2°C or as large as 30°C depending on the material.

The thermochromic behavior of cholesteric liquid crystals is the basis of many devices. In surface thermometry, such devices are used for medical diagnosis where a visual thermal map can be created and used in diagnosis of cancers and vascular disorders. There are also many decorative uses of cholesterics where they are used, in part, because of the optical quality they impart to the product.

In practice, cholesteric liquid crystals are microencapsulated. Microecapulation is a process that takes small droplets of a liquid and wraps them with a tough, resilient coating. Chiral nematics are microencapsulated for several reasons. First, chiral nematics are thick, oily, viscous fluids that are difficult to use and control. Second, the coating protects the liquid crystal from contamination by air, foreign bodies, and chemicals that can easily change the color-temperature behavior of the liquid crystal. Third, it is good economic practice; an encapsulated droplet need only be 20% liquid crystal by weight to have optimal reflectance properties. Furthermore, droplets smaller than 2-3μm in diameter have reflectance properties similar to those of bulk material. Fourth, formation of solid crystals is inhibited by the droplets. Lastly, microcapsules with different thermochromic ranges may be mixed to obtain a wide variety of optical effects.

14.4 Liquid crystal information displays

The conversion of information from electrical signal levels to a visual format is the purpose of an electronic display. There are many different types of displays, each with characteristic strengths and weaknesses. In this section, we will discuss liquid crystal displays (LCDs). First, we will discuss some general features of many different types of liquid crystal displays beginning with the dynamic scattering display. This type of display has been replaced by better types of displays; however, many of the ideas that were originally implemented in the dynamic scattering display continue to be utilized. The polymer dispersed liquid crystal (PDLC) display, another display based on scattering, will then be described. The bulk of this chapter will discuss various types of nematic displays. The final section will present some of the basics of ferroelectric LCDs. This is by no means a complete list of LCDs, but is sufficient for an introduction to this ever expanding technology.

To begin, we must define several features of information displays. First, what do all displays, even the printed page, have in common? They all control the brightness of light and its color via either reflection, as with the keyboard on which I'm typing or a printed page, transmission, as with a slide projector, or emission as in cathode ray tubes (CRT) displays or televisions. The eye has a spatial resolution that depends on many external and physiologic factors. These include the distance between the observer and object being observed, the age of the observer and the overall light intensity. Thus, there is no need for a display to display features smaller than can be resolved by the user.[2] This resolution information can be used to define the minimum sub-area needed for a display. A display can then

[2]Of course, often the smallest feature is larger than the typical person can resolve. For example, the smallest feature that can be resolved by a person in standard TV is larger than the average person's spatial resolution. This has lead to the development of high definition TV.

be constructed from a large number of these small sub-areas. In electronic displays, a display is composed of small sub-areas that are called *pixels*. My computer screen has a desk top that has an area 1024 x 768 pixels. I am able to see the words because the display electronics control which pixels are bright and which are dark. It also controls the colors of these pixels.

Rather than begin by focusing on a large-scale display, we will begin by focusing on two smaller and simpler ones. The first is the *seven-segment numeric display* used on many clocks. This display, shown in Fig. 15.2, can be made to produce any number by lighting the appropriate segments. This is a rather coarse display, and while you can, with some trickery, form letters, the letters often look poor and are difficult to discern. For many applications, both numbers and letters are needed and more "visually pleasing" shapes are desired. For this purpose the *five-by-seven dot matrix display* is often used. This is also illustrated in Fig. 14.2.

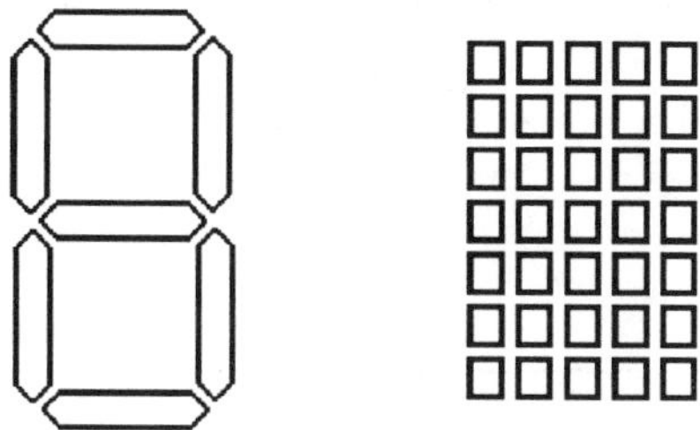

Fig. 14.2 The pixel patterns for a seven-segment numeric display and a five-by-seven dot matrix display.

In the five-by-seven dot matrix display, the brightness and possibly color of each pixel may be controlled independently, allowing all letters and digits to be produced with reasonable clarity.

All displays work on the same principle of controlling the brightness and color of light reaching the observer's eye. The light that is controlled may have two different origins. The light may originate within the display. The cathode-ray tube (CRT) on many desktop computers and TV screens is such a device. You can read a CRT or watch TV in the dark without any external light. This is contrasted with displays that work by controlling the reflection or transmission of ambient light or a light that is part of the device. The simplest example of this type of display is the liquid crystal watch display. This display is readily visible in ambient light. However, to read it in a darkened movie theater one must push a button that turns on a light. Displays that use illumination from the side or back consume electric power to generate the light. A big advantage to liquid crystal devices is that they can be made to rely on ambient light, which greatly reduces the electrical power consumption. This is why liquid crystal displays are so popular in battery operated equipment.

All non-CRT type displays rely on one of two methods of addressing the pixels. When there are relatively few pixels the choice is not serious. However, for larger displays, the method used is an important design consideration. The electrical addressing of LCDs can be divided into two broad categories or schemes called *active matrix* and *passive matrix*.

In most active matrix LCDs, a discrete, non-linear semiconductor device, such as a diode or a transistor, is associated with each pixel. The pixels are then arranged in a rectangular array or matrix. The most common active matrix LCD driving semiconductor is the thin film transistor (TFT). A schematic view of the display level electronics for a TFT display is shown in Fig. 14.3.

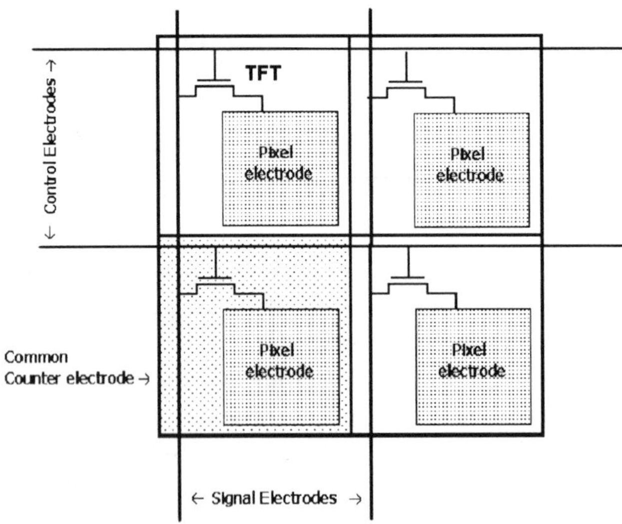

Fig. 14.3 The electrodes, transistors, and pixels for an active matrix display.

Note that there are four types of electrodes in this scheme: the control electrodes, the common counter electrode, the pixel electrodes, and the signal electrodes. An electrode is simply an electrical interconnect or plate. In this type of addressing scheme, the transistors are used as switches. A pixel changes state from dark to transmitting or vice versa when the signal voltage is applied through the transistor to the pixel electrode. There will then be a voltage between the pixel electrode and the common counter electrode that is situated on the other side of the liquid crystal layer. The transistors are controlled by voltages on the control electrodes. In this scheme, each pixel is directly driven through its own individual TFT, and so is remarkably similar to methods used to address memory in a computer.

The second type of addressing scheme is called passive matrix addressing. In this scheme there are no semiconductor devices. Once more the pixels can be arranged in a matrix. Part of a passive matrix LCD is shown schematically in Fig. 14.4. In this type of addressing, there are only two types of electrodes: row electrodes and column electrodes. Every pixel is defined by the overlap of a row electrode and a column electrode. Thus a rectangular passive matrix array that has NxM pixels needs only N+M electrodes (This is also true of active matrix displays). The voltage across the liquid crystal determines the optical state of the pixel. Since the row electrode is on one side of the liquid crystal and the column electrode is on the other side, the optical state of a pixel is determined by the voltage difference between the corresponding row and column electrodes. These row and column electrodes serve the same purpose as the pixel electrodes and the common counter electrode in active matrix addressing schemes. Note that in this type of addressing scheme, all the pixels in a row or a column see a common voltage.

A display need not be configured as a matrix in order to use matrix addressing. A common example of this is the clock display in Fig. 14.5. This is also called a 3 1/2 digit display, the half referring to the one. This clearly does not look like a rectangular matrix. To illustrate how this display can be matrix addressed (and the advantages of such addressing) one begins by considering direct addressing where every pixel has its own connection or lead and there is a common substrate for all pixels. Individually addressing each segment of the three seven-segment displays, the two pixels that form the one (which are always off or on together) and the colon requires twenty-three lines. However, by combining segments in an appropriate manner, an electrical and logical matrix can be formed with some interconnected segments on the top substrate and some interconnected seg-

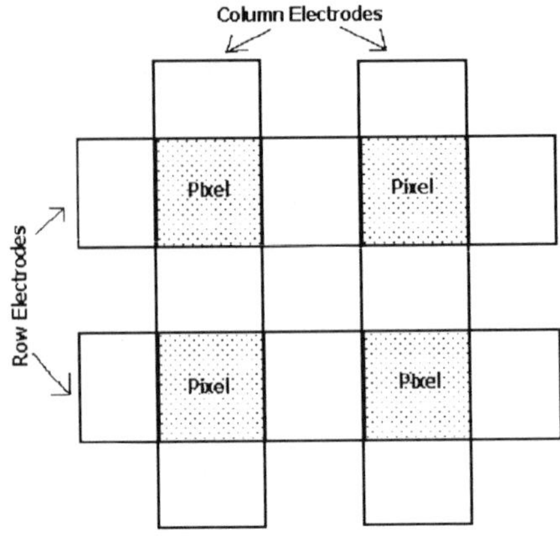

Fig. 14.4 Electrodes and pixels for a passive matrix display.

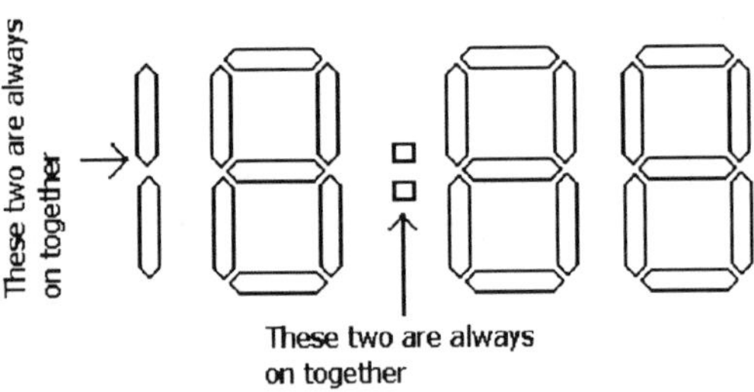

Fig. 14.5 The pixel pattern for a 3 1/2 digit display.

ments on the bottom substrate. This matrix is clearly not rectangular. However, the "equivalent matrix" may be drawn in a rectangular form. By utilizing such a scheme, one can address all 23 segments by using a 4x6 or 5x5 matrix addressing scheme with one or two addresses left over. The difference between ten and twenty-three lines is significant. For a large display, the difference in the number of lines can easily be a factor of 100.

This address line saving feature will be explicitly illustrated by considering two seven-segment displays. To directly address each segment requires 15 lines (seven for each display and 1 for the common electrode). If one can make an equivalent matrix, this number could be reduced to 8 lines (4x4 or 3x5). To see this, consider Fig. 14.6. The displays are numbered

1 and 2 and the segments are labeled a through g. The bottom electrodes are situated on the bottom or common substrate and the top electrodes on the top or segment substrate. Notice that in this 3x5 addressing scheme, each column consists of five connected segments, except for column three which has four. Each row consists of three connected segments, except for row three which has two. This approach leaves the address at row three, column three unused. In a 4x4 addressing scheme, each column and each row would be connected to four segments and there would be two addresses that are not used. Also, while the two seven-segment digits do not form a matrix, the addressing does form a matrix as shown in the equivalent matrix in Fig. 14.6.

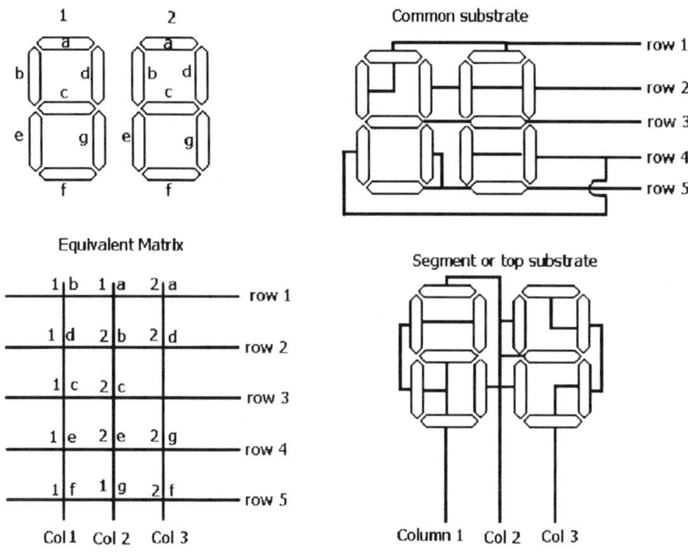

Fig. 14.6 Matrix addressing of two seven-segment displays.

For larger displays, it is impossible to keep a voltage continuously on a row or column electrode. Here, liquid crystal displays have another advantage over emissive displays such as CRTs. The electro-optic response of LCDs is determined by an average voltage called the RMS (root mean square) voltage across the liquid crystal. Because the electro-optic response is temporally accumulative, a constant voltage need not be applied between the two electrodes. Thus, displays may be temporally multiplexed (recall this was discussed earlier!).

Multiplexing does cause some difficulties. This is because multiplexing requires that the necessary voltage cannot be applied to all pixels at the same time. The normal addressing scheme has the column signals carry the pixel information. The row signal, in combination with the column signal, causes the desired state in the pixel. Suppose that the display has N rows. When the multiplexer applies the appropriate voltage to row 1 and the desired signal voltages are applied to each column electrode, the first row will have all pixels set to the desired state. The multiplexer then switches the output to row 2, and the desired signal voltages are again applied to the columns. This process repeats until all N rows are addressed, and means that each pixel has the desired addressing voltage across it only $1/Nth$ of the time. Since the difference in the average voltage between the on and off states decreases as N increases, the number of rows in a display is limited. There are some ways to reduce this problem that will be discussed later.

Finally, multiplexing and switching is not instantaneous. It takes time to switch the electronics to address a given row and apply the appropriate voltages to all the columns. For N rows, the time is longer than for say, $N/2$ rows. If this time becomes too long the display will appear to flicker. Fortunately, the eye does not respond to temporal variations much faster than about 30 variations per second or Hertz (Hz) . Incandescent lights go lighter and darker (they don't go out) 120 times a second. The eye averages this out. Similarly, the refresh rate on a typical CRT computer screen is 50-70 Hz. We normally don't observe this variation. However, if one waves ones hand rapidly back and forth in front of a CRT screen, a strobe effect, where the hand appears frozen in various positions, is observed. A similar effect is observed when watching rain fall when a porch light is on. The drops seem to have length and look like falling cylinders.

14.5 Liquid crystal displays that utilize light scattering

14.5.1 *Dynamic scattering displays*

The first commercial liquid crystal displays utilized the scattering of light that occurs in thin nematic liquid crystal devices that contain charged impurities. Here is how they work. A fairly large electric field is applied to the liquid crystal. Because liquid crystal molecules have either a permanent or an induced dipole, the liquid crystal will tend to align in the field.[3] The electric field also applies a force to the charged impurities.[4] This causes the impurities to move and destroy the orientational order of the liquid crystal imparted by the electric field. The impurities are in constant movement because the electric field is an AC field and its direction is constantly changing. Thus, there are temporal changes in the orientation of the liquid crystal. Since the two indices of refraction of the liquid crystal are not the same, these oreintational fluctuations will lead to optical inhomogeneities that, in turn, will scatter light and make the liquid crystal film appear whitish.

This type of display was usually used in ambient light and had a reflector on the back. There are several features of this display that are common in other displays. For this reason, the physical aspects of this display will be discussed in some detail. The cross-section of a typical passive LCD is shown in Fig. 14.7.

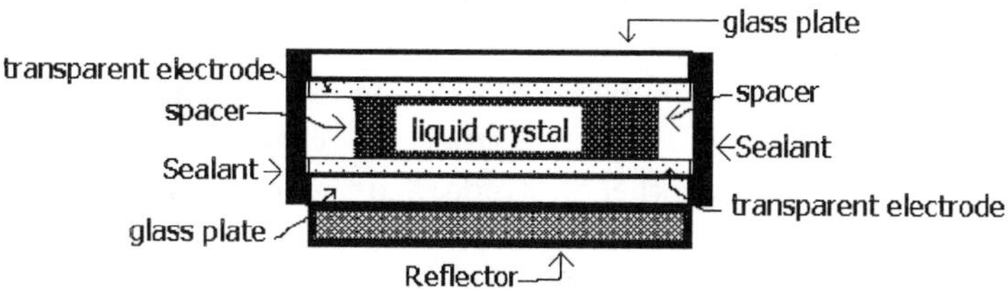

Fig. 14.7 A generic reflective passive LCD.

Several features of this display should be noted. First, the reflector is only included when the device is used in the reflective mode. Typically, the top and bottom of the LCD are glass. Glass that is extremely flat and of very low birefringence is used, and because of the volume of the LCDs manufactured, readily obtained. The glass is coated on one side with a "transparent" electrode. This is a thin layer of a material that is an electrical conductor and allows most of the incident light to pass through it. Common examples are indium tin oxide (ITO) and tin oxide. The

[3]This was discussed in the chapter on forces and fields, chapter 4.

[4]This force has a magnitude $F = qE$, where q is the charge on the impurity and E is the magnitude of the electric field. Negative charges particles are attracted towards the sources of electric field lines, while positive charges are repelled.

transparent electrodes form the column and row electrodes discussed earlier. The liquid crystal film is sandwiched between these two glass plates, and the separation of the plates is kept constant by spacers. A typical spacer is 4 to 10μm thick (0.16 to 0.40 thousandths of an inch). In order to hold the display together, prevent impurities from getting into the display, and keep the liquid crystal from leaking out, the edges of the display are sealed.

The voltage necessary to drive a dynamic scattering display is about 10-20V. This is higher than most other LCD displays. When this high voltage requirement is coupled with the low contrast of this display, it is easy to see why this display is seldom used in present day applications.

Before leaving the dynamic scattering display, there is one other design feature of this early LCD that has continued to be utilized in other LCDs. The electric field between the electrodes is uniform (except very near the edges) and its magnitude is given by $E = V/d$, where V is the potential difference or voltage between the electrodes, and d is the spacing between the top and bottom electrodes. This is in fact, a parallel plate capacitor such as discussed earlier. Suppose one applies a continuous voltage of the same sign between the two electrodes. The force on the charged impurities will always be in the same direction. Thus the charged impurities of one sign will all be swept toward one electrode, the charged impurities of the other sign will move to the other electrode, and the scattering will decrease. To avoid this, the polarity of the voltage across the display is constantly changed. The typical rate of such a change is a few thousand Hz. The way this is done is rather ingenious. Figure 14.8 shows the common (row) signal voltage, the (column) voltage for pixels that are selected and that are non-selected, and their total voltages as a function of time.

Note that when a pixel is non-selected there is still a time dependent voltage applied to both the top and the bottom electrodes but the sum of these two voltages is zero. By introducing a phase shift to the select electrode voltage waveform, the voltage across an active pixel alternates sign, has an average value of zero, but has an RMS value that is nonzero. While the voltage waveforms have changed in recent years, this basic idea continues to be applied.

14.5.2 *Liquid crystal dispersion displays*

While conventional LCDs have not been discussed, a few of their weaknesses will now be noted in order to understand why dispersion displays have been developed. Large area conventional LCDs are difficult to manufacture. The

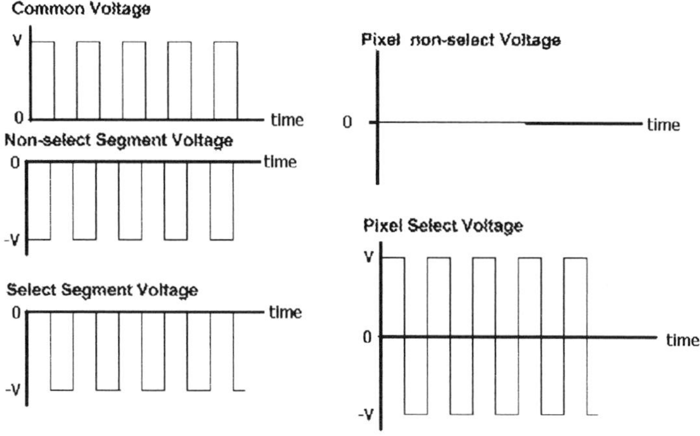

Fig. 14.8 Drive voltages for a dynamic scattering device.

optically active materials in most LCDs, nematic liquid crystals, flow like viscous liquids. Furthermore, the flatness and thickness of glass plates, as well as the spacing between them are difficult to keep uniform over very large areas. A second problem, which we will explore in greater detail later, is the relatively low light efficiency of conventional LCDs. Because these devices utilize polarizers, typical light efficiencies are about 35%. Also, since glass is not flexible, it is difficult to manufacture displays in non-flat geometries.

Dispersions of liquid crystal droplets in solid materials have been developed for high light efficiency, large area, and non-flat displays. Just as in the dynamic scattering LCD, the sample is held between two plates that are coated with a transparent conducting electrode such as ITO. The electric field causes the liquid crystal within the dispersed droplets to align and so the index of refraction matches that of the surrounding clear polymeric solid. This type of display typically utilizes nematic liquid crystals, but new types using other liquid crystalline phases are being developed.

Two different droplet technologies have been developed for dispersive displays. The first encapsulates the nematic liquid in a polymer film and is generically known as NCAP (nematic curvilinear aligned phase). The second technology creates liquid crystal droplets in a polymer binder by phase separation from a homogeneous mixture and is known as a polymer dispersed liquid crystal (PDLC). Both of these technologies are characterized by liquid crystal droplets randomly dispersed in a polymer matrix. The droplets are relatively uniform and are densely packed. The droplets tend to be aspherical and randomly shaped. This leads to a random orientation of the liquid crystal droplet directors. The following discussion will focus on PDLC devices.

The director pattern within most of the droplets in a PDLC is not random or uniform but assumes a special configuration called a bipolar droplet. The details of the director pattern in the droplets are unimportant; it is sufficient that you know that optically it is similar to the uniaxial optic material discussed earlier. Because the droplets of liquid crystal are randomly oriented in the (polymer) matrix, there is almost always a mismatch of index of refraction between the polymer matrix and the liquid crystal droplet. Since the droplets are randomly positioned in the matrix and have an average index of refraction different from that of the polymer, the medium is **not** optically uniform and light scattering results. The application of an electric field causes an aligning force on the liquid crystal molecules. By picking the correct materials, the liquid crystal molecules will align parallel to the field except very near the edge of the droplets. By careful design, the index of refraction of the polymer matches $n_\perp$ of the nematic liquid crystal, and, for normally incident light, there is very little scattering. The operation of of such a display is shown schematically in Fig. 14.9.

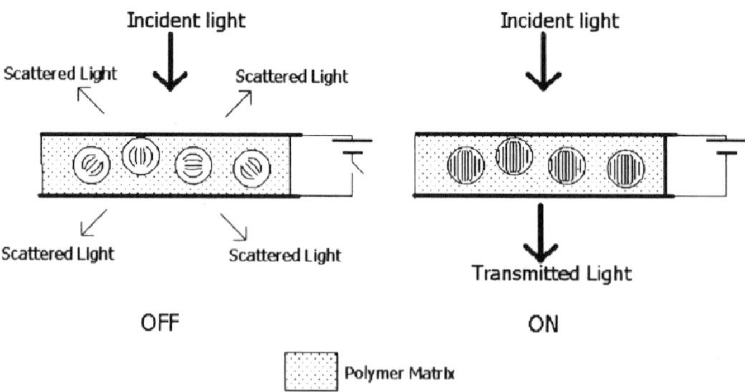

Fig. 14.9 Schematic drawing of how a PDLC display operates.

In this figure, the polymer matrix is represented by the dotted region. The droplets are shown with lines in them. that indicate the approximate direction of the director within the droplets. The figure on the left shows the field-off configuration, (the switch is open). The figure on the right shows the field-on configuration (the switch is closed). Note that the liquid crystal within the droplets can move and reorient, but the droplets do not move. An NCAP operates in a similar manner, only need not have a polymer matrix.

The electro-optics of PDLC materials can be tailored to specific applications. For example, they are commonly used in privacy windows where it is important that the off state is very highly scattering and the on state is very clear and transmitting. The driving voltage and switching speed are less important. In a display application, the difference between the on and off states is still important. However, it is critical that the driving voltages are small enough and the switching speeds fast enough. The optimization of these parameters sometimes requires a trade-off. For example, smaller driving voltage comes at the expense of longer turn off times. Similarly, increased off state scattering can reduce on state clarity. The details of these tradeoffs are beyond the level of this discussion.

Light is strongly scattered both by dynamic light scattering devices and by dispersive scattering devices. This is why light scattered from them appears white in white light. In highly scattering media, light is scattered many, many times before it leaves the sample. Thus, even though long wavelengths are scattered less efficiently than short wavelengths, there is so much scattering that the difference is negligible, and the sample appears white. The same reasoning applies to clouds and milk. You can verify this behavior by taking a few drops of milk and sequentially diluting them with water. At first the solution will appear white. After sufficient dilution it will take on a bluish hue. Finally, with further dilution, it will hardly scatter at all.

14.6 An overview of twisted nematic displays

The twisted nematic (TN) LCD was developed in the early 1970s. It was soon the most popular and successful liquid crystal display. TN-LCDs continue to dominate the LCD market because of the large number of improvements in their characteristics that have occurred in the intervening years. Because of the continuing importance of a TN-LCD, and since it brings so many ideas together, we will discuss the simplest TNs in great detail. Other, more recent extensions of this technology, including super twist nematics and double super twist nematics will also be discussed.

14.6.1 *Some properties of nematic liquid crystal devices*

The basic science of nematic devices can be explained and understood by realizing that the unusual optical properties of nematic liquid crystals occur in what is essentially a fluid phase. Fluids flow rather easily, and because liquid crystals are anisotropic, these molecules respond to small external fields. To begin, all the directors (the average local direction of the long axes of the molecules) that describe the liquid crystal molecules need not be pointing in the same direction. When all the molecules point in the same direction the free energy is a minimum. This ideal minimum free energy configuration is not always compatible with the constraints imposed by the surfaces enclosing the liquid crystal or by external fields. Typically, the configuration of the liquid crystal will be described by the minimum free energy compatible with the constraints. The free energy of an elastically distorted liquid crystal in an external electric field is given by $F_{total} = F_{elastic\ distortion} + F_{electric\ field}$. Both the elastic contribution and the electric field contribution depend on the director. The elastic energy depends on spatial variations of the director, while the electric field contribution depends on the angle between the director and the electric field.

The surfaces of the enclosing volume generally anchor the molecules so strongly that they impose *boundary conditions* for the director. For simplicity, we will discuss volumes with only plane interfaces that separate an external medium from the nematic. At such interfaces, the direction of the director will correspond to the direction that minimizes the surface free energy. Furthermore, we will only concern ourselves with *homogeneous alignment*. This alignment corresponds to all the liquid crystal molecules being parallel to the surface and pointing in the same direction. Homogenous alignment can be achieved by coating a glass substrate with a polymer and then buffing the polymer in one direction as is shown Fig. 14.10.
In this figure, the molecules are represented by heavy, short lines. It has been assumed that there are no bulk deformations. Typically, surface anchoring is so strong that the molecules at the interface remain parallel to the interface even when large deformations occur in the liquid crystal, such as when a large electric field is applied.

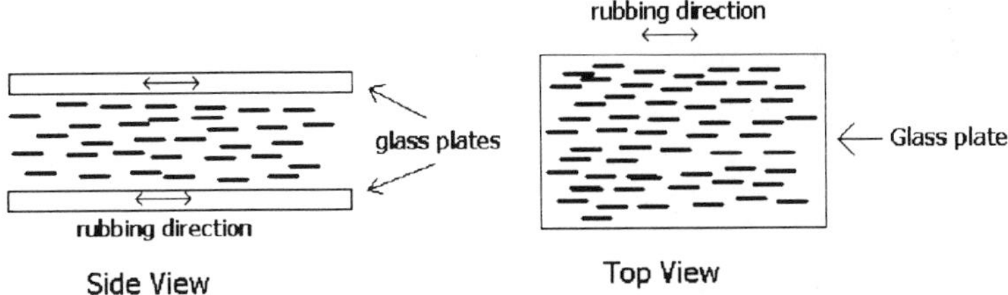

Fig. 14.10 A model of the molecular alignment when homogeneous alignment is imposed on the liquid crystal molecules by the alignment layers.

You may recall that liquid crystals are anisotropic and optically uniaxial, characterized by two indices of refraction, $n_\perp$ and $n_\parallel$. Since electromagnetic waves contain electric fields, you might suppose that the low frequency electric properties of liquid crystals are also anisotropic. This is correct, and explains why liquid crystal molecules can be manipulated by electric fields. The electrical properties of a liquid crystal are characterized by two dielectric constants. The dielectric constant parallel to the long axis of the molecule is called $\epsilon_\parallel$, while the dielectric constant in all directions perpendicular to the long axis of the molecule is called $\epsilon_\perp$. The electric field contribution to the free energy is proportional to $(\epsilon_\parallel - \epsilon_\perp)E^2 f(\theta)$, where E is the electric field and $f(\theta)$ is a function of the angle θ between the applied field and the local director. Thus the ability of an external electric field to orient molecules requires that the two dielectric constants be different. Typically, $\epsilon_\parallel - \epsilon_\perp$ has a magnitude of about 1-10 at the frequency of the drive electronics. It decreases to about 0.1 at optical frequencies. It may be either greater than zero, favoring the director being parallel to the electric field, or less than zero, favoring director alignment perpendicular to the electric field. (You will demonstrate this in Exercise 9.)

14.6.2 *Characterizing electro-optics of LCDs*

The light transmission of a typical LCD is controlled by applying a voltage to it. There are two ways to characterize the light transmission as a function of voltage. The first is to give the percent light transmitted through a display as a function of applied voltage. This will usually vary from about 1-2% in the minimum transmission state to about 35% in the maximum transmission state of a TN. The other way is to use the brightness, zero brightness corresponding to minimum transmission and 100% brightness to maximum transmission. Both characterization methods are used in practice. We will use brightness. A schematic graph of brightness versus applied voltage is shown in Fig. 14.11.

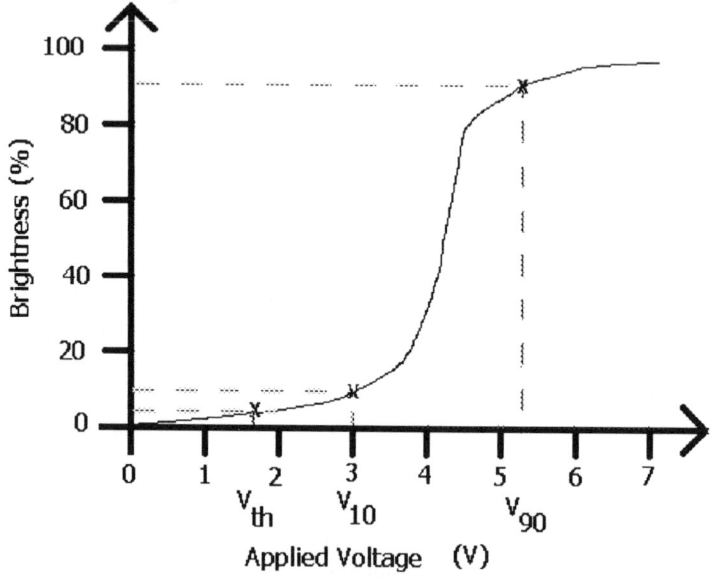

Fig. 14.11 A schematic graph of brightness vs. applied voltage. Three important voltages that are discussed in the text are shown.

There are three important voltages labeled on this graph. The first is V_{th}, the threshold voltage. This is the voltage required to achieve a detectable change in brightness. The second, V_{10}, is the voltage at which the brightness is 10%. The third is V_{90}, the voltage that corresponds to 90% brightness. Very often, the actual values of V_{10} and V_{90} are not as critical as the difference between them. The smaller V_{90}-V_{10}, the sharper the response. This is desirable in bright-dark displays, but is not useful in *gray scale* displays that utilize intermediate brightness levels. The ratio of these two voltages, V_{90}/V_{10}, can be related to the contrast and number of rows of a passive matrix display.

Another important set of display characteristics is the switch-on and switch-off times, T_{ON} and T_{OFF}, respectively. These times are related to the time for the director to reorient, which, in turn, depends on the viscosity of the liquid crystal. Since the OFF time is more dependent on the *orientational viscosity* than the ON time, T_{ON} and T_{OFF} are usually not the same. Typical T_{ON} values are approximately 40 milliseconds, but can be made faster by using higher voltages. A typical T_{OFF} is somewhat longer, 60 to 80 milliseconds.

14.7 The twisted nematic LCD

We now have enough background information on nematic devices to discuss twisted nematic LCDs. The basic geometry of a TN cell is similar to that of the dynamic scattering cell **except** that two polarizers, oriented at right angles, are placed outside each glass plate, as shown in Fig. 14.12.

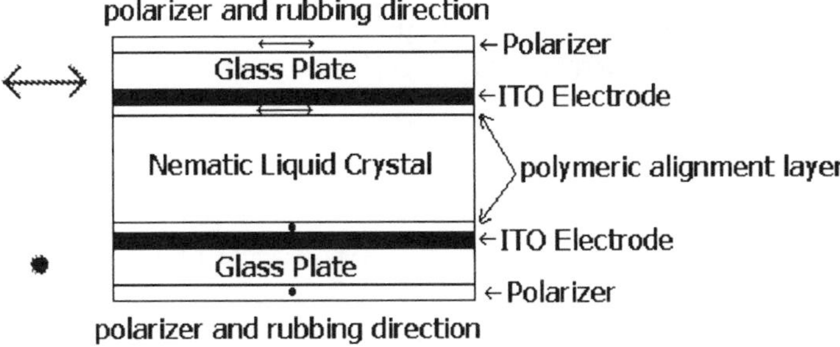

Fig. 14.12 A cross-section view of the major components of a TN cell.

Both the top and bottom plates in this cell have the rubbing direction of their respective alignment layers and the transmission direction of the affixed polarizer parallel. For the top plate, the rubbing direction of its alignment layer and the transmitting direction of the polarizer are parallel to the page. For the bottom plate, the alignment layer and polarizer are oriented perpendicular to the page. Thus, the two rubbing directions and corresponding polarizer directions are perpendicular or crossed. In Fig. 14.13, all labels have been removed from the figure to clean things up a bit, but the directions of the polarizers and alignment layers remain. This figure includes some representative liquid crystal molecules and shows the transverse nature of the light by representing it as a broad ribbon.

There are several features of this TN cell that are important to under-

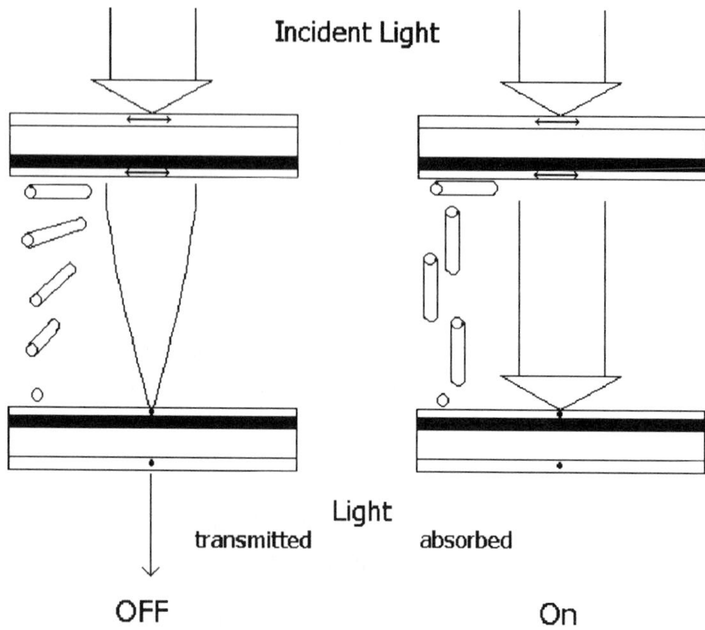

Fig. 14.13 An illustration of how a normally white TN-LCD works.

standing its operation. First, it is being used in the transmission mode. Second, note that the OFF state ($V = 0$) is the transmitting state, and that the ON state ($V \neq 0$) is the non-transmitting state. Such a TN device is said to be *normally white.* Third, the alignment layers lead to strong surface anchoring in this device. The liquid crystal molecules right next to the alignment layers do not change orientation under the action of the electric field. Fourth, when $V = 0$, the molecules perform a 90 degree twist from the top plate to the bottom plate, and the polarization of the light follows this twist. Thus, light initially polarized parallel to the page is polarized perpendicular to the page after traversing the liquid crystal. It is then transmitted by the second polarizer. Fifth, the ON State places a voltage across the electrodes and creates an essentially uniform electric field. This electric field causes the liquid crystal molecules to reorient parallel to the field. The polarization of the light is now **not** rotated as it passes through the liquid crystal. The second polarizer's transmission direction is perpendicular to this unchanged polarization. Hence, no light is transmitted (assuming perfect polarizers).

The advantages of the TN-LCD are that it is low cost, has a low threshold voltage (a few volts), and reasonable contrast. The downside? Recall that for unpolarized light, a polarizer transmits, at best, 50% of the light.[5] When one includes the absorption of the polarizers, the actual transmission may be as low as 30% or 40%. The molecules in this cell rotate by 90° and leads to rotation of the polarization of the light. While this is great for straight on viewing, it tends to produce a poorer display when the "viewing angle" is not small.[6]. Lastly, as discussed earlier, typical T_{ON} or T_{OFF} for a TN display is about 0.05 seconds; much too slow for use in TVs and other rapidly changing display applications.

There is another TN problem that has been solved, but at the same time is indicative of the subtleties of liquid crystal deformations. Recall that a nematic is non-chiral and has no tendency to twist of its own accord. The TN-LCD uses boundary layers to force a twist upon the liquid crystal director, and the only requirement of this twist is that it be 90° between the top and bottom. If the alignment is totally in the plane of the substrates, this twist can occur in either the clockwise or the counterclockwise direction and, there is no energy difference between these two deformations. Thus, both twists are equally likely. The existence of two deformations of the same energy leads to defects and reduced performance of TN-LCDs. This problem may be solved in two ways. The first is to add a very small amount of a chiral dopant to the nematic. This will make the two states energetically non-equivalent, and so only the lower energy state will occur. The second solution uses polymer alignment layers that cause the molecules to have a slight tilt of about 1° to 3° with respect to the plane of the glass plate. Once more, this means that one handedness of rotation will have a higher energy than the other, and the lower energy state will occur.

A TN-LCD with a 90° twist performs well for direct viewing of a fairly small number of lines. However, the contrast tends to decrease in large displays with many multiplexed lines. This can be somewhat alleviated by going to higher twists. The most common device uses alignment layers that cause pre-tilt of the liquid crystal molecules and, ideally, have a director rotation angle of 270° from one glass plate to the other. The basis of operation of this *supertwisted nematic LCD (STN-LCD)* is essentially the same as the 90° twist cell. In practice the twist angles are closer to 180° to 240°. The advantage of STN-LCDs is that their transmission versus voltage characteristics are significantly sharper than for regular TN-LCDs.

[5]See the chapter on light waves, chapter 5.

[6]Straight-on viewing corresponds to small viewing angle (approximately 0°).

This modification also leads to less viewing angle distortion. The disadvantage of an STN-LCD is that interference colors degrade the performance of the display. For example, orienting one polarizer at $+60^\circ$ to the rubbing direction of the alignment layer and the other -30° with respect to its alignment layer leads to a black when off and yellow when on. A rotation of one of the polarizers by 90° leads to white and blue as the colors. If the on state is to appear white in white light and the dark state black, this is clearly not the desired performance.

There are various ways to overcome these problems including birefringent surface films, thin cells, and double layer supertwist (DST) displays. In DST displays, two STN-LCD cells are stacked one behind the other. One cell is a standard STN-LCD. The other cell functions as an optical compensator and is not electrically driven. The two cells are identical except that their respective twist directions point in opposite directions and thereby cancel out birefringence interference colors. The resulting contrast is higher than that obtained with a single STN-LCD, and the effective viewing angle is wider than that of a single STN-LCD. A DST, of course, costs more than a single STN device because it uses two STN cells.

14.8 Ferroelectric liquid crystal displays

One of the recent developments in liquid crystal displays has been the *surface stabilized ferroelectric liquid crystal* (SSFLC) display. This display utilizes the unique electrical properties of the smectic C* phase and the ability of strong surface forces to cause alignment in thin cells. Recall that the smectic C* phase occurs in chiral molecules that have a permanent dielectric polarization[7] perpendicular to the long axis of the molecule. Normally, this leads to a helical structure. However, in sufficiently thin cells the helix can be unwound.

In SSFLC displays, the cell geometry is the standard two glass plate, two external polarizer geometry utilized in other displays. The smectic C* molecules are aligned parallel to the surfaces of the glass plates which are spaced 1-2μm apart. This narrow spacing allows the boundary forces to unwind the intrinsic helix of the smectic C* liquid crystal. The director points in a single direction and the dielectric polarization is perpendicular

[7] In this section, the adjective dielectric has been added when referring to the liquid crystal's molecular polarization. This should help reduce confusion about optical polarizers and dielectric polarization of molecules.

to the glass plates. A small voltage applied between the glass plates can cause the direction of the dielectric polarization to change from up to down or vice versa. The change in polarization direction causes the molecules to rotate. In the ideal case this rotation is 45°. By placing perpendicular polarizers outside the cell with one transmission direction parallel to the director, as shown in Fig. 14.14, this switching can be observed optically.

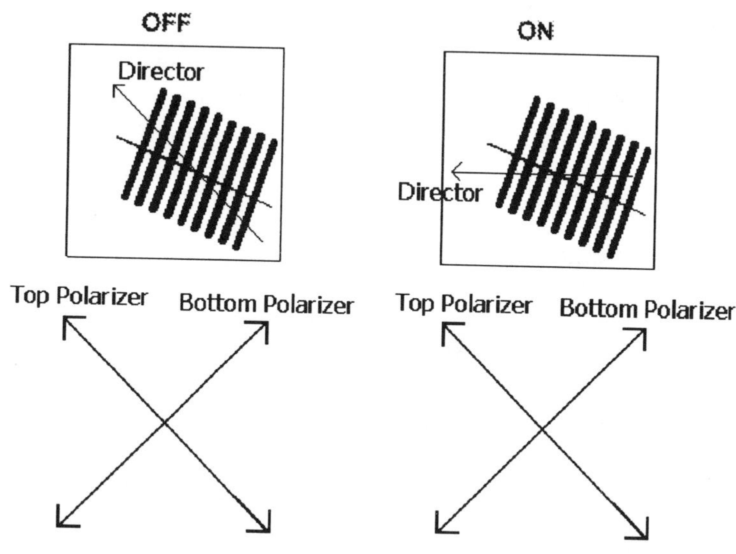

Fig. 14.14 The polarizers, smectic layers, and director in an SSFLC display. This figure is after one in Collings and Patel, pg. 12.

In this figure, the orientation of the smectic layers is indicated by the heavy lines. The director is indicated by the thin line with the arrow, and the layer normal by the thin line through and perpendicular to the heavy lines. The angle between the director and layer normal is 22.5°. The polarizers are located above and below the cell. The off-state corresponds to the director being parallel to the top polarizer. The light enters the cell with a given polarization direction and does not change. It is absorbed by the second polarizer. The on-state has the director at an angle of 45° to both polarizers. The thickness of this cell must be optimized so that there is maximal transmitted light in the on-state. This is because, in this configuration, the incident light can be described as two perpendicular polarizations that propagate through the liquid crystal at different speeds. Thus by varying the thickness of the liquid crystal, the polarization of the light entering the second polarizer can be varied. By changing the direction of the light's polarization by 90°, maximum light transmission occurs.

An SSFLC device possesses two stable states, with the director in either of the two directions shown. In the on-state, the *dielectric polarization vector of the molecules* is pointing up, while in the OFF-state it is pointing down. The director changes direction by 45° during the change in molecular polarization direction. These two states have the same energy, and hence this device is bistable. The energy of intermediate states is somewhat higher. This cell can be switched 100 to 1000 times faster than TN-LCDs.

14.9 References

The references listed in the chapter on liquid crystals (Chapter 10) are all relevant to this chapter also. Additional references tend to be more advanced. With this caveat, the following are all quite good. **Liquid Crystals Applications and Uses**, Vol. I-III, Ed. B. Bahadur, World Scientific, 1991, summarizes the research situation in the early 1990s and still provides insights today. **Optics of Liquid Crystal Displays**, by P. Yeh and C. Gu, Wiley, 1999, is both detailed and mathematical. **Handbook of Liquid Crystal Research**, Eds. P. Collings, and J. Patel, Oxford University Press, 1997, has chapters by experts in many subfields of liquid crystal, including displays.

14.10 Exercises

(1) The pitch length versus temperature of a chiral nematic is plotted on the graph below. The average index of refraction, $\bar{n}$, of the liquid crystal is 1.5, and the wavelength of light **in the liquid crystal** is $\lambda_{liquid\ crystal} = \lambda_{vacuum}/\bar{n}$.

(a) Complete the following table:

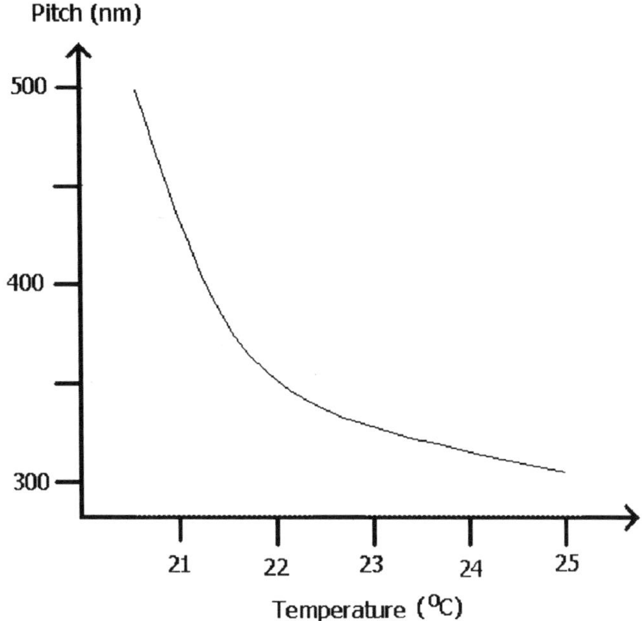

λ_{vacuum} (nm)	$\lambda_{liquid\ crystal}$ (nm)
700	
625	
600	
550	
525	
475	
425	

(b) When the wavelength of light in the liquid crystal is equal to the pitch, reflection will occur. Find the temperature that corresponds to each of the vacuum wavelengths in the table in part (a)

(c) What colors correspond to what temperatures for this thermochromic liquid crystal?

(2) Color or darken the appropriate segments on the figure on the next page to generate the numbers 0 through 9.

(3) Develop a passive matrix scheme using four (4) columns and four (4) rows to address two (2) seven-segment numerical displays. (This is similar to the example in this chapter.)

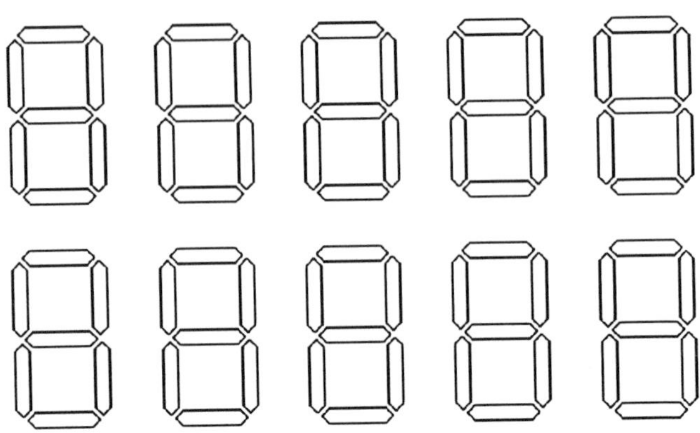

(4) Suppose a constant electric field (constant in magnitude and direction), such as that between capacitor plates, is applied to a mixture of positive and negative charges. Describe in words and pictures the behavior of the charged particles.

(5) For droplets that are not too large compared to the wavelength of light and that exhibit weak scattering (such as observed in the on state of a PDLC) the total scattering is proportional to

$$(\Delta n)^2 * r^6/\lambda^4.$$

Here r is the radius of the droplet and Δn is the difference between the index of refraction of the polymer and $n_\perp$ of the droplet, and λ is the wavelength of the light. $n_\perp$ for a particular liquid crystal has the following values at the indicated wavelengths.

$n_\perp$	λ (nm)
1.532	436
1.523	546
1.519	589
1.516	633

The index of refraction of the polymer may be designed to make Δn any number. What is a good value of $n_{polymer}$? Explain your reasoning.

(6) The transmission of light through a highly multiple scattering medium may be modeled in the "two stream" approximation by the following form:

$$T = \frac{I_{Transmitted}}{I_{Incident}} = \frac{1}{1 + 50x}$$

where x is the distance the beam propagates, and $x \geq 1$. Graph T vs. x for x between 1 and 5. Note the magnitude T. When x is doubled, by what amount does T change?

(7) For weakly scattering materials:

$$T = \frac{I_{Transmitted}}{I_{Incident}} = e^{-25x}.$$

This expression is only valid for very small x, and corresponds to the case of no multiple scattering. Graph T vs. x for x between 0 and 0.05. Within this range, when x is doubled, by what amount does T change? Let $x = 1$ and compare T in this exercise to that in Exercise 6 for $x = 1$. The reduction in transmission in multiple scattering is less than that predicted by the single scattering theory. Why do you think this is the case?

(8) The ratio of the scattering by a nematic liquid crystal to that of an isotropic liquid crystal is shown by de Gennes to be approximately

$$\frac{\text{scattering in nematic}}{\text{scattering by isotropic}} \approx \frac{1}{(qa)^2},$$

where q is roughly 6 divided by an optical wavelength and a is a typical diameter of a liquid crystal molecule. Estimate $1/(qa)^2$. Give your values of q and a.

(9) The electric field contribution to the free energy is given by the expression

$$F_{electric} = -\frac{1}{8\pi}\epsilon_a E^2 \, f(\theta),$$

where $\epsilon_a = \epsilon_\parallel - \epsilon_\perp$. $f(\theta)$ is given in the following table.

θ (in degrees)	$f(\theta)$
0	1
45	1/2
90	0
135	1/2
180	1
225	1/2
270	0
315	1/2
360	1

(a) For PAA, ϵ_a= -0.3. What θ gives a minimum electric field contribution to the free energy?
(b) For another liquid crystal, ϵ_a= +0.5. What value of θ minimizes the electric field contribution to the free energy? The actual functional form is $f(\theta) = \cos^2(\theta)$.

(10) Consider the TN-LCD discussed in this chapter. Is ϵ_a positive or negative? Explain. If alignment layers that align liquid crystal molecules perpendicular to the surface are available, can you design a TN-LCD for liquid crystals with ϵ_a of the opposite sign? Explain.
(11) How would you modify the TN-LCD discussed earlier so that it transmits in the ON-state?
(12) Add a reflector after the TN-LCD discussed in the chapter. Does this system function as a reflective display? Explain.

14.11 Research questions

(1) Consider an SSFLC display. Explain in terms of the optics of polarizers and uniaxial materials why light is transmitted at two different speeds in the ON-state.
(2) It was stated that an STN-LCD has a sharper brightness versus voltage curve than a TN-LCD. Using pictures of the molecular orientation, explain why this is the case.
(3) Modify the SSFLC discussed in the chapter so that the first picture corresponds to the off-State with transmitted light and the second figure the on-state.
(4) Discuss what and how compensation films are used in TN and STN displays.
(5) Discuss the "iron law of multiplexing" LCD displays that was developed by Alt and Pleshko. Why is it important?

Chapter 15

Putting It All Together - The Laptop Computer

15.1 Overview

In this chapter, we reach our first goal. We can now really understand a laptop computer. From our work in the previous chapters, we understand the materials that make up the various parts of a laptop whether they are polymer, metal, liquid crystal, crystalline or amorphous, pure or doped. We also understand how the semiconductors inside work and how the basic logic gates that make up the computer function. Thus, we can see how many modern technologies and their underlying sciences make the laptop computer possible.

This chapter discusses a specific laptop and connects the terminology found in the owner's manual with that discussed in earlier chapters. The goals are twofold: First, to see and understand the connections between the science, the technology and the end product. Second, to understand the terminology used when talking about computers.

Almost any laptop computer will be "outdated"[1] in six months to a year due to the ever increasing complexity and computational power of microprocessors. Thus, the laptop that has been chosen is merely illustrative, and provides a concrete example for the discussion. It was top-of-the-line when this chapter was originally drafted.

15.2 Introduction

We will now take all of the components discussed in the earlier chapters and put them together as a single high-tech consumer product. Many con-

[1]Still, like used cars, such laptops have a long service life if cared for properly.

sumer products incorporate semiconductors and integrated circuits, digital electronics and microprocessors, electric conductors and insulators, high impact plastic cases, and liquid crystal displays: pagers, telephones, stereos, walkmans, VCRs, televisions, medical devices such as insulin pumps, and laptop computers. Our modern high-tech life would be impossible without them, and they are all dependent on man made materials and components.

The laptop computer was chosen for more detailed study because it influences our lives more than any other high-tech device. It also is the most general purpose microprocessor-based device in use today. A calculator, the electronic ignition of an automobile, and the timer on a microwave oven are also computers – specialized, single-purpose computers. The personal computer is a general purpose computer because it may be programmed to perform a variety of different tasks including calculating, timing, and typing.

The simplified specifications of a Dell® Inspiron® 7500 laptop computer are listed below. You will observe that, at first, it appears to be a confusing jumble of abbreviations and specifications. Upon closer inspection, you'll see that you can understand these components, and that, in general, they are simply combinations of circuit elements or ideas that we have already discussed. Also note that they are collectively known as *hardware* since they are real physical devices.

The programs that operate and run a computer are called *software.* The most fundamental piece of software is the *operating system.* The operating system "supervises" all applications. It also sets the rules for use of memory, disk and CD drives, and other parts of the computer. The computer starts to run the operating system in a very round about manner. When the computer first starts or "boots up," the BIOS (**B**asic **I**nput **O**utput **S**ystem) runs. The BIOS is a very simple program that is stored in non-volatile RAM (typically in an EEPROM chip), performs very basic diagnostics, (sometimes with seemingly inane messages such as "Keyboard not detected, push any key to continue.") and searches for the operating system on the floppy, CD, and hard drive of the computer. Once the operating system is found, the boot program loads the necessary system files into RAM and the operating system becomes operational. This two-stage boot sequence has two advantages over a permanent operating system embedded in the computer. First, upgrades, bug-fixes, and adding new features are simpler this way. Second, it allows users to choose those parts of a software package they want. It also permits the user to pick the operating system. Some will select Microsoft® Windows® and others may pick Linux. Regardless of the choice, both operating systems will run on the same hardware.

15.3 Laptop computer specifications

The user's manual for our laptop lists the following specifications.

(1) Processor – Intel®Mobile Pentium®II 400 MHz processor with 32 KB Internal Cache (L1) and a 66 MHz external bus frequency.
(2) Display – 15 inch XGA TFT active matrix display with 1024 x 768 resolution, or a 15 inch SXGA+TFT active matrix display with 1400 x 1050 resolution.
(3) Memory – 32 MB SDRAM standard, upgradable to 512 MB.
(4) Hard Disk Drive – Ultra ATA: 4.8, 6.4, 10, 18, or 25 GB.
(5) Floppy Disk Drive Combo (with CD-ROM or DVD-ROM) – 24x variable-speed CD-ROM drive with 3.5-inch, 1.44 MB floppy disk drive; or 6x max variable DVD-ROM with 3.5", 1.44MB floppy disk drive.
(6) Card Slots – PCMCIA.
(7) Power – "Intelligent" Lithium ion battery, 12-cell, 79 W-hr.
(8) Keyboard – 87 keys, Integrated numeric keypad, 12 function keys.
(9) Pointing Device – PS/2™compatible Touchpad pointer, 20 points/mm (500 points/inch) resolution.
(10) Analog Modem – Internal 56K.
(11) Ports and Connectors –

    (a) 9-pin serial connector.
    (b) 25-pin parallel connector.
    (c) Audio jacks: headphones (same as line-out).
    (d) External microphone - in.

(12) AC Adapter – Input voltage: 90-264 VAC, Output power: 70 W.
(13) Dimensions and Weight (with CD-ROM) and 15-inch XGA display

    (a) Height: 63.5 mm (2.5 inch).
    (b) Width: 327.7 mm (12.9 inch).
    (c) Depth: 266.7 mm (10.5 inch).
    (d) Weight: 9.1 lbs.

15.4 The parts

The microprocessor is the first part to specify in any computer. The microprocessor includes the computer's CPU and is the boss, brains, and message control for the whole computer. All of the other components, including several other types of coprocessors exist to make communication between the user and the microprocessor possible. It is important to realize that a PC contains many coprocessors that are used for various special tasks such as generating graphics and sound. There is even a special processor inside the keyboard that generates the appropriate signals when a key is pressed.

The processor specified is an Intel Pentium II operating at a clock frequency of 400 MHz. At one time, the Pentium II was the highest performance processor available. It's computational power is more than sufficient for almost all common uses of laptops. It contains 7.5 million transistors and is designed to get data into and out of the processor as speedily as possible. An important improvement in this processor is that, unlike the generic processors we discussed earlier, it has two arithmetic logic units (ALUs). These ALUs are used only for integer mathematics. There is a separate calculational unit, the floating-point unit, FPU, which is optimized for floating point numbers (numbers with decimal fractions.) Pentium processors also have multiple storage areas called *caches* that can quickly store and retrieve information and thereby reduce computation time.

A few details of the Pentium II processor are useful to know. First, the RAM communicates with the rest of the microprocessor over a 64-bit bus. This means that there are 64 switchable lines. The speed of this bus is 66 MHz. The "Bus Interface Unit" (BIU) within the microprocessor receives data and instructions from the RAM. It sends data to one cache memory (in this case, a 32 kilobit, L1 or Level 1 cache) and instructions to another, identical cache memory. These two cache memories help ensure faster operation of the microprocessor. The control of such a microprocessor is rather complicated and utilizes a circular buffer, and a predictive unit called the "Branch Target Buffer" (BTB) to speed up operation. The Pentium II also has a sub-unit that improves the performance of multimedia and graphics software. Referred to as MMX, this unit has about 60 special instructions required for the highly repetitious operations needed in multimedia applications. The MMX unit uses the registers of the floating-point unit. The FPU has 80 registers, each of which can hold 80 bits, although the MMX unit only uses 64. Graphics and video often use 64 bits so that, provided one doesn't try to generate live video of the results of a complex mathematical calculation, these registers are usually sufficient for an application. Audio data usually consists of 16 bit groups, so four groups can fit into one 64-bit register and can be executed in parallel.

The laptop's display is a color, active matrix, 15-inch diagonal measure LCD. It has thin film transistors as the driving elements. There are two types of displays available. The first has a resolution of 1024 x 768 pixels with 256 colors and follows a display standard called e**X**tended **G**raphics **A**rray, or XGA. The other display has a resolution of 1400 x 1050 pixels and follows the **S**uper e**X**tended **G**raphics **A**rray (SXGA) standard. A block diagram of the electronics in such an LCD is shown in Figure 14.1.

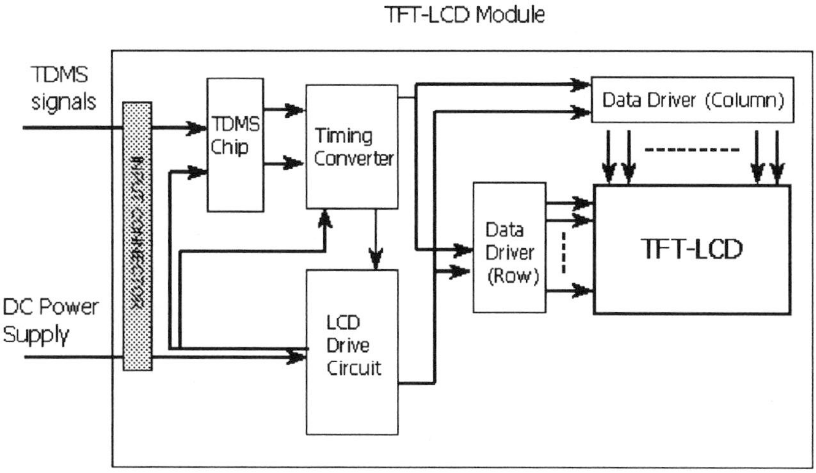

Fig. 15.1 The functional block diagram of a TFT-LCD (Thin Film Transistor - Liquid Crystal Display) module. TDMS stands for Terminal Display Management System.

The LC display has associated with it many electronic components which transform the 0s and 1s from the microprocessor into visible signals. The drive circuitry is non-trivial and critically important, but you already know about row and column lines, and understand how they are driven.

The next component specified is memory. A common rule of thumb (that does not always work) is "the more RAM the better." The rational for this rule is that the more data and/or programming information that is stored in RAM, the less time the computer has to go to slower access time devices such hard drives, floppies or CD-ROMs. Realistically, one buys as much RAM as one can afford consistent with the intended application. For the laptop discussed 32-64 MB (Megabytes) is sufficient. Like many changes, newer laptops often need and are sold with significantly more RAM. An 8-bit sequence of binary digits (bits) is called a *byte*. This term originates in the early days of digital computers. One occasionally sees the term *nibble*, which is half a byte, or 4 bits.

The specification lists SDRAM which means **S**ynchronous **D**ynamic **R**andom **A**ccess **M**emory (also called Synchronous Dynamic Read and Write Memory). This type of RAM was discussed in Exercise 17 of chapter 13. To review, DRAM is a form of semiconductor memory in which the information is stored in capacitors on an integrated circuit chip. Each bit is stored as an electrical charge on the capacitor. Due to leakage, the capacitor discharges gradually and the memory cell loses its contents (forgets its state). Therefore, to preserve the information (the charge state), this RAM has to be refreshed periodically and hence is said to be d dynamic RAM, DRAM. Despite this inconvenience, DRAM is a very popular memory technology because of its high density and low price. SDRAM is a form of DRAM that adds a separate clock signal to the control signals. SDRAM chips can contain more complex circuitry that allows them to support "burst" access modes that clock out a series of successive bits very rapidly. The periodic refreshing is typically done by reading all rows at regular intervals. A 16 MB SDRAM chip needs to be refreshed every 32 milliseconds. The 66 MHz external bus frequency mentioned in the Processor specification means that it takes approximately 15 nanoseconds to read one bit of information out of the laptop's SDRAM.[2]

The next specification is for the hard disk drive. You can think of a hard drive as a type of **non-volatile** memory that takes longer to access than SDRAM. Good disk drives have access times of about 8 milliseconds, are inexpensive, and store large quantities of data. A hard drive has several disk-shaped platters, that have a magnetic coating approximately 3 millionths of an inch thick. The platters are attached to a motor that rotates very rapidly. Rotational speeds as fast 7200 revolutions per minute, above the redline of many automobile engines, are not uncommon.

A hard drive is truly an engineering masterpiece. Each time the hard drive is accessed, the read/write *heads* are very precisely positioned at the correct position over the platters. The gap between the platters' surface

[2]The time necessary to read one bit is approximately the reciprocal of the clock frequency in Hertz (Hz).

and the heads is significantly smaller than the diameter of a human hair or a small dust particle. For this reason, hard drives are fabricated in cleanrooms and then sealed in a metal case. The read/write heads are at the ends of movable arms that position the heads over the platters. The write head writes data to a platter by aligning the magnetic fields of the particles in the magnetic coating. The read head reads the data by detecting the magnetic field of the aligned particles.

The Windows operating system and associated programs use large amounts of disk storage space. A hard drive with a capacity of at least 3-4 *gigabytes* (GB) is required. The minimum size that one may purchase with this computer is 4.8 gigabytes, or $4.8 * 1024$ megabytes. Compare this to the 32 megabytes of RAM that is standard on this computer. It is believed that the storage capacity of hard drives will continue to increase and the cost will decrease.[3]. On the downside, keep in mind that this is a mechanical and electrical component, and, as such, is one of the first candidates for failure.

The "Ultra ATA" referred to in the hard drive specification is a data transfer interface. The strength of this interface is that it allows "fast burst" transfer rates of data. Since it's introduction, the ATA interface has been improved and it has gotten faster. The slower interface (ATA/33) runs at 33.3 MB/s (33 megabits per second), while the Ultra ATA/66 has a transfer rate of 66.6 MB/s. Thus, data can be transferred from this hard drive rather quickly.

Why then is reading from or writing to a hard drive such a slow process? The reason is that a single file may be strewn over hundreds of different locations on every platter in the hard drive. Two adjacent portions of the file could be separated so part is on one platter and the next portion on the last platter a few degrees before the previous part of the file.[4] In this case, it would take one rotation of the platters, which takes 8.3 milliseconds (at 7200 RPM) plus reading time to read these two parts of the file.

This problem is minimized by *defragmenting* or *optimizing* the hard disk. This process puts all the parts of a given file adjacent to each other. In this way the drive does not have to waste as much time moving the heads about reading all the scattered fragments of the file. Defragmenting

[3]This is indeed the case. In late 2001, it was difficult to find a replacement drive smaller than 10 GB, and the drive in a standard laptop is at least this big.

[4]This somewhat chaotic distribution of file parts occurs because a hard drive will always write to the first bit of free disk space it can find. As many files of differing sizes are created and erased, the free disk space can become quite spread out on the platters.

programs are part of operating systems and should be used regularly.

The laptop also includes a floppy disk drive. Like the hard drive, a floppy is a form of non-volatile RAM. Data in the form of programs or results can be stored on these discs. The drive cited in the specifications is a combination CD-ROM (or DVD-ROM) floppy drive. A floppy drive reads and writes to thin Mylar disks coated with a magnetic material. The heads in a floppy drive use tiny electromagnets to change the polarity of the magnetic particles that are embedded in the magnetic material. The data can be written on the top or the bottom of the disk. 3 1/2 inch diameter floppy disks are the present standard size, but there are several data *densities*. The most common formats can store 1.44 MB or 2.88 MB. The floppy drive in this system reads and writes a maximum of 1.44 MB. Floppy disks are relatively inexpensive, dependable, and portable.

A CD-ROM drive is an optical, non-volatile, random access, read-only memory (ROM) device. The CD drive in a computer accepts small plastic discs similar to music CDs. The data is read with a laser, hence the term optical. CD-ROM is popular for distribution of large databases, software, and especially multimedia applications. The maximum capacity of a standard CD-ROM is about 640 megabytes. CD-ROM drives are rated with a speed factor relative to the speed of a music CD (1x or 1-speed which has a data transfer rate of 150 kilobytes per second). 12x drives allow higher data transfer rates, up to 12 times the "1x" rate. Some of the earlier CD drives did not have the data transfer rate increase linearly with the increase in speed. This is no longer the case. A Samsung Electronics 32x CD-ROM has a maximum data transfer rate of 4,800KB/sec which is 32 times the 1x speed. The average access time for this drive is 80 milliseconds.

A CD drive is fundamentally different from the other drives discussed so far. First, it is optical rather than magnetic, as are the hard drive and floppy. Second, the motor that rotates the disk is a variable speed motor. The rotational speed of the disk varies so that the surface of the disk moves by the detector lens at the same linear speed no matter what portion of the disk is being read.

A CD is a sandwich of many materials, but, basically, it is a reflective metal film or foil covered by plastic. The reflective foil alternates between *lands* and *pits*. Lands are flat areas of foil that reflect the light from the reading laser into the detector. Pits are tiny depressions in the reflecting area that tend to scatter light, thus reflecting less light into the detector. For commercial CDs, the pits are created by mechanically pressing a plastic disk with a die, just as records used to be pressed, and then depositing metal over the pressed surface. Recordable CDs have a photosensitive layer that is transparent or scattering depending on if it has been written to or "burned." Thus, 1's and 0's are represented by pits and lands. The reflected light strikes a photodiode that converts the light pulses into electrical pulses that are processed and used by the computer.

Another type of optical ROM is the DVD. DVD stands for **D**igital **V**ersatile **D**isk. A DVD can hold up to 8.5 GB (gigabyte) of data. The first drives used a single-layer disc of with a capacity of 4.7GB. In 1997, dual-layer discs increased the disc capacity to 8.5 GB. Double-sided, dual-layer discs will eventually increase the capacity to 17 GB. The one-sided dual layer disk consists of four layers. The bottom layer is a thick plastic that is the structural foundation for the other layers. On top of this is a thin reflecting layer. The third layer is a transparent layer. The fourth and final layer is a clear protective plastic layer. The data on a DVD is also stored as pits and lands, and once more the light reflected from these is converted into electrical pulses. The pits on a DVD are smaller than those on a CD. This explains part of its improved data storage capacity. The other portion of this increase is because the laser used to read this data has a shorter wavelength and can be focused electronically. The laser is focused so sharply that the data stored on the transparent layer and that stored on the opaque layer can be read separately, thereby doubling the capacity of the disk from 4.7 GB to 8.5 GB. The other doubling comes from using both sides of the structural base.

The next specification is a bit cryptic: "Card Slots – PCMCIA." Since most computers eventually need extra memory or new or different features as the user's interests and needs change, this laptop has slots for installing PCMCIA cards. PCMCIA stands for the group that developed the design standards for these cards. These cards are barely bigger than your driver's license, only weigh a few ounces, and can perform all sorts of tasks including memory cards, I/O cards for connection to networks, FAX/modems, hard drives, and digital photography.

Since this laptop is meant to be portable, it has a built in power source, a rechargeable battery. As specified, the battery, an "Intelligent" Lithium Ion battery, has an energy storage rating of 79 W-hr. This is slightly less energy than a 60 watt light bulb uses while it is on for 80 minutes.

Rechargeable batteries produce electricity via a reversible chemical reaction, known as an oxidation-reduction reaction. The battery is designed so that electrons produced by this reaction can only flow through an external circuit. As the electrons flow, the component materials in the battery change. These materials return to their initial states upon recharging the battery, a process where the electron flow through the battery is reversed. In a laptop, recharging is automatically accomplished when the AC adapter is plugged in. There are several common rechargeable batteries. The lead storage battery in automobiles uses lead, lead oxide, and sulfuric acid. Another common battery is the Nickel Cadmium (NiCd) battery. These tend to be inexpensive and bulky. They are used for portable radios, cellular phones, video cameras, power tools and some biomedical instruments. Sealed Lead Acid (SLA) batteries are also used for biomedical equipment, wheelchairs, and other heavier applications where energy-to-weight ratio is not critical and low battery cost is desirable. Lithium Ion (Li-Ion) batteries have a higher energy per unit mass than NiCd, but at a higher cost. Li-Ion batteries are presently limited to fairly low current applications. Most batteries are made of several cells connected in series in order to increase the voltage output. A car battery usually has six cells, and the lithium battery used in this laptop has 12 cells.

The choice of a Li-Ion battery was made largely on the basis of energy storage capacity for a given mass. Some characteristics of NiCd and Lithium ion batteries are compared in the following table.[5] A Li-Ion battery currently costs about twice as much as a NiCd battery.

Property	NiCd	Li-Ion
Energy density (W-hr/kg)	50	100
Cycle life, typical. (How often can it be recharged?)	1500	300-500
Fast-Charge Time (hours)	1 1/2	3-6 hrs
Self-Discharge	medium	very low
Cell Voltage	1.25 V	3.6 V
Load Current	very high	high

Note that though the Li-Ion battery has a higher energy density, it has some drawbacks – shorter cycle life and lower current capacity than NiCd batteries. This laptop will operate for 3.5 to 4 hours on a full charge. The batteries will charge in three to six hours when the computer is off.

The keyboard and pointing device are standard. The AC adapter is also

[5]This summary was found on the web at http://netbox.com/powersource.

fairly standard. The audio out and external microphone jacks are similar to those on a desktop computer or stereo. The modem is the subject of a Research Question at the end of this chapter.

The serial and parallel ports are also standard, but they need to be discussed, because they represent two basic methods of transferring data between hardware components. The serial port has been described as the jack-of-all-trades among computer interfaces. It is very simple with one line to send data, one line to receive data, and several lines to "handshake" or "direct traffic." The functioning of a serial port may be thought of as a single line of students trying to get money at the ATM on a Friday night. Only one student at a time can use the ATM, and the ATM carefully checks every person both before and after the transaction before proceeding to the next person. The parallel port, also called the printer port because most printers use it, gets its name from its use of many "parallel" lines to send and receive data simultaneously. Typically, parallel data is sent one byte (8 bits) at a time down eight separate lines. This is analogous to watching a good marching band. A row of the band moves at the same time and in the same direction. Sending data in parallel is much faster than in serial, provided the electronics employed are similar.

15.5 Exercises

(1) The bus speed of a computer is 66 MHz.

  (a) What is the bus speed in Hertz (Hz)?

  (b) How much time does it take for a bit of data to be transferred at this bus speed? Assume 1 bit/cycle.

(2) If a microprocessor operates at 400 MHz, how does the bus speed limit the rate at which data can be processed?

(3) A CD-ROM rotates at 7200 RPM (revolutions per minute).

  (a) How many revolutions are there per second?

  (b) Assume that two pieces of data that are needed sequentially by a program are offset by 1/2 of a rotation, how long does it take to access and read the two pieces of data?

  (c) An effective clock speed for the CD-ROM may be obtained by calculating the reciprocal of the time required for one rotation of the disk. What is this time? Why is it desirable not to have to read from a CD-ROM while running a program?

(4) A microprocessor operates at a frequency of 1 GHz. The reciprocal of this frequency is approximately the time duration of for one bit of information to be formed.

  (a) How long, in seconds, does it take for one bit of information to be formed in this computer?
  (b) Assume that this bit (a voltage pulse) travels in the computer at 95% of the speed of light. How far does this bit travel in one clock pulse? Recall distance = speed*time.

(5) List the different materials that are used in a laptop computer, and discuss their type (liquid, metal, polymer, etc.). Could a PC easily be made of all natural materials?

15.6 Research questions

(1) What are CISC and RISC processors?
(2) What is the RS-232 serial line standard?
(3) How does a modem work?
(4) Using catalogs, distributors, or the Internet, design a computer different than the one described in this chapter. Try to include new features and capabilities. Discuss why you chose certain options and types of devices. Define any new terms.
(5) What is a CRT monitor? How does it work?
(6) What is Moore's law? How does it explain why computers continue to become less expensive and more powerful? Can Moore's law be true forever?
(7) Some laptop computers are now being sold without a floppy disk drive. Explain why this is done.

Chapter 16

Liquid Crystals in the Visual Arts

16.1 Overview

Scientists have demonstrated that much about biological molecules and life itself can be explained by the laws of physics and chemistry. For this reason, this book began with several chapters on basic science. From there, we looked at various special types of matter. Since our goal at that stage was to show the connections between basic science and the laptop computer, the primary focus was on non-living matter. However, as discussed in the Introduction of this book, there is more to liquid crystals than displays, and this chapter and the next look beyond displays. This chapter focuses on liquid crystals and the visual arts. The reader will see how colors are produced in paintings and how liquid crystals can be used to create interesting color effects. I hope that the reader is curious about art, since this chapter will answer the following questions.

(1) How does one describe colors?
(2) How do pigments work?
(3) What are interference colors?
(4) Why does a thin layer of oil on water produce colors?
(5) How do cholesteric liquid crystals produce colors?
(6) Why do these "cholesteric" colors change?

16.2 Introduction

Liquid crystals have been used as a medium for the visual arts for almost thirty years. They have unique properties that make them an interesting and exciting medium to study. One of the prominent artists utilizing this medium is David Makow of Ottawa, Ontario, Canada. The sections of this chapter on artistic applications of liquid crystals are largely based on a review article he wrote a number of years ago.

This chapter has four main sections. We begin by introducing color theory, review the many ways colors are produced, and provide some cement to bind the rest of the discussion together. The second section will compare colors produced by "normal" pigments with those produced by cholesteric liquid crystals. The ways in which cholesteric liquid crystals are used as an artistic media will discussed in the third section. The final section will take a careful look at the details of light reflection from cholesteric liquid crystals.

16.3 Colors and color mixing

We were introduced to the basics of color theory at a very early age when we mixed watercolors or other paints together. Later, we learned that by mixing together appropriate amounts of the three "primary colors," blue, yellow, and red, we could produce "any" color. If you pursued art to a higher level, you found that this rule was a simplification, but a good starting point. This is not quite enough of a foundation to carry us through this chapter, so this section will discuss how various colors are produced.

Color theory dates back to Isaac Newton and the years just after 1666. During this time, he experimented with prisms. He used a prism to separate a beam of white light into its colored components or spectrum. He than separated one color from this spectrum, say red, and passed it through a second prism. He found that the red beam could not be further separated into other colors. These experiments demonstrated that white light could be divided into *monochromatic* (single color) components or beams. Newton continued his experiments and found that by mixing two appropriately selected monochromatic beams, a new color was perceived. For example, by combining green light and red light, yellow was perceived. So we see that yellow can be produced in *at least* two ways: either as a single monochromatic beam of light or as the combination of two different monochromatic beams of light. That a single color may be derived from different processes is very significant, and points out that **color is a perception**. When we say something appears blue, in the absence of instruments to study this "blueness" more carefully, we are saying nothing about how this "blue" perception occurred.

Since color, unlike wavelength, is a perception, we must rely on people to describe it. This makes quantifying it exactly, the scientist's dream, impossible[1] The best we can do is to compare it to other colors or to our own experience. For example, I cannot look at my wife's violet sweater and say that is "violet 635-23." However, I can say it is less violet than the flowers I saw in a painting or the socks I saw at the store. In order to describe colors, we need experience with colors and some basic concepts.

Psychological studies indicate that there are three attributes necessary to specify a perceived color: *hue*, *saturation*, and *brightness*. Hue is the attribute that most closely matches the color of an object. Blue-green, orange, and violet are all examples of hue. Saturation refers to the **lack of whiteness** of a color. For example, a cobalt blue is a very **saturated** blue and pink is a very **unsaturated** red. Brightness describes perceived intensity. The sun provides a good example of all three attributes.[2] At noon, the sun appears "yellowish-white". Thus, the hue of the sun is yellow, it is very unsaturated and very bright. At sunset, the sun can appear red. Here, its hue has shifted to a red, moreover, it is more saturated and less bright than at noon.

In this scheme, gray represents a completely hueless, unsaturated color. Variations in gray occur because of changes in brightness. Thus, one can properly think of black and white as extremes of the same "non-color," gray. Black represents a total lack of brightness, and white a very bright source. You will get some practice in using these terms in the Exercises.

There are two basic ways to combine colors to obtain other colors – color addition and color subtraction. A common example of color addition occurs when two stage lights fitted with different colored filters are simultaneously directed to the same spot on stage to produce a third color. Another common example of color addition is the full range of color we perceive on the CRTs in televisions or computer terminals. In CRTs, color mixing depends on the colored regions being so small that they are not independently visible (resolvable) by a viewer's eye. The colors are fused together by the viewer's eye-brain system. The primary colors for CRTs are essentially red, green and blue, but they are not monochromatic colors. Thankfully, the rules of color addition work whether the light is monochromatic or not. An examination of these rules and some examples will be left for the Exercises.

[1]By the use of chromaticity diagrams, color can be quantified to a degree. However, there are still an infinite number of ways to achieve a non-spectral color. A non-spectral color is a color that can not be obtained from the spectrum of white light.

[2]This example is from *Light and Color* by Williamson and Cummins.

Interestingly, the color effects of adding paints and pigments together are best described by color subtraction. Color subtraction describes how colors are created when light is passed through colored filters such as stained glass, or the colored coating of a Christmas light. These types of filters remove part of the spectrum via absorption, which changes both the brightness and the spectrum of the transmitted light. This absorption explains how mixing paints together is considered color subtraction, since the perceived color is from light reflected by the paint. The primary colors for color subtraction are magenta (loosely reddish, but it contains some violet), yellow, and cyan (loosely blue, but it includes violet and green). The Exercises and the following section provide additional examples of color subtraction.

16.4 Color – pigments and liquid crystals

The observed colors of **most** objects are caused by one of three basic mechanisms. The simplest mechanism is reflection of some of the incident light. This occurs when a portion of the incident "white" light is reflected while the rest is somehow separated or suppressed. The common methods of suppression are selective absorption and reduction or elimination by destructive interference. A second mechanism for generating colors is interference. In this case, certain wavelengths of light add together to produce large effects, while other wavelengths are suppressed through destructive interference. A third mechanism is dispersion, which occurs when light passes from one medium to another. Dispersion produces the colors observed through prisms and rainbows, and is also responsible for the colors observed when light refracts off spider webs or other thin fibers. There are other sources of color, such as scattering, which will not be discussed.[3]

Most pigment colors are based on selective absorption while most liquid crystalline colors are due to interference. We will now explore this important distinction.

16.4.1 *Colors by selective absorption or reflection*

Pigments are the most common materials used to produce colors and work by selective absorption. All matter absorbs light to some extent, and different colors arise based on how different wavelengths of light are affected by the material. Recall that white light consists of "all" wavelengths in the

[3]Nassau (see references) discusses fifteen causes of color in his book.

visible spectrum, which ranges between approximately 400 and 700 nm. When white light is incident on a non-transparent material, some of the light will be reflected and some will be absorbed. If all wavelengths of the incident "white" light are reflected more or less equally (total reflection), then the object will appear white. When there is some reflection and some absorption, again, more or less equally for all wavelengths, the object will appear gray. When all wavelengths of the incident light are absorbed approximately equally and completely, the object will appear black. A pigment produces color by selectively reflecting more strongly some wavelengths of the incident light. That is, a pigment reflects some portion of the incident wavelength spectrum and absorbs the rest. For example, a mouse pad appears blue because it absorbs the yellow, orange, and red (longer wavelength) portions of the spectrum while reflecting the bluish (shorter wavelength) portions of the spectrum. Most plants appear green because they absorb both longer and shorter wavelengths and reflect green light with a wavelength near 510 nm.

One of the unique features of pigments is that the absorption and reflection of light does not generally occur at sharply peaked wavelengths, but rather as broad peaks, sometimes separated by valleys. A graph of the reflectance spectrum of a typical bluish pigment is shown in Fig. 16.1. There are, of course, many varieties of pigments available. Different media

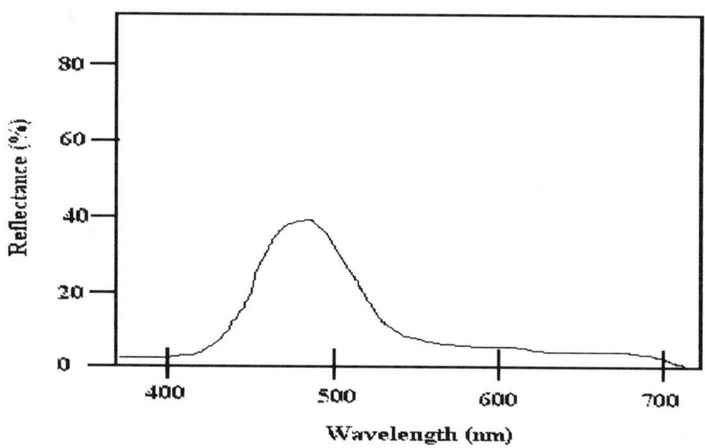

Fig. 16.1 The reflectance spectrum of a typical blue pigment.

such as watercolor, oils, dyes, and acrylic require pigments with different chemistries. The basic optical science of all of these pigments is as we have discussed.

16.4.2 *Interference colors*

Colors produced by optical interference exist because of the structural details of the medium at length scales comparable to the wavelength of light. The colors obtained from such interference may have nothing to do with the color of the medium itself. Makow refers to interference colors as *structural colors*. These occur in layered media where the incident light must penetrate into the media and be reflected at two or more layer interfaces. Two very common examples of interference color are the colored fringes observed in soap bubbles and in thin oil slicks.

In both situations, interference occurs when light is split into two or more paths that recombine to form the color. To see how this works, consider a light beam to be represented by the sine wave shown in Fig. 16.2. If

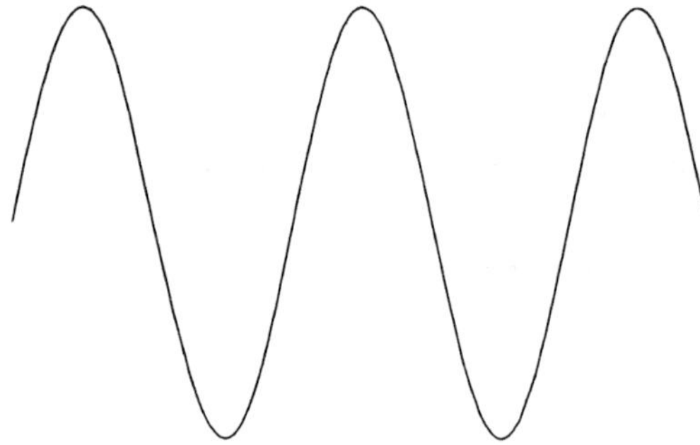

Fig. 16.2 A light wave represented by a sine wave.

another light beam represented by another sine wave of the same amplitude and frequency as the first sine wave, but shifted by an integer number of wavelengths, is added to the first sine wave, a sine wave of twice the original amplitude results. This example of *constructive* interference is shown in Fig. 16.3.

Finally, suppose that the second light beam is a sine wave of the same amplitude and frequency as the first sine wave, but is shifted by an odd number of half wavelengths ($1/2$, $3/2$, $5/2$, ...). Then the two light beams will be exactly out of phase so that one will be most positive when the other is most negative and the net effect will be no displacement. This complete

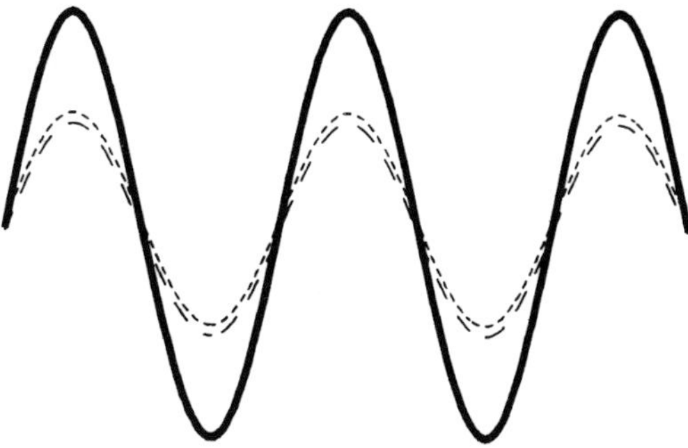

Fig. 16.3 Constructive Interference – The sum of two identical sine waves produces a sine wave of twice the original amplitude. The two original sine waves have been displaced slightly for clarity, and the heavy line represents their sum.

destructive interference is shown in Fig. 16.4. The addition of waves to

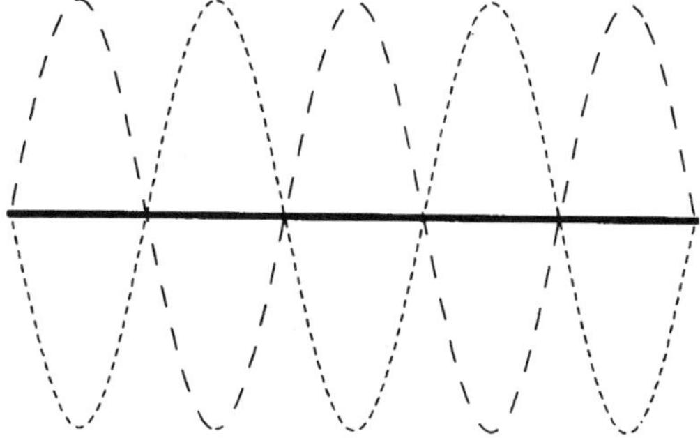

Fig. 16.4 Destructive Interference – The sum of two identical sine waves that are out of phase by an odd number of half wavelengths produces a sine wave of zero amplitude. The heavy solid line represents this sum.

produce zero displacement is called *destructive interference.*

To see how interference can occur in layered materials, consider an observer looking at the light reflected from a thin, flat soap film. The geometry of the situation is shown in Fig. 16.5.
Notice that the incident light is reflected off both the front surface and the rear surface of the soap film. Such behavior is experimentally verified using lasers and many films of varying thicknesses. From the figure, it is clear that the light reflected from the back surface of the film has to travel a longer distance than the light reflected from the front surface. This extra

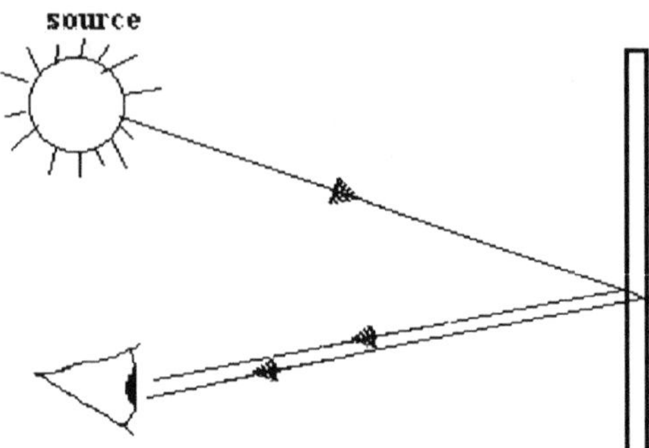

Fig. 16.5 How interference from a thin soap film occurs and is observed.

distance or "optical path length" is all in the film. Thus, we have two beams of light traveling different distances and reaching the observer's eye where they will add together and produce interference effects.

Does reflection off of thin films always result in interference? No. How does one accurately predict interference? Carefully. There is some subtlety in this process, since predicting interference requires the phases of the reflected light to be known. This is complicated because a phase shift[4] of 1/2 wavelength **may or may not** occur upon reflection. The rules of reflection are simple, yet in analyzing some applications, it is easy to make errors. There are two rules for understanding reflection of light.

- The first rule of reflection states that light is always reflected when there is a difference in refractive index between two adjacent media. Generally, the medium the ray is initially traveling in is called medium 1 and the medium on the other side of the interface is called medium 2.
- The second rule of reflection states that there is a phase shift of 1/2 wavelength in the reflected light when the index of refraction of medium 2 is greater than that of medium 1.

Taken together, the rules for addition of sine waves and the rule for phase shift upon reflection give the rules for interference from thin films. From the addition of sine waves, constructive interference occurs when the

[4]Phase shift can be expressed as a length, as we do in this chapter, or as an angle. Since one wavelength (λ) of phase shift corresponds to a phase shift of 360° or 2π radians, we can easily convert between phase shift as length and phase shift as angle.

phase shift between the two waves is an integer multiple of the common wavelength, λ. Thus, the rule for interference from thin films, where the incident light beam is split in two, may be stated as follows:

Constructive interference occurs when the total phase shift, the phase shift due to reflection plus the phase shift due to the extra path length traveled by one beam, is an integer multiple of the wavelength of light *in the film*. Destructive interference occurs when the total phase shift is an odd integer multiple of half the wavelengths of light in the film. If a range of wavelengths are present in the light, the following analysis can be carried out for each wavelength.

To simplify the mathematics, we assume that the light is incident perpendicular to the film and that it originates from directly behind the observer. In this geometry, the extra distance traveled by the wave reflected from the back of the film is simply twice the film thickness, t. Then, for **constructive interference** to occur, we must satisfy the following relationship:

$$2t + (\text{change in phase upon reflection}) = k\frac{\lambda}{n}, k = 1, 2, 3, \ldots$$

Similarly, for **destructive interference** to occur, we must satisfy:

$$2t + (\text{change in phase upon reflection}) = \left(\frac{2k-1}{2}\right)\frac{\lambda}{n}, k = 1, 2, 3, \ldots$$

In both cases, λ is the wavelength of the light in vacuum and n is the index of refraction of the medium of the film. λ/n is the wavelength of light in the film.

Example 16.1 Consider a soap film of index of refraction $n = 1.3\overline{3}$ and thickness $t = 150$ nm suspended in air. By the first rule of reflection, light will reflect at both the front and rear surfaces of the film. Since the index of refraction of air is 1.00, there will be a phase shift of $\lambda/2n_{\text{film}}$ at the front surface, but no phase shift at the rear surface.Given these phase shifts, the respective equations for constructive and destructive interference become:

$$2t + \frac{\lambda}{2n} = k\frac{\lambda}{n}$$

and

$$2t + \frac{\lambda}{2n} = (2k-1)\frac{\lambda}{2n}$$

Simplifying, one obtains:

$$t = \left(\frac{2k-1}{4}\right)\frac{\lambda}{n}, \quad k = 1, 2, 3, \ldots \qquad \text{constructive interference}$$

$$t = \left(\frac{k}{2}\right)\frac{\lambda}{n}, \quad k = 1, 2, 3, \ldots \qquad \text{destructive interference}$$

By letting $k = 1, 2, 3$, and so forth, one observes that constructive interference will occur when $\lambda = 4nt, 4nt/3, 4nt/5, \ldots, 4nt/(2k-1)$. Similarly, destructive interference will occur when $\lambda = 2nt, nt, 2nt/3, \ldots, 2nt/k$. Substituting the numbers for the soap film, we see constructive interference occurs at $\lambda = 800$ nm, 267.7 nm, and so forth. Destructive interference occurs for $\lambda = 400$ nm, 200 nm, and so forth.

If the light is not perpendicularly incident to the film and the observer is not between the film and the light source, geometric factors come into play, and somewhat different colors will be seen at different positions within the film. The variations of colors in bubbles and oil slicks are also caused by variations in the thickness of the film. In the exercises, a modification of this problem is used to discuss anti-reflection coatings on glass.

Given the intricacies of interference from thin films, it is not surprising to learn that structural colors vary with both viewing angle and ambient lighting. Structural colors are not limited to soap films or oil slicks, they are also observed in nature in some insects, birds, and fishes. Such colors are often referred to as being iridescent. You may also be familiar with them from some cosmetics that use thin mica plates to achieve this effect. Another example is iridescent plastic wrapping film which is made up of over one hundred layers of two different types of polymers.

Of course, the primary reason for this discussion is that interference effects have been observed in nematic, smectic, and cholesteric liquid crystals. The origin of these colors is slightly more complex than it is for thin films, and will be discussed later. The strongest interference effects are observed with cholesteric liquid crystals. Since cholesterics have been used in art, we will focus our attention upon them.

16.5 Cholesteric liquid crystals as an artistic medium

As a quick review of cholesteric liquid crystals, recall that the molecules are rod-shaped and locally point in a common direction designated by the director. As one moves through the material, the director rotates or twists in a manner analogous to a spiral staircase. Cholesteric liquid crystals are also thermotropic, meaning, in this case, that the twist changes with temperature.

Cholesteric liquid crystals are used as an artistic medium in several different ways. We will discuss three: (1) free cholesteric liquid crystals, (2) encapsulated cholesteric liquid crystals, and (3) polymeric liquid crystals. Each use has advantages and disadvantages that will be discussed.

16.5.1 *Free cholesteric liquid crystals*

A typical cholesteric at room temperature acts like a thick, oily, viscous liquid that does not dry. Thus, when a drop of such material is placed on a substrate (canvas, wood, *etc.*) it must be covered by a clear protective coating. Typically, these coatings are made of Mylar, plastic, or glass, and are sealed to the substrate using techniques similar to those employed in the construction of liquid crystal displays.

For this technique to be effective, the substrate should be black. The reason for this is clear from the earlier discussion. The cholesteric liquid crystal transmits most of the incident light; the exception is a narrow band of reflected color. The transmitted light must be absorbed at the liquid crystal-substrate interface or the reflection off the substrate would wash out the color reflected from the liquid crystal.

The colors produced by free cholesteric liquid crystals appear very pure and highly saturated when viewed on a black substrate. These reflected colors also exhibit a shift towards shorter wavelengths as the viewing angle changes from straight on. The viewing angle is taken to be the angle between the viewer and the "normal" to the art work, the line perpendicular to the substrate. Thus "straight-on" would be zero degrees, while viewing from the extreme right- or left-hand side corresponds to a large viewing angle.

In practice, one often mixes several cholesterics together to obtain the desired range of colors under normal room temperature changes. Since the reflected wavelength is temperature dependent, this can lead to very striking effects. Interestingly, when cholesterics are mixed the resulting cholesteric liquid crystal mixture has a single pitch and hence strongly reflects only a single wavelength of the incident light. Another effect may be obtained by putting separated layers of different cholesterics in consecutive layers. In this case, each of layer's individual reflectance peaks remain and the perceived color is an additive mix of the individual colors. Further information can be found in Charnay's paper that discusses some of the techniques he used to produce paintings using free cholesteric liquid crystals. (See the references.)

Makow states that, "Paintings with free liquid crystals sparkle with rainbow colours and appear like nothing ever seen before in painting." He observes that the blues and greens are very striking since they are highly saturated, quite unlike the corresponding colors obtained using ordinary pigments. Furthermore, he believes that paintings that allow self-mixing and alignment of the liquid crystal will be the most effective. Lastly, he notes that it is very time consuming to produce a large painting with liquid crystals.

16.5.2 *Encapsulated liquid crystals*

The difficulty of working with free liquid crystals, adding a protective covering and sealing the entire assembly, prompted efforts to encapsulate the liquid crystal in small transparent droplets that could be dispersed and handled like regular paint. The most common technique for encapsulation makes droplets 10 to 40μm in diameter from an aqueous mixture of gelatin and gum arabic. These capsules are then dispersed in polyvinyl alcohol and water to form a whitish slurry. This slurry may be applied using a variety of techniques. The colors appear after the material has dried. Once more, a black substrate is essential for obtaining vivid colors. These paintings do not need a protective coating because the liquid crystal is protected by the capsule. However, Makow recommends an acrylic coating that contains a UV absorber since UV light can degrade the liquid crystals.

The colors of encapsulated cholesteric liquid crystals are not as striking as those obtained with free cholesteric liquid crystals. Blues and greens are still quite good, but the reds tend to be duller than those obtained with oil or acrylic paints. This dullness is due to the variation in alignment of the liquid crystal within the droplets. This weakness may be reduced by painting the substrate red or mixing some red into the encapsulated liquid crystals when a more vivid red is desired. Furthermore, the variation of the reflected colors with viewing angle is not as strong as in free liquid crystals.

The temperature variation of the pitch is the same in both techniques.

16.5.3 *Liquid crystalline polymers*

A third technique for employing liquid crystals, based on polymeric liquid crystals, has more recently been introduced. These cholesteric polymer liquid crystals exhibit a cholesteric phase at moderate temperatures, roughly 40-50°C, and undergo a glass transition between these temperatures and room temperature. Once the material is cooled below the glass transition temperature, it turns into a plastic coating, the liquid crystal alignment is "frozen in," and brilliant reflected colors are observed. These iridescent coatings are durable and adhere to many substrates. They can also be used in transmission on a transparent substrate to create a new type of stained glass. Since these materials are solids or powders at room temperature, they must be melted before being applied. However, it may be possible, with a suitable combination of solvents, to dissolve the polymer for application and then let the solvent evaporate.

16.6 The physics of color production by cholesterics

Now that we have discussed the origin of structural colors and some of their uses as artistic media, we will discuss the physics of the selective reflection in cholesteric liquid crystals. The mechanism responsible for the observed colors is selective reflection, previously discussed in the chapter on liquid crystal displays. The following discussion will analyze this material from a somewhat more mathematical perspective.

Since a cholesteric liquid crystal is a layered structure that is characterized by a "repeat distance" equal to one-half of the pitch, the twist of the helical axis, one anticipates interference effects. Since the layers are of the same material as the surrounding medium, there should be no phase change upon reflection. For light incident parallel to the helical axis, one expects **constructive interference** when

$$2\left(\frac{p}{2}\right) = k\frac{\lambda}{n},$$

where p is the pitch of the cholesteric, n is the average index of refraction of the cholesteric, λ is the strongly reflected wavelength, and k is an integer (1, 2, 3, ...). From this equation, one expects that strongly reflected light will occur for incident wavelengths given by

$$\lambda = \frac{np}{k}, k = 1, 2, 3, \ldots$$

This is essentially correct. Experimentally, for light incident parallel to the helical axis, only a small band of wavelengths centered about the $k = 1$ wavelength is reflected. The higher order reflections do not exist. The existence of only one order is indicative of the subtlety of this problem. A detailed mathematical theory is in agreement with the experimental data.

The wavelength variation about the mean wavelength is typically about 20-30 nm, and is the result of the variation in index of refraction of the molecules. You might recall from our discussion of nematic liquid crystals that these molecules can be characterized by two indices of refraction, $n_{\|}$ and $n_{\perp}$, and that the difference in these indices of refraction is about 0.05. The general expression for the width of the wavelength range, $\Delta\lambda$, is given by:

$$\Delta\lambda = \left|(n_{\perp} - n_{\|})\right| p.$$

This equation can be confusing. The difficulty arises because $n_{\perp}$ need not always be greater than $n_{\|}$. To allow for the general case, and to have $\Delta\lambda \geq 0$, it is necessary to take the absolute value of $n_{\perp} - n_{\|}$.

The dependence of color on illumination angle and viewing angle can be found by considering Fig. 16.6. In this figure, a cholesteric is illuminated at an angle θ with respect to helical axis and viewed at an equal angle θ on the other side of the axis. Since the cholesteric has a periodic structure perpendicular to this axis, distances separated by the same director can be thought of as being in different layers; hence the picture as a layered structure.

The extra distance traveled by the lower ray in reaching the second molecular layer in the figure is shorter than it was in our previous example when $\theta = 0°$. This extra distance, $t_{effective}$, can be found from geometry, and is given by:

$$t_{effective} = \frac{p}{2}\cos(\theta).$$

The *total* extra distance traveled by the lower ray once it leaves the layered material is twice this distance or $2t_{effective}$, and corresponds to the

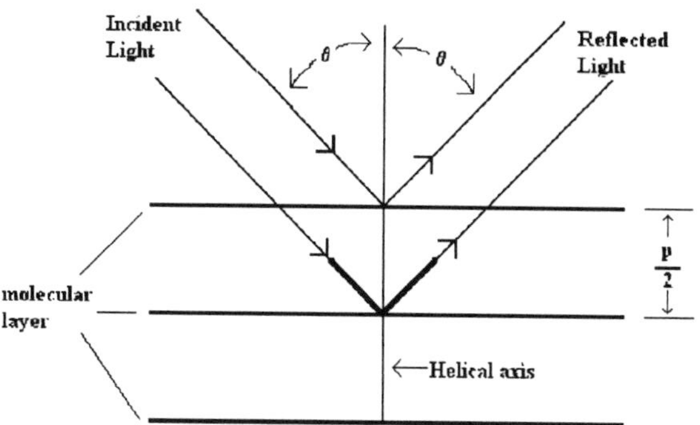

Fig. 16.6 Light incident at an angle θ on a layered material. The extra distance traveled by the lower ray is indicated by the heavy line.

heavy line if Fig. 16.6. Then, using the same arguments as before, and noting that now all k correspond to allowed reflections, the wavelengths of light that are strongly reflected is given by

$$\lambda = \frac{np}{k} \cos\theta, k = 1, 2, 3, \ldots .$$

When the incident and viewing angles are different, a more complicated expression is obtained, but the essential result is not changed too much.

The important effect contained in this equation is that increasing the viewing angle θ decreases the wavelength of the maximal reflection. The actual situation is generally more complicated because our discussion has assumed that the cholesteric is perfectly aligned. Nevertheless, the general observation is that increasing viewing angle decreases the reflected wavelength and shifts the color towards the blue (shorter wavelengths), as our model predicts.

16.7 References

A primary reference for the applications of liquid crystals in the arts is "Liquid Crystals in Decorative and Visual Arts," D. Makow; in *Liquid Crystals: Applications and Uses*, Vol. 2, Ed. B. Bahadur, (World Scientific, New Jersey) 1991. *Light and Color in Nature and Art*, S. J. Williamson and H. Z. Cummins (John Wiley and Sons, New York) 1983, has a very good discussion of color mixing and is very readable. Lastly, *The Physics and*

Chemistry of Color, K. Nassau, (John Wiley and Sons, New York) 1983, is a more advanced, but very understandable book.

An interesting review article on colors in the biological world is "Nano-optics in the Biological World: Beetles, Butterflys, Birds and Moths," M. Srinivasarao, Chemical Reviews, **99**, 1935 (1999).

Charnay's paper referenced in the text is in *Leonardo*, **15**, 219 (1982).

16.8 Exercises

(1) Pick a color picture from a color magazine and describe the colors in this picture to a friend without showing the friend the picture. Then have the friend look at the picture and compare their direct perceptions of the colors with their perception based on your description.

(2) Find examples of various colors and describe them in terms of their hue and saturation.

(3) Review Chapter 5 on light and Chapter 10 on liquid crystals.

(4) Lenses are often coated with a thin transparent coating of magnesium fluoride, MgF_2, that acts to reduce the effects of reflected light. (This also makes the lenses appear purplish in reflected light). The situation is sketched in Fig 16.7.

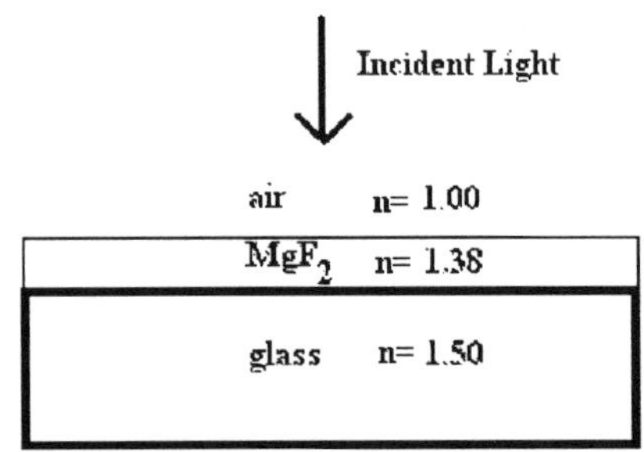

Fig. 16.7 Anti-reflection coatings.

Notice that the index of refraction of the anti-reflection layer is between that of glass and air. When answering the questions, **ignore the glass-air interface at the bottom of the figure.**

(a) At which interface(s) do you expect a phase shift upon reflection?

(b) Assume that the average wavelength of visible light is $\lambda = 550$ nm, and write the equation describing destructive interference between the ray reflected from the air-MgF_2 interface and the MgF_2-glass interface. This equation will be in terms of the wavelength of light, λ, the index of refraction, n of the MgF_2, the thickness of the MgF_2 layer, t, and the integer k.

(c) Put in the numbers and solve for the thickness, t.

(d) t should have an infinite number of solutions. In practice, the smallest value is the one used. What is this value of t? Why pick the smallest value?

(e) Why do lenses coated in this manner appear purple? (Hint: What wavelengths are reduced the least in reflection?)

(5) White light is incident on a thin soap film. The reflected light appears red ($\lambda = 630$ nm). What is the thinnest possible film for which this observation may be valid?

(6) White light is incident on a well-aligned cholesteric liquid crystal that has its helical axis parallel to the incident light. The pitch of the liquid crystal is 450 nm and its average index of refraction is 1.65.

(a) What is the repeat distance between identical planes in the cholesteric liquid crystal?

(b) What wavelength is selectively reflected?

(c) Assume that the incident angle and viewing angle now change to $\theta = 60°$. What is the wavelength of the selective reflection now?

(d) Make a graph of the reflected wavelength versus viewing angle for this material. Let the illumination angle and the viewing angle be the same as in the text. Let θ vary from $0°$ to $60°$.

16.9 Research questions

(1) Write a paper on how color addition works. Include examples. Define complementary colors, primary colors, and secondary colors.

(2) Discuss subtractive color mixing in greater detail than in this chapter.

(3) Investigate interference colors in nature.

(4) Is there a difference between violet and purple? If so, what is it?

(5) In 2001 BASF started to market iridescent colorants. These colorants contain liquid crystal and a chiral dopant. How do these colorants work? Are they encapsulated liquid crystals? BASF has a reference to these products at http://www.basf.com/newsinfor/features1.html.

Chapter 17

Lyotropic Liquid Crystals and Life

17.1 Overview

Life as we know it requires matter to enter and leave cells. This "transport" is accomplished in a number of ways, and central to this is the flexibility of cell membranes. This flexibility arises because, structurally, cell membranes are liquid crystalline phases. This chapter explores why living organisms use liquid crystalline phases as the structure of choice to achieve some of their necessary functions and how serious diseases may result when these liquid crystalline phases or their chemical components are modified. It will introduce the relationship between lyotropic liquid crystals and life. The following questions will be addressed:

(1) Why should liquid crystalline phases be necessary for life?
(2) Where do liquid crystalline phases exist in cells?
(3) How are some diseases related to liquid crystalline phases and phase transitions?

17.2 Introduction

Life, by its very nature, puts constraints on the structures that will allow it to exist. For example, our bodies are largely composed of water; yet our overall shape and size changes slowly except under catastrophic conditions. Our skin is elastic but doesn't stretch too much, and our bones are strong and rigid. Moreover, even in specialized cells such as red blood cells, which are in an essentially aqueous environment, there is still separation between the inside contents of the cell and the surrounding, outer fluid. Detailed studies of cells and the components of cells reveal ordered regions in these objects. What structures are simultaneously rigid and flexible, able to provide an orienting substrate, and act as membranes that separate and enclose various regions? Liquid crystalline structures have all these properties.

This chapter will first introduce some basic biological structures and how liquid crystals are incorporated in them. Then various examples of biological molecules that form liquid crystalline phases will be discussed. The next several sections will discuss how liquid crystals affect biophysics and biological processes. The last section will discuss liquid crystals and diseases.

17.3 Biological structures: advantages and disadvantages

All biological structures, on Earth at least, form in environments that contain large amounts of water. While this allows molecules to diffuse easily, it could also be a serious handicap. How, for example, does bone, which is quite rigid, form in this fluid environment? The rather surprising answer to this question is a combination of ordered protein structures and inorganic precipitation.

Collagens are a family of proteins that are present in all multicellular animals and are generally fibrous in nature. It is the most abundant protein in mammals, typically forming approximately 25% of total protein mass. Collagens can form insoluble fibers that have a high tensile strength. Extensively cross-linked, it is the major fibrous element in skin, bone, tendons, blood vessels, and teeth. It also serves to hold cells together in discrete packages and has a role in developing new tissue.

The basic structural unit of collagen is called *tropocollagen*. This rod-shaped aggregate is roughly 300 nm long and 1.5 nm in diameter and consists of three polypeptide chains. The composition of the collagen chains depends on the type and purpose of the collagen. For example, Type I collagen has two different types of chains and Type II has three identical chains. Collagen fibers are formed from staggered arrays of tropocollagen molecules that can be modeled as staggered rows. Each tropocollagen molecule is separated from it neighboring molecules by a distance, called a hole, of 40 nm. Adjacent rows are displaced by 68 nm. The 40 nm hole is important for cross-linking and may play a role in bone formation.

Bone consists of two phases: an organic phase that is almost entirely collagen and an inorganic phase that is calcium phosphate. It appears that the holes in the collagen rows act as nucleation sites for the calcium deposition. Since bones form inside a living organism, a precipitate must form in an aqueous environment. Chemists know that it is easy to combine two water-soluble chemicals and obtain a water insoluble precipitate. However, doing it exactly where one wants to in a precisely controlled manner is a different ball game, yet nature has solved this problem.

We couldn't stand up straight if it didn't happen.

We're clearly not made 100% of bone, as wonderful as bone is. Other biological structures exist, in part, because there must be a fluid component to allow proteins, enzymes, and other essential biological molecules to move about easily and interact. Solids do not allow molecules to diffuse[1] very easily and large particles do not readily diffuse in liquids. A further requirement of life is that these biological molecules may have to be manufactured in one cell and then leave that cell and enter another cell. This requires the possibility of breakable self-healing structures.

From our earlier discussion of lyotropic liquid crystals, it would appear (quite correctly) that these phases are ideal candidates for solving these problems. Many of the molecules of life are amphiphilic and can self-assemble into structures that have an inside and an outside and are self-healing. Proteins that act as "gates" and "switches" can be incorporated into these membranes. Furthermore, our study of lipid bilayers showed that they are fluid rather than solid and let water into and out of the structure. And while it is true that a lipid bilayer is not very rigid, many of the filaments of proteins that are associated with these cells are rather rigid and provide the necessary support.

17.4 Liquid crystal structures and biological molecules

This section will discuss some of the liquid crystalline phases that are observed in biomolecules, in various cellular structures, and even in some higher forms of life.

Many molecules extracted from biological structures exhibit liquid crystalline behavior in bulk. Actually, the list of molecules that form liquid crystalline phases is rather large. Broadly speaking, members of all four types of biological molecules discussed earlier form liquid crystalline phases.

[1]A typical self-diffusion coefficient for a metal at room temperature is on the order of 10^{-24} m^2/s, for a liquid, 10^{-9} m^2/s, and for a gas 10^{-4} m^2/s. The diffusion coefficient of a particle in solution depends on concentration, particle size, and other factors. For dilute solutions of spherical particles of radius r (in nm) in water at room temperature: $D \approx 2 * 10^{-10}/r$ m^2/s.

For example, many different types of lipids have been observed to form liquid crystalline phases. In fact, the lamellar, hexagonal and cubic phases have all been observed in solutions containing lipid molecules. Similarly, purified protein and polypeptide solutions exhibit liquid crystal behavior. The same is true of DNA and RNA, and some polysaccharides.

At a somewhat higher level of biological functionality, it has become clear that living cells must have both mobility and structural order. Liquid crystals fill the bill, they possess the mobility of a liquid and the structural order of a solid. At its simplest, a cell is a dynamic entity that is enclosed by a membrane. This membrane has structural elements and is composed of lipids and proteins.

At this point, it is useful to separate cellular organisms into primitive and advanced forms. The distinction is based on the existence of membranes within the cell. Primitive cellular organisms such as blue-green algae and bacteria do not have membranes surrounding their nucleus and have genetic material throughout their structures. This class of cells is called *prokaryotes.* Cells in higher organisms have many membrane-enclosed structures within the cell, including the nucleus and other *organelles* such as the *golgi apparatus.* This class of cells is called *eukaryotes.* Not surprisingly, the organelles in plants and animals are different.

Another cellular component with liquid crystalline structure is the exoskeleton of many cells which is made of a polysaccharide polymer called chitin. Chitin is also part of the exoskeleton of higher animals such as anthropoids and crustaceans. Microscopic examination of chitin indicates that the fibers trace a helical pattern that is highly reminiscent of cholesteric liquid crystalline phases. Mitochondria, a critical cellular structure, contain the respiratory enzymes of the cell. Mitochondria are similar in both plants and animals and up to 50% of their dry weight is lipid. They consist of two membranes, an outer, limiting membrane and an inner membrane. Both membranes are well modeled as bilayers with proteins. The inner membranes have been observed to undergo a reversible lamellar-to-twisted state transition. This may be related to a liquid crystalline phase change.

A living organism responds to external stimuli. In order to respond, an organism needs transducers to convert external stimuli to chemical signals. Of particular interest are those transducers that convert energy. As an example, we will briefly describe the structure of chloroplasts, the receptor *structure* or organelle for photosynthesis. The receptor *molecule* for photosynthesis is chlorophyll which resides in a pigment-protein complex in the cell membranes of chloroplasts.[2] Chloroplasts have a number of specialized names. They are called *chromaphores* in photosynthetic bacteria, *plastids* and chloroplasts in higher plants. We will use the generic term chloroplasts to describe all of these organelles. Typically, chloroplasts are approximately ellipsoids of revolution with major axes of 1 to 10 μm and minor axes of 1 to 5 μm. They exhibit optical birefringence and appear to be composed of regularly spaced lamellar membranes, clearly a structure with liquid crystalline properties. The analogy can be further extended. It appears that the chlorophyll molecules are spread as monolayers on the surfaces of the lamellae. One anticipates that since photosynthesis takes place in the lamellar membranes, the liquid crystalline structure of the lamella is critical to photosynthesis.

Photoreceptors or eyes are another important class of transducers. In all vertebrate photoreceptors, the rods and cones responsible for vision are embedded in membranes that consist of lipids and proteins. These membranes are separated from each other by aqueous media and are equally spaced. Closer examination suggests that these lamellae are actually broken up into flat bilayer vesicles. The existence of many bilayers accounts for the sensitivity of the eye. Invertebrates photoreceptors are a bit more complicated. In these life forms, the relevant membranes form microtubules that arrange into a hexagonal phase analogous to lyotropic liquid crystals.

Nerve impulses are carried along nerve cells, neurons, by electrical impulses. Neurons consist of ends that are rather diffusely connected by a highly elongated section called an axon. This axon is often surrounded by a sheath. The purpose of the sheath is to isolate the nerve cell from its surroundings. In many ways, it is analogous to insulation on an electrical wire. Nerves that are myelinated have a much higher conduction velocity than those that are not. Microscopic examination shows that the sheath is actually a membrane that is wound around the axon. Once more, a liquid crystalline structure is adapted to a biological function.

On a still larger scale, muscle cells are similar in many ways to nematic and smectic liquid crystals. Muscle contraction is caused by the sliding of interdigitating filaments of two proteins, *myosin* and *actin*. Vertebrate muscle that is under voluntary control (a leg or arm muscle) has a striated appearance when observed under an optical microscope. It consists of cells that are surrounded by an electrically excitable membrane. The muscle cell contains many parallel *myofibrils* that are approximately 1 μm in diameter.

[2]There are various types of chlorophylls, a complication that we shall ignore.

A myofibril contains two different types of protein filaments or straight polymers each with different diameters. The thicker ones have a diameter of approximately 15 nm and are composed primarily of the protein myosin. The thin filaments have a diameter of approximately 7 nm and contain several proteins, most importantly actin, tropomyosin, and troponin.

A muscle shortens by as much as one third of its initial length when it contracts. This behavior is explained by the "sliding-filament model" The basic features of this model are: a) the lengths of the filaments do not change during contraction of a muscle, b) the contraction is due to an increase in the overlap of the two types of filaments, c) the force of contraction arises because of this active movement. This is illustrated in Fig. 17.1.

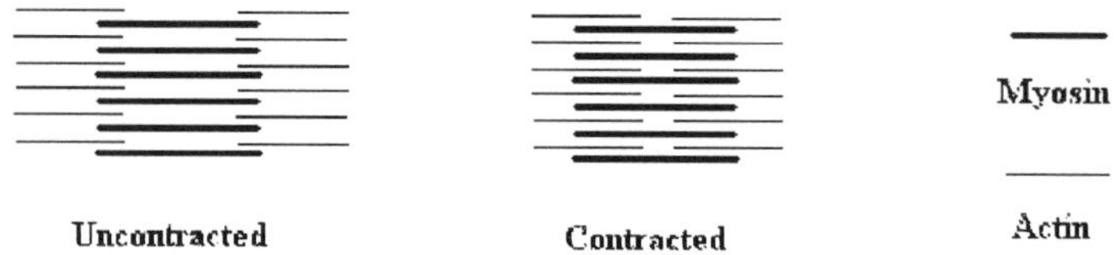

Fig. 17.1 The sliding-filament model of muscle contraction. The filaments exhibit liquid crystalline behavior.

This muscle structure is very much like a smectic liquid crystal. The region where there are thick filaments shows hexagonal ordering of both the thick and the thin filaments. The region where there are only thin filaments shows no positional order. Thus part of of a muscle is a layered structure with in-plane hexagonal order and the rest is a smectic phase. The situation is different in involuntary muscles (the heart, for example). In this case, the filaments are staggered and the arrangement is more analogous to a nematic liquid crystal. The question of how much of a living structure may truly be liquid crystalline has yet to be determined. Nevertheless, as this discussion illustrates, both "large" and "small" structures are liquid crystalline, so the answer might well be significantly larger than you originally imagined.

17.5 Diseases and liquid crystalline phases

Liquid crystalline phases are implicated in a number of diseases. The four most prevalent are hardening of the arteries (artherosclerosis), gallstones, sickle cell anemia, and multiple sclerosis. Liquid crystalline phases have also been implicated in a number of rather rare diseases, and possibly in some aspects of cancer. These are all active research areas, and the present discussion merely scratches their surfaces. It is just meant to help the reader appreciate the subtle role of liquid crystalline states in disease. A more complete understanding of these diseases, their treatments, and bio- and physical-chemistry is left for more detailed texts and primary references.

17.5.1 *Atherosclerosis*

Atherosclerosis[3] or hardening of the arteries is a disease in which a local thickening of the walls of the arteries that supply blood to vital organs obstructs the flow of blood. The thickening is caused by the built-up "gunk" on the arteries that tends to be somewhat localized in the form of lesions. These lesions have been found to contain a significantly higher concentration of cholesterol compounds than healthy arteries. Excess cholesterol is toxic to cells, so the elimination or *sequestering* of excess cholesterol is essential to good cellular health.

The chemical makeup of these arterial lesions is striking. They are almost totally (about 95%) cholesterol esters[4] and free cholesterol (2%), and contain very little phospholipid (about 1%). This is in contrast with non-lesion areas that contain approximately 40% cholesterol esters and phospholipids, and about 20% free cholesterol.

How do liquid crystals tie into this disease? Cholesterol is insoluble in water and hence precipitates out of aqueous solutions. However, a three component system of water, phospholipids, and cholesterol (or cholesterol esters) can form lamellar liquid crystalline Microscopic studies of diseased artery walls confirm that the cholesterol esters are stored in a liquid crystalline state. This is actually not surprising since a large number of cholesterol compounds form liquid crystals at body temperature.

The more serious question of why cholesterol esters build up in arteries is not known. It is clear that cholesterol is soluble in cholesterol esters and

[3]Medical terminology can be even more confusing than the decidedly obtuse liquid crystal terminology. *Athero*sclerosis is a disease of the large arteries of the body and doesn't greatly affect vessels smaller than about 2 mm. *Arterio*sclerosis describes a different group of arterial diseases.

[4]An ester forms when an acid and an alcohol combine. Water is also produced. Because of its terminal OH^- group, cholesterol is an alcohol (called a *sterol*, which belongs to a subgroup of compounds called steroids that also includes many hormones). "Fatty acids" form when *triglycerides* (which are the fats we eat) are acted on by *lipase* (an enzyme) and are transformed into *monoglycerides*, fatty acids, and water as part of the digestive process.

that fairly large amounts of cholesterol may be dissolved in cholesterol esters without the cholesterol precipitating out. This may be a way to get excess cholesterol out of the blood stream without precipitating (crystallizing) the cholesterol out of solution. This is reasonable since cholesterol crystals are observed in very advanced cases of artherosclerosis.

It appears that the various lipids in an arterial lesion are essentially in equilibrium. This suggests various ways to help combat this disease. By reducing the amount of cholesterol in one's body (through diet, for instance), the equilibrium between cholesterol in the blood stream and the lesions may change and the extent of the lesions could decrease. As long as a lesion is in the liquid crystalline phase, the time scale of this decrease may be fast. In laboratory experiments, the addition of phospholipids can help solubilize the liquid crystalline phase. The time needed for a cholesterol crystal to dissolve is much slower.

17.5.2 *Gallstones*

Gallstones form when components of the digestive liquid known as bile hardens into a stone-like piece of material. Bile is made in the liver, stored in the gallbladder, and used to help digest fats. The major components of human bile are water, cholesterol, phospholipids (the second most abundant component), and bile salts (the most abundant component). Bilirubin, a breakdown product of hemoglobin is required for the digestion of fat, is also found in bile. Bile salts are structurally related to cholesterol and, when ionized in solution, act as surfactants that are critical to the digestion of fats. Gallstones result when there is too much cholesterol, bile salts, or bilirubin in the bile.

There are three types of gallstones: cholesterol stones, that vary in color from whitish to to yellowish to green and are essentially crystals of cholesterol, black pigment stones and brown pigment stones, both of which are dark colored and consist of bilirubin, cholesterol and other minor constituents. The cholesterol stones account for approximately 75% of the gallstones in Western populations. Cholesterol stones are defined as stones with greater than 50% cholesterol, though it is not uncommon for these stones to be 80% or greater cholesterol. Black pigment stones occur approximately 20% to 25% of the time. These stones are usually small, but exist in large numbers and are composed primarily of calcium bilirubinate and may contain up to 30% cholesterol. Generally though, the cholesterol content is less than 10%. Brown pigment stones represent roughly 5% of the stones in the West, but are more common in the Far East. They tend to form in chronically diseased settings. Gallstones come in a variety of sizes. Some can be as big as a golf ball and others the size of a grain of sand.

Cholesterol gallstones could occur in everyone, because the the bile of both healthy and diseased individuals is supersaturated with cholesterol. The formation can be understood by considering the phase diagram of the four major components: water, lecithin (a phosopholipid), bile salt, and cholesterol. Cholesterol-rich (with molar ratios of up to 2:1 cholesterol:lecithin) vesicles are used to transport cholesterol, and can exist as an equilibrium phase. Given sufficient time in the gallbladder, equilibrium **may** be obtained. If equilibrium is obtained, then the issue is where in the phase diagram does this equilibrium occur. If equilibrium occurs in a multiphase region containing cholesterol crystals, cholesterol stones will form. If the equilibrium is in region without crystals, then stones will not form. Thus, gallstone formation has both a thermodynamic and a kinetic component. Actually, the issue is even more complex than this because mucus, which is secreted by the gallbladder, is an integral component of both cholesterol and pigmented stones, and could serve to accelerate neucleation of cholesterol crystals in bile.

Several treatment options depend on the thermodynamic and kinetic nature of the formation of these stones. The most common treatment is surgical removal of the stones or the entire gallbladder. Removing the gallbladder allows the bile to continuously drain into the intestine, solving the problem "kinetically." There are currently three well-known non-surgical treatments which are used only for cholesterol stones: oral dissolution therapy, contact dissolution therapy, and extracorporeal shockwave lithotripsy. Oral dissolution therapy uses drugs made from bile salts to dissolve the gallstones. This may take several months to be effective and has some side effects. The contact dissolution technique injects a drug directly into the gallbladder. It can dissolve stones in one to three days but is quite toxic. Extracorporeal shockwave lithotripsy uses shock waves to break up the stones. It is hoped that the little pieces can pass out of the gallbladder without difficulty. The success of this technique is not too high. In all cases, cholesterol stones reoccur about half the time after treatment.

Different phases of human bile fluid have been observed under the microscope and liquid crystal structures have been identified. It is believed that the structure is lamellar. While the role of this liquid crystal is somewhat speculative, it appears that one of its major roles is to solubilize the cholesterol so that it does not precipitate out of solution and into gallstones. It has been verified in laboratory experiments that water and bile salts solubilize cholesterol, so this is a reasonable model. Once one thinks of bile as having a liquid crystalline phase, then the three modes of treatment seem reasonable. While surgically removing the gall bladder seems extreme, the gallbladder is basically a holding tank, and letting the bile drip continuously into the intestine prevents the build-up of cholesterol deposits. The method of adding more bile salts and/or lecithin should shift the system towards more solubilized cholesterol and should reduce the size of the stones. The time frame for this to occur is rather long, and one might have to take these drugs for extended periods. The direct injection of a drug that dissolves the stone also seems reasonable.

The long-term goal, preventing the stones from forming in the first place, may also be studied using this liquid crystalline model, as it suggests once more that too much cholesterol is a bad thing. Of course, so is too little. Cholesterol may well be the most peculiar of all biological molecules – you can't live with too much or too little of it.

17.5.3 *Sickle cell anemia*

Sickle cell anemia is an inherited, chronic disease in which one amino acid residue, located on two of the four polypeptide chains that form the hemoglobin molecule, is mutated. This causes a change in the shape of red blood cells. This genetic disease affects primarily people of African descent. The incidence of this disease appears to be somewhat dependent on the source consulted. One reference stated that it affects 1 in 400 African-Americans while another text stated the incidence is 1 in 250. Regardless, this is a significant public health problem.

Sickle cell anemia is genetically transmitted. Children who receive the gene from one parent but not the other have the "*sickle cell trait.*" These individuals are usually not symptomatic, and only about 1% of their red blood cells is sickled, as contrasted with about 50% sickled cells in those with the disease. Unfortunately, even just having the trait is problematic. Red blood cells from an individual with the trait will sickle in vitro upon reduction of oxygen.

Early experiments by Linus Pauling and his co-worker showed that sickle-cell hemoglobin, also know as hemoglobin S, is more highly charged than normal hemoglobin, making sickle cell anemia a "molecular disease." The change in the amino acid sequence substitutes a non-polar residue for a polar residue. This residue markedly reduces the solubility of deoxygenated hemoglobin. In fact, the solubility of deoxygenated sickle-cell hemoglobin in blood plasma is only 4% that of normal deoxygenated hemoglobin, a huge effect. Fortunately, it has little effect on the solubility of oxygenated hemoglobin.

This molecular mutation also produces a sticky patch on the hemoglobin. This patch is present on both oxygenated and deoxygenated sickle cell hemoglobin. There is also a complementary site on both deoxygenated normal and S hemoglobin. Sticky and complementary sites on neighboring deoxygenated hemoglobin S molecules can join and form fibers. Normal hemoglobin has this complementary site masked so that fiber formation does not occur. Thus, one expects fiber formation to occur only in high concentrations of the deoxygenated form of sickle cell hemoglobin. This is illustrated by Fig. 17.2, adapted from Stryer.

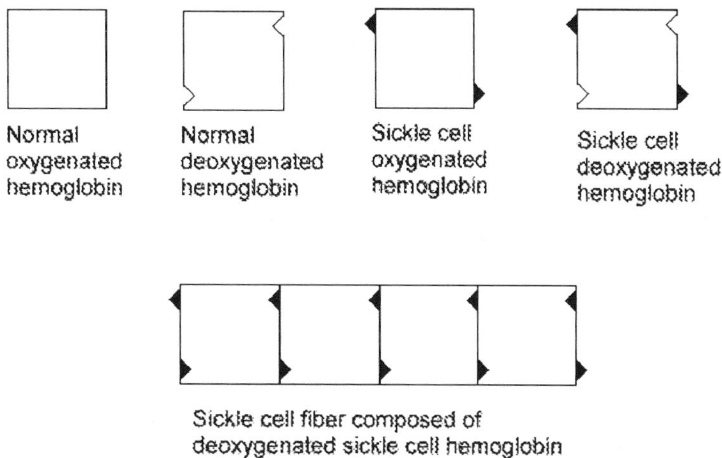

Fig. 17.2 Schematic models of oxygenated and deoxygenated normal and S hemoglobin. The sticky sites are indicated by the protruding filled triangles and the complementary sites by the triangular indentations. The bottom of the figure illustrates how long strands can form when there is a high concentration of deoxygenated hemoglobin S.

The sticky patches are denoted by the filled triangles (two per hemoglobin S molecule) and the complementary site by the open triangle notches. In the deoxygenated state, both normal and S hemoglobin have the complementary site, but only sickle cell hemoglobin has the sticky sites. The key and lock fitting of sticky and complementary sites allow the fibers to form as shown in the bottom of Fig. 17.2.

The precipitated fibers that form from deoxygenated sickle cell hemoglobin deform the red blood cells and give them their characteristic sickle shape. These fibers appear to have a diameter of approximately 21.5 nm and are formed from a fourteen-strand helix of hemoglobin S molecules. The formation of these fibers is a highly cooperative venture, and the rate of fiber formation is a function of the tenth power of the concentration of deoxygenated hemoglobin S. In fact, the rate of formation of the nucleus is the slow step in this process. This strong concentration dependence of the formation rate seems to explain why those who carry the trait are largely asymptomatic.

How do liquid crystals help us understand this disease? Certainly, the size and shape of blood cells is controlled by the liquid crystalline structure of the cell membranes. A normal red blood cell is disk shaped and thinner in the center than at the edges. This structure is well suited for the rapid exchange of gases between the hemoglobin inside the membrane and the outside. The resulting flexibility of these membranes is also useful since red blood cells must be forced through capillaries that are very similar in size to a red blood cell.

The liquid crystalline nature of the red blood cell membrane suggests treatments. For example, could the membrane be strengthened so that it does not collapse? While such an investigation would be interesting, the shape of the cell is secondary to the distortion caused by the formation of fibers. These fibers of deoxygenated hemoglobin S adopt a helical geometry. Thus, at least within the sickled cell, a cholesteric liquid crystal structure occurs. This suggests avenues for further research. For example, can the formation of fibers be reduced by changing the interactions between the molecules? Also, since the formation of a nucleus of hemoglobin S molecules is the slowest step in the process, can the intermolecular forces responsible for this binding be sufficiently reduced without harming the person? Other questions and answers will become clearer as we learn more about this disease.

17.5.4 *Multiple sclerosis*

Multiple Sclerosis (MS) is a chronic disease of the central nervous system. It generally affects young adults, roughly one person per 1000 in the United States and Europe, and is characterized by demyelination of some parts of the axons. It is not known if MS is viral or auto-immune in nature, as both models have some weaknesses, but there is evidence of some genetic influence. The symptoms characteristic of MS are due to the loss of the myelin coating which affects the transmission of electrical nerve impulses along the nerve axons. Research has discovered that the central nervous system has limited capacity to regenerate normal myelin.

The myelin sheath that surrounds a nerve is actually a membrane that is wound several times around the axon. In cross-section, it is rather like a roll of paper towels, and can be modeled as a lamellar phase. The breakdown of the myelin sheath can therefore be thought of as the destruction of this lamellar liquid crystal. The relative inability of the body to repair this damage suggests that an understanding of what causes this phase change or destruction of the lamellar is a good mode for attacking this disease.

17.5.5 *Other diseases involving liquid crystalline phases*

Many other diseases have been identified with the destruction or formation of liquid crystalline phases. Fortunately, the occurrence of these diseases is rare. A partial list includes (a) Krabbe's Disease, (b) Fabry's Disease, (c) Gaucher's Disease, (d) Wolman's Disease and (e) Tangier's Disease. These five all have a genetic origin. The first four are caused by an enzyme deficiency that leads to the accumulation of metabolic products that cause the destruction of liquid crystalline structures and hence the membranes.

17.5.6 *Cancer and liquid crystalline phases*

The study of cancer and liquid crystalline phases is in its infancy and this section is rather more speculative than some of the previous sections. In cultures of normal epithelial (skin) cells, the cells are in contact with each other. In cultures of malignant cells, gaps appear between the cells and the cell-to-cell adhesion is greatly reduced. Many factors may affect this adhesion. The cell membranes of malignant cells contain more cholesterol than non-malignant cells and the proteins in malignant cells are more clustered. These two changes will have an effect on the liquid crystalline characteristics of the cell membrane. Thus, a disturbance in the normal liquid crystal characteristics of the cell membrane occurs with the onset of the cancer. Of course, does the change in structure cause the cancer or does the cancer cause the change in structure? The answer is not obvious but may be found in the significant fact that these changes are occurring in a liquid crystalline environment as opposed to an aqueous environment.

17.6 Summary

In recent years, the role of liquid crystalline structures in disease has been discovered and its exploration has begun. It is clear that many diseases alter the liquid crystalline structures of cells, membranes, and organelles. It is also becoming increasingly clear that many critical biological reactions occur in structured liquid crystalline environments and not the the largely aqueous and structureless environments that make up the bulk of organisms. The relationship between liquid crystals and life remains one of the truly exciting frontiers of science.

17.7 References

Two particularly good discussions of this material are found in: "Liquid crystals and life," J. W. Goodby, Liquid Crystals, **24** (1998) 25-38; and **Liquid Crystals and Biological Structures**, G. H. Brown and J. J. Wolken, Academic Press (New York) 1979. **Biochemistry** by Lubert Stryer, W. H. Freeman and Company, NY, 1988 was very useful for several sections, as was **Biochemistry and Molecular Biology**, by W. H. Elliot and D. C. Elliot, Oxford University Press, 1997. The texts on liquid crystals were helpful in some cases. In particular, there is an interesting section in Collings' book. A number of medical books have very good discussions of certain aspects of this material. **Clinical Diagnosis and Management by Laboratory Methods, 19th ed.** by J. D. Henry, W. B. Saunders 1996, and the classic **Robbins Pathologic Basis of Disease, 6th ed.** by Cotran, Kumar and Collins, W. B. Saunders, 1999, **Textbook of Medical Physiology, 10th ed.** by Guyton and Hall W. B. Saunders, 2000 were all of great help. The diffusion coefficients in the footnotes are in part from **Physical Chemistry** by Berry, Rice and Ross Wiley 1980, p. 1111.

17.8 Exercises

(1) A footnote in the text states that the diffusion coefficient of a spherical particle in water is $D \approx 2 * 10^{-10}/r$ m^2/s, when r is measured in nm. Assume that this expression is exactly true and that hemoglobin and oxygen can be modeled as spheres of radius 2.68 nm and 0.36 nm, respectively. Calculate their diffusion coefficients.

(2) In three dimensions, the mean distance, d, a particle diffuses in a time

t, is given by the expression $d = \sqrt{6Dt}$, where D is the diffusion coefficient. Assume the diffusion coefficient of a particle in the solid phase is $D_{solid} = 10^{-28}$ m^2/s, and that in the liquid phase, $D_{liquid} = 10^{-11}$ m^2/s. Calculate the time for a particle to diffuse 10^{-6} m (1 μm), about the size of a cell, and 1 mm in both phases. Discuss why, even in the liquid state, diffusion is inefficient for transporting materials over distances much greater than cellular dimensions, and why diffusion in a solid is inefficient even over cellular dimensions.

(3) Chloroplasts appear to be composed of regularly spaced lamellar membranes. What liquid crystalline phases are consistent with this structure? Explain.

The next four exercises explore some aspects of blood flow. In all exercises, blood will be treated as an incompressible fluid and blood flow as continuous. In the first three exercises, assume that blood has zero viscosity. These simplifications do not affect the qualitative conclusions of these exercises.

(4) Conservation of blood mass, artery area, and blood flow velocity.

  (a) By conservation of mass, the relationship between mean blood velocity and cross-sectional area of an artery is $A_1v_1 = A_2v_2$, where A_1 and A_2 are the cross-sectional areas and v_1 and v_2 are the corresponding velocities at two points. When a person is at rest, the average speed through an aorta of cross-sectional area 2.5 cm^2 is 33 cm/sec. The total cross sectional area of the small arteries is 20 cm^2. What is the mean speed of blood through these small arteries?

  (b) The cross-sectional area of all the capillaries is 2500 cm^2. What is the average speed of blood through one of these capillaries? (Assume all capillaries have the same cross-sectional area.)

  (c) A typical capillary is 0.3-1.0 mm long. On average, how long does a given volume of blood remain in a capillary? Is this long enough for a molecule the size of hemoglobin to diffuse through the capillary wall? (Assume that the capillary wall does not affect diffusion.)

(5) One effect of atherosclerosis is the formation of *atheromatous plaques* that can become so large that they reduce or even stop blood flow through an artery. This exercise will explore how *atheromatous plaques* affect blood flow.

(a) Suppose an artery has a cross-sectional area of 0.5 cm^2 and a mean blood flow velocity of 8 cm/s. A plaque reduces the cross-sectional area to 0.1 cm^2. What is the flow velocity past the plaque?

(b) Invoking conservation of energy, and ignoring the viscosity of blood and gravity, the velocity and pressure in a flowing fluid are related by Bernoulli's equation:

$$P_1 + \frac{1}{2}\rho v_1^2 = P_2 + \frac{1}{2}\rho v_2^2.$$

Here, P_1 and P_2 are the pressures and v_1 and v_2 are the velocities at points 1 and 2, respectively, and ρ is the density of blood. Take the density of blood to be 1.06 dyne/sec^{2}cm^4 and calculate the pressure difference $P_2 - P_1$ using your answer to part (a). This is the difference between the pressure where there is no plaque and the pressure where there is plaque. What does a negative pressure mean? What will be a result of a sufficiently large negative pressure?

(6) A positive pressure difference, the opposite of Exercise (5), occurs in an aneurysm. An aneurysm is is an outward bulge caused by a weakening of the arterial wall, and makes the cross-sectional area of an artery larger.

(a) Suppose that a healthy section of an artery has a cross-sectional area of 0.5 cm^2 and a mean blood flow velocity of 8 cm/sec. An aneurysm increases the artery cross-section to 1.0 cm^2. What is the flow velocity of blood past the aneurysm?

(b) Use Bernoulli's equation from Exercise (5)(b) to calculate the pressure difference between the aneurysm and a healthy section of artery. What does this positive pressure mean? While there is a rather small increase in pressure in this case, exercise will increase the velocity and the corresponding rise in pressure may burst the artery.

(7) A further effect of an increase in blood flow velocity is the development of turbulence. In turbulent flow, not only does the blood flow along the vessel, it also flows across the vessel creating flow patterns similar to the whirlpools seen in a rapidly flowing river. The propensity to form turbulence depends on the smoothness of the vessel wall, the branches in the vessels, and many other factors. The *Reynolds' number*, *Re*, is a

measure of the tendency of a fluid flow to form turbulence: $Re \equiv \rho v d/\eta$, where v is the velocity (in this problem, the maximum velocity), d is the diameter of the vessel, η is the shear viscosity (roughly 0.03 poise for blood), and ρ is the density of blood. At arterial branches, when the Reynolds' number is greater than 200-400, turbulent flow will develop and then die out. When Re is greater than about 1000 to 2000 turbulent flow will occur and persist everywhere along the artery.

(a) Assume that the maximum velocity through an aorta of diameter 1.7 cm is 66 cm/sec. Calculate Re. Do you except the flow to be turbulent?

(b) The maximum velocity through an artery of diameter 0.5 cm is 15 cm/sec. If this flow occurs far from an arterial branch do you expect the flow to be turbulent? Explain.

(8) Gallstones are the result of a complex series of events whose net result is the precipitation of various components of bile into insoluble "stones" in the gallbladder. Cholesterol gallstones are often roughly spherical and made of a large number of filament shaped cholesterol (monohydrate) crystals cemented together by other biomolecules. These filaments are generally oriented perpendicular to the surface of the stone, similar to the quills of curled up porcupine. Since all healthy individuals have bile supersaturated in cholesterol, people who form cholesterol stones have a higher propensity to nucleate stones. Nucleating and non-nucleating agents have been found in bile. This exercise explores the rate at which the filament shaped crystals grow. Assume that very small crystals have formed.

(a) The rate at which material is added to a filament, $\Delta n/\Delta t$, depends on the area, A, over which it is added (or removed), the rate at which cholesterol is incorporated, k_f, the concentration of cholesterol, c, and the rate at which cholesterol is released or dissolved from the stone, k_d. Then, assuming A, k_f, k_d, and c are greater than zero, we can write:

$$\Delta n/\Delta t = A(k_f c - k_d).$$

In the steady state, $\Delta n/\Delta t = 0$, and the filament is not growing or shrinking. By setting $\Delta n/\Delta t = 0$, show that $c_{ss} = k_d/k_f$, where c_{ss} is the steady state concentration.

(b) Use your answer to part (a) to rewrite the equation for $\Delta n/\Delta t$ as $\Delta n/\Delta t = Ak_f(c - c_{ss})$. Show your work.
(c) If $c > c_{ss}$, is the filament growing or shrinking?
(d) If $c_{ss} > c$, is the filament growing or shrinking?
(e) The number of moles of cholesterol in the filament, n, may be expressed as the volume of a filament times the number of moles of cholesterol per unit volume. This second factor may be written as the mass density, ρ, divided by the molecular weight of cholesterol, M. ($\rho/M = 1/376$ mole/cm^3.) Assume the filament is a cylinder of length l and radius r with a volume of $\pi r^2 l$. As the filament lengthens, which quantity changes, r or l? Explain.
(f) If the growth of a filament is diffusion limited, then k_f can be related to the diffusion coefficient, D by the equation $k_f = D/r$. Take A as the area of the end of a filament, include parts (b) and (e), and show that:

$$\Delta l/\Delta t = \left(\frac{M}{\rho}\right)\left(\frac{D}{r}\right)(c - c_{ss}).$$

Solve this equation for Δl.
(g) The formation of cholesterol stones is associated with the existence of a viscous fluid called *biliary sludge*. Biliary sludge causes the diffusion coefficient for cholesterol to be smaller than it is in water. Take the diffusion coefficient, D, of cholesterol in this sludge to be $2 * 10^{-8}$ cm^2/s. Further assume that $r = 0.01$ cm and $M/\rho = 376$ cm^3/mole. Show that

$$\Delta l = 7.52 * 10^{-4}(c - c_{ss})\Delta t.$$

(h) Assume the cholesterol filaments grow for 10 years ($3.15 * 10^8$ s) and that $(c - c_{ss}) = 10^{-5}$ moles/cm^3. What is Δl? (For reference, $c_{ss} \approx 2.5 * 10^{-5}$ moles/cm^3.) Clearly, longer times and different excess concentrations can lead to different size stones.

(9) The formation of polymers of sickle cell hemoglobin is proportional to c^{10}, where c is the effective concentration of the deoxygenated sickle cell hemoglobin. This is a very rapid rate of increase. This exercise explores this functional behavior.

(a) Assume that c varies between 0 and 1, and graph c^{10} for $0 \leq c \leq 1$.

(b) How does c^{10} change when c doubles form 0.5 to 1? Similarly, how does c change when c doubles from 1 to 2? The answers you just obtained will be true of any doubling of concentration when a process is proportional to c^{10}. With this in mind, why would a successful strategy for minimizing the effects of sickle cell anemia be to reduce the concentration of sickle cell hemoglobin in the blood?

(10) Since demyelination causes the serious medical conditions associated with Multiple Sclerosis, you might wonder why myelinated axons exist at all. This exercise explores some aspects of this question.

(a) There are examples of non-myelinated axons in nature, such as those in the giant squid. Axons from the giant squid can have a diameter of roughly 238 microns, and are easily removed from the squid and kept functional for up to a day. Theoretical models predict and experimental studies confirm the speed of axon conduction, u, is proportional to the square root of the axon diameter, d: $u \propto \sqrt{d}$. Of course, the area of the axon is proportional to d^2. Suppose you could make an axon of any diameter and you wanted to double the speed of conduction. By what factor does the diameter have to increase when u is doubled? By what factor does the area increase? (The volume of the axon will increase by the same factor.) Suppose an organism needs a large number of axons. Can all the axons increase in size without completely filling the organism with axons?

(b) For an unmyelinated axon, the conduction speed is $u = \sqrt{2d}$, where u has units of m/s and d is measured in microns. Predict u for $d = 238$ microns and $d = 10$ microns. How long would it take for a signal to propagate 2/3 m (about the length of an adult leg) for both sizes of axons? Discuss which speed is better for walking and which is better for running.

(c) The approach seen in vertebrates that keeps both the axon volume and conduction speed reasonable is to allow "normal conduction" to occur only over a very small fraction of the total length of the axon. In vertebrates, the axon is almost everywhere covered by an electrically insulating coat of *myelin*, and only at very small nodes called *the nodes of Ranvier*,[1] spaced roughly 1 mm apart, does the axon conduct "normally." This physiological change actually alters the propagation properties of the axon and allows the diameter of an axon to be as small as 10 microns and still have a speed similar to that of an uncoated axon. Conduction in myelinated axons occurs in a step-wise manner similar to a row of dominoes falling over. This type of conduction initially seemed to jump across the nodes of Ranvier and the effect became known as "*saltatory conduction.*" For saltatory conduction, the conduction speed is given by $u_s = D/T$, where D is the distance between adjacent nodes and T is the "delay time." The delay time is the time for a signal propagate between the nodes of Ranvier. To a first approximation, $T = RC$, where R is the internal resistance between two nodes on a single axon, and C is the capacitance of the cell membrane of the axon between two nodes. A myelinated frog axon has a diameter of 14 microns, $D = 0.002$ m, $R = 28 * 10^6\ \Omega$, and $C = 3.1 * 10^{-12}$ F. Calculate u assuming that the axon is not mylinated using the formula in part (b). Then calculate u_s for such a myelinated axon.

(d) By carefully modelling the dependence of the resistance and the capacitance on axon diameter, and by using the numbers from (c), it can be shown that $u_s = 1.65d$, where u is measured in m/s and d in microns. (There is variability in the data and Tasaki quotes $u_s = 2.5d$.) Graph u (from part (b)) and u_s (part(c)) on the same scales for $0 < d \leq 100$.

(e) In vertebrates, it is common for many axons to run close together in a bundle typically called a "nerve." Based on the preceding discussion, what will happen in a nerve when demyelination occurs? (Recall that the myelin also acts as an insulator between neighboring axons.)

17.9 Research questions

(1) Describe and discuss Fabry's disease,
(2) Describe and discuss Krabbe's disease
(3) Describe and discuss Tangier's disease.
(4) Sphingomyelin is prevalent in the myelin sheath of nerve cells. What is its chemical structure?
(5) What is the chemical structure of lecithin (phosphatidylcholine)?

Appendix A

Review of Mathematics

A.1 Overview

The language of physics and engineering, and increasingly of biology and chemistry, is mathematics. This does not mean that you must be a math whiz or a math major to understand the material in this book. It does require you to have an open mind, a willingness to learn, and a desire to understand new material and sometimes to re-examine old material in a new way. This appendix reviews the mathematics that is needed in this book. All of this is covered in a high school algebra class, but since you may have not used it since then, you may have forgotten it.

We will try to answer the following questions in this appendix:

(1) What is a proportionality?
(2) What is an equation?
(3) Why do we care about proportionalities and equations? Why are they useful?
(4) What is a graph? Why and how do scientists use them?
(5) How does one solve a simple equation? Why would one want to?
(6) How does one write very large and very small numbers economically (without writing too many zeros)?
(7) What are "units" and how do they relate to measurement. Why are units important?

A.2 Relationships, proportionality, equations, and graphs

A.2.1 *Relationships*

As we have seen, quantitative science seeks to discover relationships between various physical quantities. For example:

(1) How does the volume of a block of material depend on temperature and pressure?
(2) How does the amount of light transmitted through a liquid crystal cell depend on the voltage applied across the cell?

We can find the answers to these questions by performing experiments. In an experiment, we strive to vary one parameter at a time while keeping all other parameters constant, and measure how another quantity changes. This process is not very efficient since every material, temperature, and control parameter requires a new experiment. This inefficiency is one more reason why we invent and use models. The simplest mathematical model we may invoke is that of proportionality.

A.2.2 *Proportionality*

Proportionality is most easily explained by example.

Example A.1

How does the perimeter of a square depend on the length of one of its sides?

We know that a all four sides of a square are of the same length and that a square with large sides has a larger perimeter than one with small sides. Thus we may state that the perimeter of a square is proportional to the length of one side. Mathematically we could write this as: $P \propto l$, where P is the perimeter, l is the length of a side, and $\propto$ means is proportional to. Stated another way, this means that P increases when l increases, and that P decreases when l decreases. This behavior is called the functional dependence of P on l. Often, such functional dependencies are the important information in the phenomena that we focus on.

Example A.2

How does one relate $^\circ F$ to $^\circ C$ (Fahrenheit to Celsius)?

By driving by the bank on several days, we make the following observations from the time and temperature sign:

$^\circ F$	$^\circ C$
14	-10
32	0
86	30
59	15

We note that when the temperature measured in $^\circ F$ increases, the temperature measured in $^\circ C$ increases, and vice versa. This behavior makes excellent sense since we are just using different "rulers" ($^\circ F$ and $^\circ C$) to measure the same property (temperature). Thus, we can state that ($^\circ C \propto {}^\circ F$).

A.2.3 *Equations*

Note that proportionality does not tell us the actual numbers. The question, "How fast does P increase with l?," cannot be answered by a proportionality. To answer this question, we must find the actual equation. Empirically, the equations which describe the two examples given above are $P = 4l$, and $^\circ C = 5/9(^\circ F - 32)$.

Most of the equations that we solve in this book are linear and can be described graphically by a straight line. If x is the independent variable, the variable whose value we may arbitrarily pick, and y is the dependent variable, then the equation for a straight line is $y = mx + b$. In this expression, m is the slope (how rapidly y increases with a unit change in x), and b is the y-intercept, (the value of y when the value of x is zero). So, in the equation $^\circ C = 5/9(^\circ F - 32)$, $^\circ C$ is the dependent variable, $5/9$ is the slope, and $(-5/9 * 32)$ is the intercept, the temperature in $^\circ C$ when it is 0 $^\circ F$. For any value of $^\circ F$, a value of $^\circ C$ may be calculated. This illustrates an important point about equations. The independent variable need not be given the symbol "x," the slope need not be represented by the symbol "m," the intercept need not be represented by "b," and "y" need not be the symbol of the dependent variable. We are free to, and indeed should, pick symbols that have explicit meaning in a given situation.

A.2.4 *Graphs*

It is extremely important for you to realize that not all relationships can be represented by an equation of this simple, linear form. An important tool in determining the form of any relationship, linear or not, is a graph.

Suppose a scientist constructed a liquid crystal cell and then measured the amount of light transmitted through the cell as a function of applied voltage. The results of such a hypothetical experiment are shown in the following data table:

% Light Transmitted	Applied Voltage (V)
0	0
1	1
4	2
9	3
16	4
25	5
36	6
50	7
63	8
82	9
100	10

From the data presented in this table, you can tell that a higher voltage leads to more transmitted light, but not too much else. A better understanding of this data may be obtained by making a graph of it. A graph may be thought of as a picture of a relationship between variables. This data is plotted in the graph in Fig. A.1.

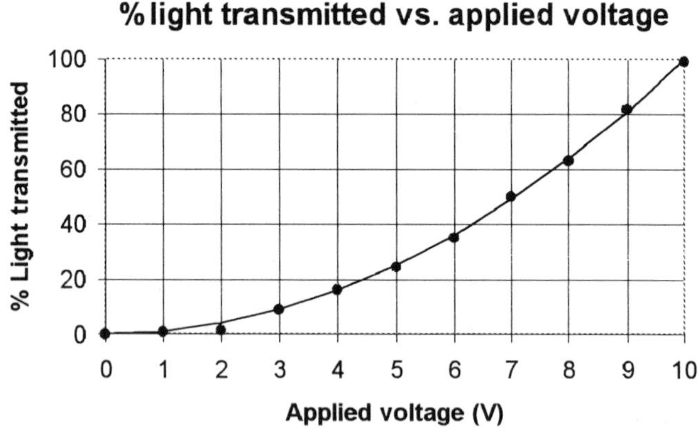

Fig. A.1 Percent light transmitted vs. applied voltage

Observe that "Percent light transmission vs. applied voltage" means that the dependent variable, the "Percent (of) light transmitted" is plotted on the vertical (y) axis, and that the independent variable, "applied voltage," is plotted on the horizontal (x) axis. This graph allows us to construct a simple model by drawing a smooth, continuous curve through the data that can be used to make some predictions about this liquid crystal cell. For example, from this graph one can determine the voltage necessary to make the cell transmit 75% of the light. Similarly, one can determine the transmission at an applied voltage of 4.5 volts. For those who need a short reminder of how to do this, we can proceed as follows. To find the voltage needed for 75% transmission, find 75% transmission on the vertical axis, and then move horizontally until you intercept the curve. Finally, move straight down and estimate the applied voltage. The correct answer is 8.5 volts. Estimating the percent transmitted when the applied voltage is 4.5 volts, one obtains a value of about 20%. Note also that the data points on this graph are clearly drawn as small circles. This indicates that they are experimental points and have some uncertainty associated with their values.

Sometimes the data is not as ideal as that shown above. The data is said to be noisy and may not appear to fall on a single smooth curve. In this case, the correct scientific graphing technique is to draw the "best" curve through the data. While many spreadsheet programs simply connect the data points, this is not proper scientific practice. You should **never** connect the dots if you want to properly analyze the data. This **incorrect** technique is illustrated Fig. A.2, and the **correct** method for graphing the data is shown in Fig. A.3

It is important to note that the correct graphing technique represents a model of one aspect of the phenomena under investigation. This is one reason why graphs are such powerful modeling tools.

A.3 A wee bit of algebra

From time to time, the exercises in this book require that simple equations be solved. This will generally require no more than simple algebra. You will not be asked to reproduce all intermediate steps in such calculations.

An algebraic equation has at least one unknown. The unknowns are generally represented by letters such as x, y, and z but any letter may be used. It is often useful to pick a letter that has some further meaning such as P for perimeter. As an example, consider $2x + 7 = 15$. This is an algebraic equation in which we need to solve for the unknown x. An

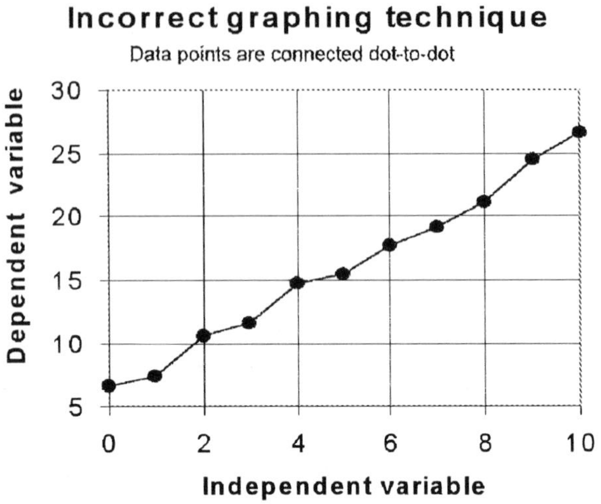

Incorrect graphing technique
Data points are connected dot-to-dot
Dependent variable
30
25
20
15
10
5
0
2
4
6
8
10
Independent variable

Fig. A.2 The **incorrect** way to graph noisy data.

Correct graphing technique
Best fit line is drawn
Dependent variable
30
25
20
15
10
5
0
2
4
6
8
10
Independent variable

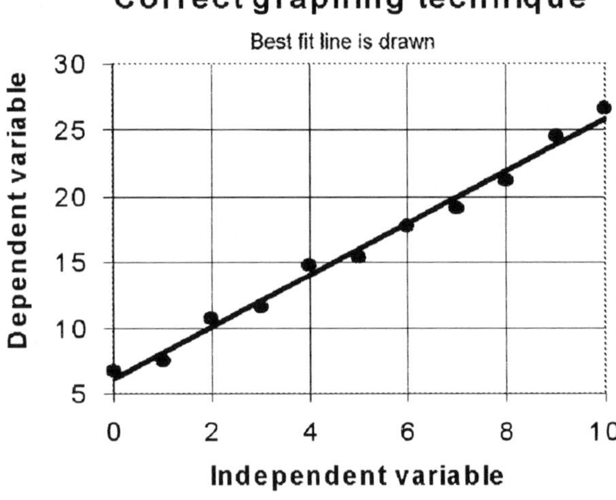

Fig. A.3 The **correct** way to graph noisy data.

equation like this may be solved by remembering two rules:

(1) Never divide by zero.
(2) Treat both sides of the equal sign the same.

The example equation is solved as follows:

Example **A.3**

Equation	**Comment**
$2x + 7 = 15$	Initial equation.
$2x + 7 - 7 = 15 - 7$	Subtract 7 from both sides of the equation.
$2x = 8$	Simplify.
$2x/2 = 8/2$	Divide both sides of the equation by 2.
$x = 4$	The solution.

A.4 Exponents and scientific notation

In our studies we find that we deal with very large and very small numbers; for example, 115,000,000 N/m^2 or 0.0000000000000523 m. Writing all these zeros is awkward, time consuming, and presents a great chance of error. For this reason, scientific notation was introduced. In scientific notation, the number of zeros to the left or right of the decimal point is indicated by a power of ten and an implied positive sign or an explicit negative sign. Our two "extreme" quantities can be written as $1.15 * 10^8 N/m^2$ and $5.23 * 10^{-14}$m.

What do 10^8 and 10^{-14} mean? The answer can be found by induction. We begin by considering $100 = 10 * 10$. We can write $10 * 10$ as 10^2 which is pronounced "ten squared." Note that there are two factors of ten and two zeros to the left of the decimal point in 100. Similarly, $10,000 = 10 * 10 * 10 * 10 = 10^4$ where are now four (4) factors of ten and four zeros. In these examples, the four or two is called the *exponent.* We say that 10 is raised to the $4th$ power or the $2nd$ power. By repeating this procedure a few more times we may conclude:

$10 * 10 * 10 * 10 * 10 * 10 * \ldots = 10^n$ n factors of $10 = 10$ raised to the nth power

Negative exponents describe numbers that are between 0 and 1. For example, $0.01 = 1/100 = 1/(10^2) = 10^{-2}$. When converting to scientific notation, one moves the decimal point to the right until one passes the first non-zero digit. For example, $0.000125 = 1.25 * 10^{-4}$. Notice that the decimal point was moved four places to the right. This is accounted for by multiplying by 10^{-4} (0.0001). Negative numbers between 0 and -1 are described by a negative number between 1 and 10 times 10 raised to a negative exponent. For example, $-0.00876 = -8.76 * 10^{-3}$. The negative three follows because the decimal point moved three places to the right, and the minus sign out front is because the number is a negative number.

There are two more points worth mentioning:

(1) Any number or symbol may be raised to a power.
(2) There are a few specially named exponents
    (a) $2 \Rightarrow$ "squared." f^2 is pronounced "f squared."
    (b) $3 \Rightarrow$ "cubed." g^3 is pronounced "g cubed."
    (c) $1 \Rightarrow$ "first power." h^1 is pronounced "h to the first power," or simply "h."

A.5 Rules for exponents

Here we will simply list the common rules for combining and manipulating exponents. In these rules, a is any non-zero number.

(1) Multiplication: $a^n a^m = a^{(n+m)}$. To multiply two **identical** numbers raised to different powers, add the exponents
(2) Division: $a^n/a^m = a^n * a^{-m} = a^{(n-m)}$. To divide two **identical** numbers raised to different powers, subtract the exponent that is in the denominator from the exponent in the numerator.
(3) **BE CAREFUL!!!** $a^n * b^m$ cannot be simplified because a and b are **not identical** numbers.
(4) Negative powers correspond to reciprocals: $1/a = a^{-1}$, $1/a^5 = a^{-5}$, etc.
(5) $a^0 = 1$ for **all** non-zero a.
(6) Fractional exponents are allowed and correspond to the "roots" of numbers. Specialized notation exists, thus, $\sqrt{a} = a^{1/2}$, $\sqrt[3]{a} = a^{1/3}$, etc. You might recall that the square root of a number, say a, is the number which multiplied times itself will give you the number. For example, $\sqrt{8} = 8^{1/2}$ so that $(\sqrt{8})^2 = \sqrt{8} * \sqrt{8} = 8$. Similarly, the *nth* root of a number multiplied by itself n-times will yield the original number again.

A.6 Measurement and units

One of the basic functions of science is to correlate measurements of physical quantities. To do this, one must define basic units of measure. Furthermore, a self-consistent set of units must be used. To illustrate this point, consider a person who is "5-8" (pronounce this "five eight") First, we must ask 5-8 what? Physical quantities need to have dimensions (basic units of measurement) associated with or attached to them. Does 5-8 mean 5 yards and 8 inches or something else? In the U.S., we can reasonably assume that the person is 5 feet 8 inches tall, but in science, we cannot assume this. We must always specify the units in which the quantity is measured.

Here is another example. We can measure speed in miles per hour (mph). If you want to answer the question, "How far can I travel in x hours if my speed is y mph?," it makes sense to measure distance in miles also. This indicates an important property of length; it is a dimensional quantity. That is, its **numerical value** depends on what system of units used. For length, we commonly use inches, feet, yards, and miles. And the same person who is 5 feet 8 inches tall is also 5 2/3 feet tall, 1.89 yards tall, 68 inches tall, and $1.07 * 10^{-3}$ miles tall. Many of the quantities we deal with are dimensional quantities, so we must always specify our basic dimensions and units of measure.

For the level of science discussed in this book, the common quantities that we use have a mechanical and/or geometrical meaning. Some examples of dimensional quantities of interest include energy, force, velocity, volume, area, and voltage. It is customary to express these and other quantities in terms of four basic dimensions. Also following custom, we will take these four basic dimensions, which for now will be written inside square brackets, as mass [M], length [L], time [T], and charge [Q]. All other dimensional quantities can be written in terms of these basic dimensions. In this way we obtain *derived* dimensional quantities such as area, which has dimension $[L]^2$, or speed, which has dimension [L]/[T]. Speed can be measured in ft/sec, kilometers/hr, miles/hr, furlongs/fortnight, all of which are a length divided by a time. Thus we say speed has dimensions of [L]/[T]. Similarly, the volume of a cube of side l is l^3. This can be generalized to other shapes, but we always find that the dimension of volume is length cubed, volume$=[L]^3$.

Having decided on our basic dimensions, we must now specify our fundamental units of measurement. The *International System of Units* (abbreviated SI) is used in science. The SI unit of length is the meter, which is 39.37 inches or 3.281 ft. The SI unit of time is the second, which you are familiar with. The SI unit of charge is the Coulomb. While Coulombs are not commonly used in daily life, this unit allows us to measure differences in electric potential in the familiar units of volts. Lastly, the SI unit of mass is the kilogram. You are probably least familiar with mass. As you are aware, objects at rest tend to resist being moved (they stay at rest). For a given set of conditions, this resistance to being moved, an attribute of matter called inertia, is quantified by measuring its mass. Mass should be contrasted with weight, which is the force of the earth's gravity on a mass.

In all cases, there are rather complicated internationally agreed upon standards for any SI unit you'd care to use to characterize a phenomenon. Finally, the SI system of units has prefixes which describe multiples or fractions of the basic unit. A table of important, common prefixes follows.

A.7 References

The material in this appendix is a standard part of introductory chemistry and physics courses. The texts for such courses will discuss this material in varying degrees of detail. **Fundamentals of Physics**, by Halliday and Resnick, published by John Wiley is a classic and was consulted during the writing of this book.

indexmathematics!metric prefixes

Prefix	Abbreviation	Meaning	Examples
tera	T	10^{12}	$2*10^{21}$ Tg $= 2*10^{33}$ grams $= 2*10^{30}$ kg ($\approx$ mass of sun)
giga	G	10^{9}	1 GHz = 1 gigahertz (A microwave frequency)
mega	M	10^{6}	89.7 Mhz = 89.7 megahertz (The broadcast frequency of WKSU, Kent, Ohio)
kilo	k	10^{3}	1 kg = 1000 gm = 10^3 gm (2.21 lb on earth)
deci	d	10^{-1}	1 deciBel = 0.1 Bel (A unit of relative intensity)
centi	c	10^{-2}	1 centimeter = 0.01 m ($\approx$ 1/30 ft)
milli	m	10^{-3}	1 milligram = 0.001 gm = 10^{-3} gm (500 mg dose of a medicine)
micro	μ	10^{-6}	1 micrometer = 1μm = 10^{-6} m (typical size of bacteria)
nano	n	10^{-9}	1 nanosecond = 1 ns = 10^{-9} sec (The time for light to travel one foot)
pico	p	10^{-12}	1 pF = 1 picofarad = 10^{-12} farad ($\approx$ the capacitance of 1 inch of twisted wires)
femto	f	10^{-15}	1 femtometer = 1 fm = 10^{-15} m ($\approx$ the size of a proton.)

A.8 Exercises

(1) Express the speed of light, $3*10^8$ m/s, in standard notation.

(2) Green light has a wavelength near 515 nm. Express this in meters (m) and in standard notation.

(3) The Bohr radius is $5.292*10^{-11}$m. Express this in nanometers (nm), picometers (pm), and standard notation.

(4) In a certain experiment the following data is obtained:

m (grams)	50	70	100	120	150	170	200
L (cm)	37.3	39.5	43.4	45.1	48.8	51.2	54.0

It is predicted that there is a linear relationship between L and m of the form $L = L_o + C * m$. Graph L vs. m and determine L_o and C from your graph. (You will need to use a somewhat larger range for L and m than the data range.)

(5) Plot the following points:

t	905	929	942	957	969	984
r	17.12	13.72	11.64	8.59	6.38	4.23

Plot r as the independent variable. Start the r axis at zero and extend the t axis to at least 1020. Note that the range of t is much smaller than t itself. In this case, it is a good practice not to start t at zero, but at a more reasonable number such as 850 or 900.

(6) Convert the following numbers into scientific notation: 1,001,000; 10,200; 0.000123; 0.000000034; 1025; 0.1450; -0.0235; -123576.

(7) For the following graph, estimate the value of y at the following values of x: x= 4.5, 6.5, 9.5, 3, 5, 6, 9.

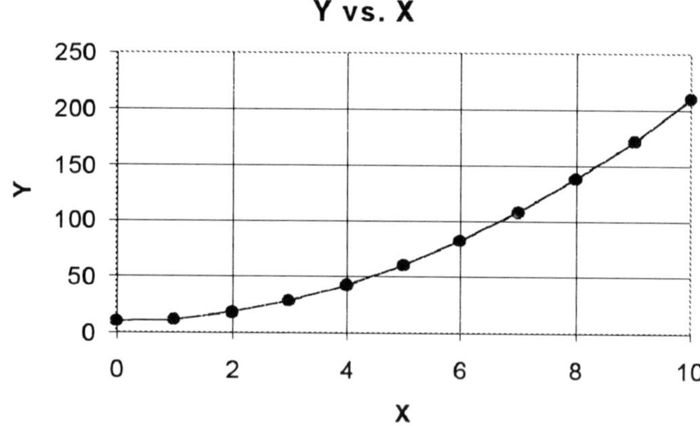

Index